Structural Fire Fighting:
Initial Response Strategy and Tactics

Second Edition

Alex Abrams
Lead Senior Editor

Veronica Smith, Libby Snyder, Michael Fox
Senior Editors

Lynn Hughes
Lead Instructional Developer

Validated by the International Fire Service Training Association
Published by Fire Protection Publications • Oklahoma State University

Cover photo courtesy of Ron Jeffers, Union City, NJ. • Title page photo courtesy of Bob Esposito.

The International Fire Service Training Association (IFSTA) was established in 1934 as a *nonprofit educational association of fire fighting personnel who are dedicated to upgrading fire fighting techniques and safety through training.* To carry out the mission of IFSTA, Fire Protection Publications was established as an entity of Oklahoma State University. Fire Protection Publications' primary function is to publish and distribute training materials as proposed, developed, and validated by IFSTA. As a secondary function, Fire Protection Publications researches, acquires, produces, and markets high-quality learning and teaching aids consistent with IFSTA's mission.

IFSTA holds two meetings each year: the Winter Meeting in January and the Annual Validation Conference in July. During these meetings, committees of technical experts review draft materials and ensure that the professional qualifications of the National Fire Protection Association® standards are met. These conferences bring together individuals from several related and allied fields, such as:

- Key fire department executives, training officers, and personnel
- Educators from colleges and universities
- Representatives from governmental agencies
- Delegates of firefighter associations and industrial organizations

Committee members are not paid nor are they reimbursed for their expenses by IFSTA or Fire Protection Publications. They participate because of a commitment to the fire service and its future through training. Being on a committee is prestigious in the fire service community, and committee members are acknowledged leaders in their fields. This unique feature provides a close relationship between IFSTA and the fire service community.

IFSTA manuals have been adopted as the official teaching texts of many states and provinces of North America as well as numerous U.S. and Canadian government agencies. Besides the NFPA® requirements, IFSTA manuals are also written to meet the Fire and Emergency Services Higher Education (FESHE) course requirements. A number of the manuals have been translated into other languages to provide training for fire and emergency service personnel in Canada, Mexico, and outside of North America.

Copyright © 2017 by the Board of Regents, Oklahoma State University

All rights reserved. No part of this publication may be reproduced in any form without prior written permission from the publisher.

ISBN 978-0-87939-623-7   Library of Congress Control Number: 2017934843

Fifth Edition, First Printing, April 2017   *Printed in the United States of America*

10   9   8   7   6   5   4   3   2   1

If you need additional information concerning the International Fire Service Training Association (IFSTA) or Fire Protection Publications, contact:

Customer Service, Fire Protection Publications, Oklahoma State University
930 North Willis, Stillwater, OK 74078-8045
800-654-4055   Fax: 405-744-8204

For assistance with training materials, to recommend material for inclusion in an IFSTA manual, or to ask questions or comment on manual content, contact:

Editorial Department, Fire Protection Publications, Oklahoma State University
930 North Willis, Stillwater, OK 74078-8045
405-744-4111   Fax: 405-744-4112   E-mail: editors@osufpp.org

Oklahoma State University in compliance with Title VI of the Civil Rights Act of 1964 and Title IX of the Educational Amendments of 1972 (Higher Education Act) does not discriminate on the basis of race, color, national origin or sex in any of its policies, practices or procedures. This provision includes but is not limited to admissions, employment, financial aid and educational services.

Chapter Summary

Chapters

1. Fire Dynamics 8
2. Prefire Planning 78
3. Managing the Incident 132
4. Size-Up: Evaluation and Assessment 156
5. Strategy 204
6. Tactics 224
7. Residential Scenarios 274
8. Commercial Scenarios 308
9. Special Hazard Scenarios 342

Appendices

A. Chapter and Page Correlation to FESHE Requirements 375

Glossary **379**

Index **387**

Table of Contents

List of Tables viii
Acknowledgements ix
Introduction 1
Purpose and Scope 1
Book Organization... 2
Terminology 2
Key Information 3
Metric Conversions 4

1 Fire Dynamics... 8

Science of Fire 11
  Physical Science Terminology.................... 12
  Fire Triangle and Tetrahedron 15
  Ignition... 15
  Combustion.. 17
    Nonflaming Combustion 17
    Flaming Combustion................................ 17
    Products of Combustion 18
  Pressure Differences 22
Thermal Energy (Heat)................................. 22
  Difference between Heat Release and Temperature.. 22
  Sources of Thermal Energy........................ 23
    Chemical Energy...................................... 23
    Electrical Energy..................................... 24
    Mechanical Energy 25
  Heat Transfer.. 26
    Conduction... 26
    Convection.. 27
    Radiation.. 28
    Interaction among the Methods of Heat Transfer 30
Fuel 31
  Gases... 32
  Liquids.. 33
  Solids.. 35
Oxygen.. 37
Self-Sustained Chemical Reaction............ 39
Compartment Fire Development.............. 41
  Stages of Fire Development........................ 41
  Incipient Stage.. 43
  Growth Stage 44
    Thermal Layering.................................... 46
    Transition to Ventilation-Limited Decay........... 48
    Rapid Fire Development 50
  Fully Developed Stage 55
    Fuel-Limited Conditions 56
    Ventilation-Limited Conditions............. 56
  Decay Stage... 58
    Fuel-Limited Decay................................. 58
    Ventilation-Limited Decay..................... 58
Structure Fire Development 59
  Flow Path 59
  Ventilation and Wind Considerations 61
    Unplanned Ventilation 62
    Wind Conditions...................................... 62
  Smoke Explosions 63
  Effects of Fire Fighting Operations 63
Reaction of Building Construction to Fire............ 63
  Construction Type and Elapsed Time of Structural Integrity 64
  Fuel Load of Structural Members and Contents 65
    Furnishings and Finishes 68
    Combustible Exterior Wall Coverings 68
    Combustible Roof Materials................... 68
  Building Compartmentation...................... 69
  Effects of Building Construction Features........... 70
    Modern vs. Legacy Construction 70
    Compartment Volume and Ceiling Height........ 72
    Thermal Properties of the Building................... 72
    Failure of Lightweight Trusses and Joists 73
  Construction, Renovation, and Demolition Hazards 74
Chapter Review 75
Discussion Questions 76
Chapter Notes... 76

2 Prefire Planning....................................... 78

Understanding Surveys 81
  Updating Prefire Plan.................................. 82
  Area Familiarization 83
Preparing for Preincident Surveys 84
  Emergency Response Considerations................ 85
  Survey Information Records 86
    Common Survey Form Data................... 86
    Field Sketches 87
    Photography, Videography, and Global Positioning 87
    Building Exterior..................................... 87

  Building Interior 88
  Life-Safety Information 90
   Building Conditions *91*
   Building Contents *92*
  Ventilation Systems 93
   Built-In Ventilation Devices *94*
   Underfloor Air Distribution Systems *95*
  Fire Protection Systems 95
Available Resources 97
  Water Supply 98
  Required Fire Flow Calculations 99
  Resource Needs 100
   Automatic Aid *101*
   Mutual Aid *102*
   Additional Resources *102*
Understanding a Building Construction 102
  Lightweight/Engineered Construction 103
  United States Construction 103
   Type I *104*
   Type II *105*
   Type III *105*
   Type IV *106*
   Type V *107*
   Unclassified Construction Types *109*
  Canadian Construction 110
  Interior Building Arrangement 111
   Open Floor Plans *112*
   Compartmentalization *113*
   Basements, Cellar, and Crawl Spaces *114*
   Attics and Cocklofts *115*
   Concealed Spaces *116*
   Structural Collapse Potential *117*
  Roof Types and Hazards 125
   Flat *126*
   Pitched *126*
   Arched *127*
   Roof Hazards *128*
  Occupancy Types 128
   Single Use *129*
   Multiple Use *129*
Chapter Review 130

3   Managing the Incident 132

Decision-Making 135
  Identify and Prioritize the Problems 139
  Determine a Solution 139
  Implement the Solution 140
  Monitor the Results 141
  Adjust Plan as Necessary 141
  Result of Indecision 141

National Incident Management System-Incident Command System (NIMS-ICS) 142
  Common Terminology 144
   Organizational Levels *144*
   Resources *146*
  Common Communications 146
  Unified Command Structure 148
  Incident Action Plan 149
  Manageable Span of Control 150
  Comprehensive Resource Management 151
  Personnel Accountability 151
  Resource Tracking 152
Chapter Review 155

4   Size-Up: Evaluation and Assessment .. 156

Incident Size-Up Considerations 159
  While Responding 159
   Time of Day *161*
   Weather Conditions *161*
   Capabilities of Your Department's Response..*162*
  Facts, Perceptions, and Projections and Probabilities 162
   Facts *163*
   Perceptions *164*
   Projections and Probabilities *165*
  On Arrival 167
  Reading Smoke 172
   White Smoke *173*
   Brown Smoke *173*
   Gray Smoke *173*
   Black Smoke *174*
   Unusual Colors in Smoke *175*
   Neutral Plane *175*
   Smoke Volume *175*
   Smoke Density *176*
   Types of Smoke Activity *176*
   Velocity *177*
   Pulsations *177*
   Smoke Movement *177*
   Heat *178*
   Flame Color *179*
  During the Incident 179
  Decision-Making 180
  Plan of Operation 180
Firefighter Survivability Approaches 181
  Occupant Survival Profiling 181
  Crew Resource Management 183
  Rules of Engagement 184
Critical Fireground Size-Up Factors 185
  Building Characteristics 186

v

  Street Access..................................*187*
  Structure Access..............................*188*
  Egress..*188*
  Metal Security Doors and Bars........*189*
  Special Hazards...............................*190*
 Life Hazard via Occupancy Condition 191
  General Considerations....................*191*
  Residential Occupancy Type.............*193*
  Business Occupancy Type *194*
  Mercantile Occupancy Type..............*194*
  Mixed Occupancy Type.....................*195*
  Industrial and Storage Occupancies...*196*
  Institutional Occupancy Type...........*196*
  Assembly Occupancy Type................*197*
  Educational Occupancy Type............*197*
  Unoccupied, Vacant, or Abandoned
   Structures *198*
 Arrival Condition Indicators 200
  Time of Day *200*
  Weather...*201*
  Visual Indications *202*
Chapter Review...................................**202**

5   Strategy 204

Incident Priorities..............................**207**
 Life Safety... 207
 Incident Stabilization.......................... 210
 Property Conservation........................ 210
Risk versus Benefit.............................**211**
Operational Strategies......................**212**
 Offensive Strategy 212
 Defensive Strategy.............................. 214
 Transitioning Strategies...................... 216
Incident Action Plan..........................**217**
 Developing the IAP 217
 Implementing the IAP 218
  Command Options............................*218*
  Resource Allocation *220*
Chapter Review...................................**222**

6   Tactics 224

Search and Rescue..............................**227**
 Search Safety Guidelines for the IC and Search
  Teams 231
 Conducting a Search........................... 231
  Primary Search *232*
  Secondary Search.............................*234*
  Victim Removal *234*
Exposures...**234**

Confinement **235**
Extinguishment..................................**239**
 Fire Attack 241
  Direct Attack....................................*241*
  Indirect Attack.................................*242*
  Gas Cooling *242*
 Transitional Attack 243
 Positive-Pressure Attack 245
Overhaul..**248**
Tactical Ventilation............................**253**
Considerations Affecting the Decision to
Ventilate **255**
 Risks to Occupants and Firefighters 256
 Building Construction and Modern
  Furnishings 256
 Basement Ventilation.......................... 258
 Fire Behavior Indicators 258
 Location and Extent of the Fire 259
 Type of Ventilation 259
 Location for Ventilation 260
 Weather Conditions 261
 Exposures... 262
 Staffing and Available Resources 263
Types of Tactical Ventilation **263**
 Horizontal Ventilation 263
  Natural Horizontal Ventilation *263*
  Mechanical Horizontal Ventilation...*264*
 Vertical Ventilation............................. 267
 Other Types of Ventilation Situations ... 268
  Basement Ventilation *269*
  High-Rise Fires *269*
Salvage/Property Conservation **271**
Chapter Review...................................**273**

7   Residential Scenarios............. 274

Note to Instructors **276**
Scenario 1...**277**
Scenario 2...**283**
Scenario 3...**289**
Scenario 4...**295**
Scenario 5...**301**

8   Commercial Scenarios 308

Note to Instructors **310**
Scenario 1...**311**
Scenario 2...**317**
Scenario 3...**323**
Scenario 4...**329**
Scenario 5...**335**

9 Special Hazards Scenarios 342
Note to Instructors 344
Scenario 1 345
Scenario 2 351
Scenario 3 357
Scenario 4 363
Scenario 5 369

Appendix A
Chapter and Page Correlation to FESHE Requirements 375

Glossary 379
Index 387

List of Tables

Table 1.1	Common Oxidizers	11
Table 1.2	Common Products of Combustion and Their Toxic Effects	20
Table 1.3	Health Effects of CO	20
Table 1.4	Health Effects of HCN Exposure	21
Table 1.5	Spontaneous Heating Materials and Locations	24
Table 1.6	Thermal Conductivity of Common Substances	27
Table 1.7	Response to Temperature of an Object Being Heated	30
Table 1.8	Representative Peak Heat Release Rates (HRR) During Unconfined Burning	31
Table 1.9	Characteristics of Common Flammable Gases	33
Table 1.10	Characteristics of Common Flammable and Combustible Liquids	35
Table 1.11	Pyrolysis of Wood and Polyurethane Foam	36
Table 1.12	Flammable Ranges of Common Flammable Gases and Liquids (Vapor)	39
Table 1.13	Fire Resistance Ratings for Type I through Type V Construction (In Hours)	64

Dedication

This manual is dedicated to the men and women who hold devotion to duty above personal risk, who count on sincerity of service above personal comfort and convenience, who strive unceasingly to find better and safer ways of protecting lives, homes, and property of their fellow citizens from the ravages of fire, medical emergencies, and other disasters

...The Firefighters of All Nations.

Acknowledgements

The second edition of **Structural Fire Fighting: Initial Response Strategy and Tactics** is designed to meet the objectives listed for FESHE courses.

Acknowledgement and special thanks are extended to the members of the IFSTA validating committee who contributed their time, wisdom, and knowledge to the development of this manual.

IFSTA Structural Fire Fighting: Initial Response Strategy and Tactics Second Edition Validation Committee

Chair
Brian Morrow
Assistant Chief of Operations
Orange County Fire Rescue Department
Winter Park, Florida

Vice Chair and Secretary
Jeff Parsons
Division Chief
Idaho Falls Fire Department
Idaho Falls, Idaho

Committee Members

Jason Balleto
Assistant Drillmaster/Instructor
New Haven Fire Department Training Division
New Haven, Connecticut

Rob Bocanegra
Lieutenant
Hutto Fire Rescue
Hutto, Texas

Steve Brisebois
Lieutenant
Flash Information
Mirabel, Quebec, Canada

Glenn Doers
Deputy Chief
Port Washington Fire Department
Port Washington, Wisconsin

Patrick Dunn
Training Captain
Kansas City Fire Department
Kansas City, Kansas

Paul Egizi
Fire Captain
Los Angeles Fire Department
Van Nuys, California

Mark Flagler
Firefighter
Cincinnati Fire Department
Cincinnati, Ohio

Tim Frankenberg
Deputy Chief
Washington Fire Department
Washington, Missouri

Ed French
Emergency Response Chief (Retired)
ConocoPhillips Canada
Soldotna, Alaska

Joseph Guarnera
Coordinator
Massachusetts Firefighting Academy
North Chelmsford, Massachusetts

Alan Hartford
Assistant Fire Chief-Operations Division
Contra Costa County Fire Protection District
Pleasant Hill, California

Thomas Hughes
Fire Chief
Jerome City Fire Department
Jerome, Idaho

Mike McLaughlin
Deputy Fire Chief
Cosumnes Fire Department
Elk Grove, California

Kevin Scheuerman
Lieutenant
Madeira & Indian Hill Joint Fire District
Cincinnati, Ohio

IFSTA Structural Fire Fighting: Initial Response Strategy and Tactics Second Edition Validation Committee

Committee Members (cont.)

John Schutt
Captain
Mesa Fire and Medical Department
Mesa, Arizona

Demond Simmons
Captain
Oakland Fire Department
Oakland, California

Keith Stakes
Research Engineer
Firefighter Safety Research Institute
Underwriters Laboratories
College Park, Maryland

Much appreciation is given to the following individuals and organizations for contributing information, photographs, and technical assistance instrumental in the development of this manual:

- Bob Esposito
- Essentials of Fire Fighting, Seventh Edition, committee members
- Ron Jeffers, Union City, New Jersey
- Chris Mickel, New Orleans (Louisiana) Fire Department Photo Unit
- Jeremy Potter, Idaho Falls Fire Department

Last, but certainly not least, gratitude is extended to the following members of the Fire Protection Publications staff whose contributions made the final publication of this manual possible.

Structural Fire Fighting: Initial Response Strategy and Tactics Second Edition, Project Team

Lead Senior Editor
Alex Abrams, Senior Editor

Technical Editor
Derek Alkonis
Battalion Chief
Los Angeles County (CA) Fire Department

Lead Instructional Developer
Lynn Hughes, Curriculum Developer

Director of Fire Protection Publications
Craig Hannan

Curriculum Manager
Colby Cagle

Editorial Manager
Clint Clausing

Production Manager
Ann Moffat

Editors
Veronica Smith, Senior Editor
Libby Snyder, Senior Editor
Michael Fox, Senior Editor

Illustrators and Layout Designer
Errick Bragg, Senior Graphic Designer

Curriculum Development
Lynn Hughes, Lead Curriculum Developer
Lori Raborg, Curriculum Developer
Beth Ann Fulgenzi, Curriculum Developer Proofer

Photographers
Alex Abrams, Senior Editor
Jeff Fortney, Senior Editor
Mike Sturzenbecker, Sales Director
Mike Wieder, Associate Director

Editorial Staff
Clint Clausing, Editorial Manager
Tara Gladden, Editorial Assistant

Indexer
Nancy Kopper

The IFSTA Executive Board at the time of validation of the **Structural Fire Fighting: Initial Response Strategy and Tactics, Second Edition** was as follows:

IFSTA Executive Board

Executive Board Chair
Steve Ashbrock
Fire Chief
Madeira & Indian Hill Fire Department
Cincinnati, OH

Vice Chair
Bradd Clark
Fire Chief
Ocala Fire Department
Ocala, FL

IFSTA Executive Director
Mike Wieder
Fire Protection Publications
Stillwater, OK

Board Members

Steve Austin
Project Manager
Cumberland Valley Volunteer Firemen's Association
Newark, DE

Mary Cameli
Assistant Chief
City of Mesa Fire Department
Mesa, AZ

Dr. Larry Collins
Associate Dean
Eastern Kentucky University
Safety, Security, & Emergency Department
Richmond, KY

Chief Dennis Compton
Mesa & Phoenix, Arizona
Chairman of the National Fallen Firefighters
Foundation Board of Directors

John Hoglund
Director Emeritus
Maryland Fire & Rescue Institute
New Carrollton, MD

Tonya Hoover
State Fire Marshal
CA Department of Forestry & Fire Protection
Sacramento, CA

Dr. Scott Kerwood
Fire Chief
Hutto Fire Rescue
Hutto, TX

Wes Kitchel
Assistant Chief
Sonoma County Fire & Emergency Services Dept.
Santa Rosa, CA

Brett Lacey
Fire Marshal
Colorado Springs Fire Department
Colorado Springs, CO

Robert Moore
Division Director
Texas A&M Engineering Extension Services
College Station, TX

Dr. Lori Moore-Merrell
Assistant to the General President
International Association of Fire Fighters
Washington, DC

Jeff Morrissette
State Fire Administrator
State of Connecticut
Commission on Fire Prevention and Control
Windsor Locks, CT

Josh Stefancic
Division Chief
Largo Fire Rescue
Largo, FL

Paul Valentine
Senior Engineer
Nexus Engineering
Oakbrook Terrace, IL

Steven Westermann
Fire Chief
Central Jackson County Fire Protection District
Blue Springs, MO

Introduction

Introduction Contents

Introduction 1	Terminology 2
Purpose and Scope 1	Key Information 3
Book Organization 2	Metric Conversions 4

Introduction

Fire fighting is commonly accepted as a hazardous profession that places personnel, regardless of their experience level, at risk throughout their daily work shift. It is also an accepted fact that many of the fatalities and injuries that firefighters suffer every year can be prevented through numerous means. Fire departments have emphasized ongoing training for personnel, strong decision-making by Incident Commanders, and no tolerance for freelancing at a scene. The fire service has also benefitted from improvements in technology as well as ground-breaking research in fire dynamics that has the potential to change long-standing beliefs about response times, extreme fire conditions, and the stages of fire. This research is presented in the first chapter of IFSTA **Structural Fire Fighting: Initial Response Strategy and Tactics**.

Although there have been numerous books on fire fighting strategy and tactics written over the past half century, **Structural Fire Fighting: Initial Response Strategy and Tactics** differs in some essential ways. First, unlike other current literature in the field, a committee of fire and emergency service professionals of all ranks from North America has reviewed and validated this manual. Peer validation means that the committee reached a consensus on the information contained in this manual.

Second, this manual takes a focused approach on strategy and tactics as they apply to the most frequently types of fires in occupancies: single- and multiple-family residential dwellings and small-to-medium-sized retail (mercantile) occupancies. This manual provides the strategy and tactics needed to control average fires in average structures.

Finally, this manual is written for the initial response and the person, regardless of rank, who must evaluate the incident, assign resources, and attack the fire. The allocation of resources begins with the first-arriving unit and evolves to an average first assignment of apparatus, usually two engines and a ladder/truck company. The initial response ends with control of the incident, transfer of command to a battalion or district chief, or expansion of the incident management system to a higher level involving additional units and agencies.

Purpose and Scope

This 2nd edition of **Structural Fire Fighting: Initial Response Strategy and Tactics** is written for the initial response and the person, regardless of rank, who must evaluate the incident, assign resources, and fight the fire. The **Purpose** of this manual is to provide the initial Incident Commander with the strategic and tactical concepts that can be applied to various situations with the resources available to them. The manual meets the requirements of the FESHE course outcomes and outlines for the Associates level noncore course, "Strategies and Tactics." This manual also broadens the knowledge of the strategy and tactics requirements found in NFPA®1021, *Standard for Fire Officer Professional Qualifications*. The manual is written for use in training agencies, college degree programs, and as a resource guide for all firefighters who are or may become unit commanders.

The **Scope** of this manual is to provide the first arriving unit commander with the knowledge, skills, and abilities to assess the situation, initiate a command structure, and deploy resources until transfer of command or termination of the incident. The target audience is all career and volunteer personnel who will be assigned to incidents involving single and multifamily dwellings, commercial occupancies, and unique target or special hazards. Nationally accepted strategies and tactics are presented and applied to hypothetical incidents throughout the manual. The overall goal of this manual is to teach Incident Commanders how to make better decisions at emergency incidents and to reduce fireground injuries and fatalities.

Book Organization

The second edition of **Structural Fire Fighting: Initial Response Strategy and Tactics** is arranged in a manner designed to assist the reader in developing the knowledge, skills, and abilities required to manage an initial attack on a structure fire. The first two chapters provide information that an Incident Commander should have knowledge of prior to arriving at a scene, such as the basics of fire dynamics and prefire planning. The next four chapters cover the different stages of managing an incident, from assigning resources, selecting a strategy and tactic, and finally performing overhaul and salvage. The last three chapters offer real-life scenarios for readers to formulate and discuss their approach to a fire, such as the size-up considerations, command options, and unusual hazards or conditions.

Terminology

This manual is written with a global, international audience in mind. For this reason, it often uses general descriptive language in place of regional- or agency-specific terminology (often referred to as *jargon*). Additionally, in order to keep sentences uncluttered and easy to read, the word state is used to represent both state and provincial level governments (or their equivalent). This usage is applied to this manual for the purposes of brevity and is not intended to address or show preference for only one nation's method of identifying regional governments within its borders.

The glossary at the end of the manual will assist the reader in understanding words that may not have their roots in the fire and emergency services. The sources for the definitions of fire-and-emergency-services-related terms will be the IFSTA **Fire Service Orientation and Terminology** manual.

Key Information

Various types of information in this book are given in shaded boxes marked by symbols or icons. See the following definitions:

Case History

A case history analyzes an event. It can describe its development, action taken, investigation results, and lessons learned.

Safety Alert

Safety alert boxes are used to highlight information that is important for safety reasons. (In the text, the title of safety alerts will change to reflect the content.)

Tie Rods

Some Type III buildings have tie rods and anchor plates that span from one side of a building to the opposite side. These rods are connected to either ornamental stars or plates on the exterior of the building. When conducting a walk around or pre-planning a structure, personnel should make note of this construction feature. When exposed to heat in a fire, the tie rod will expand and allow the exterior walls to move outward and potentially collapse.

What This Means To You

These boxes take information presented in the text and synthesize it into an example of how the information is relevant to (or will be applied by) the intended audience, essentially answering the question, "What does this mean to you?"

A **key term** is designed to emphasize key concepts, technical terms, or ideas that readers need to know. They are listed at the beginning of each chapter and the definition is placed in the margin for easy reference. An example of a key term is:

Three key signal words are found in the book: **WARNING**, **CAUTION**, and **NOTE**. Definitions and examples of each are as follows:

- **WARNING** indicates information that could result in death or serious injury to readers. See the following example:

Fuel Load — The total quantity of combustible contents of a building, space, or fire area, including interior finish and trim, expressed in heat units of the equivalent weight in wood. *Also known as* Fuel Loading.

Introduction 3

WARNING
Visible smoke may be unreliable as an indicator of the conditions inside a structure.

CAUTION
Keep a charged hoseline available when exposing a concealed space with suspected fire conditions.

- **CAUTION** indicates important information or data that readers need to be aware of in order to perform their duties safely. See the following example:
- **NOTE** indicates important operational information that helps explain why a particular recommendation is given or describes optional methods for certain procedures. See the following example:

NOTE: Your AHJ will determine who is responsible for determining fire cause and origin and when an investigator is required at the incident.

Chapter End Notes refer the reader to sources referenced within the chapter. Not all chapters have Chapter End Notes as they are only included when appropriate to do so.

1. Kerber, Stephen, "Analysis of Changing Residential Fire Dynamics and Its Implications on Firefighter Operational Timeframes," Underwriters Laboratories, Inc. 2012.

To find curriculum or study materials associated with this manual and its contents, please go to ifsta.org and use the search tool in the shop to find accompanying products. You can also search for IFSTA apps using your smartphone or tablet device.

Metric Conversions

Throughout this manual, U.S. units of measure are converted to metric units for the convenience of our international readers. Be advised that we use the Canadian metric system. It is very similar to the Standard International system, but may have some variation.

We adhere to the following guidelines for metric conversions in this manual:

- Metric conversions are approximated unless the number is used in mathematical equations.
- Centimeters are not used because they are not part of the Canadian metric standard.
- Exact conversions are used when an exact number is necessary such as in construction measurements or hydraulic calculations.
- Set values such as hose diameter, ladder length, and nozzle size use their Canadian counterpart naming conventions and are not mathematically calculated. For example, 1 1/2 inch hose is referred to as 38 mm hose.

The following two tables provide detailed information on IFSTA's conversion conventions. The first table includes examples of our conversion factors for a number of measurements used in the fire service. The second shows examples of exact conversions beside the approximated measurements you will see in this manual.

U.S. to Canadian Measurement Conversion

Measurements	Customary (U.S.)	Metric (Canada)	Conversion Factor
Length/Distance	Inch (in) Foot (ft) [3 or less feet] Foot (ft) [3 or more feet] Mile (mi)	Millimeter (mm) Millimeter (mm) Meter (m) Kilometer (km)	1 in = 25 mm 1 ft = 300 mm 1 ft = 0.3 m 1 mi = 1.6 km
Area	Square Foot (ft^2) Square Mile (mi^2)	Square Meter (m^2) Square Kilometer (km^2)	1 ft^2 = 0.09 m^2 1 mi^2 = 2.6 km^2
Mass/Weight	Dry Ounce (oz) Pound (lb) Ton (T)	gram Kilogram (kg) Ton (T)	1 oz = 28 g 1 lb = 0.5 kg 1 T = 0.9 T
Volume	Cubic Foot (ft^3) Fluid Ounce (fl oz) Quart (qt) Gallon (gal)	Cubic Meter (m^3) Milliliter (mL) Liter (L) Liter (L)	1 ft^3 = 0.03 m^3 1 fl oz = 30 mL 1 qt = 1 L 1 gal = 4 L
Flow	Gallons per Minute (gpm) Cubic Foot per Minute (ft^3/min)	Liters per Minute (L/min) Cubic Meter per Minute (m^3/min)	1 gpm = 4 L/min 1 ft^3/min = 0.03 m^3/min
Flow per Area	Gallons per Minute per Square Foot (gpm/ft^2)	Liters per Square Meters Minute (L/(m^2.min))	1 gpm/ft^2 = 40 L/(m^2.min)
Pressure	Pounds per Square Inch (psi) Pounds per Square Foot (psf) Inches of Mercury (in Hg)	Kilopascal (kPa) Kilopascal (kPa) Kilopascal (kPa)	1 psi = 7 kPa 1 psf = .05 kPa 1 in Hg = 3.4 kPa
Speed/Velocity	Miles per Hour (mph) Feet per Second (ft/sec)	Kilometers per Hour (km/h) Meter per Second (m/s)	1 mph = 1.6 km/h 1 ft/sec = 0.3 m/s
Heat	British Thermal Unit (Btu)	Kilojoule (kJ)	1 Btu = 1 kJ
Heat Flow	British Thermal Unit per Minute (BTU/min)	watt (W)	1 Btu/min = 18 W
Density	Pound per Cubic Foot (lb/ft^3)	Kilogram per Cubic Meter (kg/m^3)	1 lb/ft^3 = 16 kg/m^3
Force	Pound-Force (lbf)	Newton (N)	1 lbf = 0.5 N
Torque	Pound-Force Foot (lbf ft)	Newton Meter (N.m)	1 lbf ft = 1.4 N.m
Dynamic Viscosity	Pound per Foot-Second (lb/ft.s)	Pascal Second (Pa.s)	1 lb/ft.s = 1.5 Pa.s
Surface Tension	Pound per Foot (lb/ft)	Newton per Meter (N/m)	1 lb/ft = 15 N/m

Conversion and Approximation Examples

Measurement	U.S. Unit	Conversion Factor	Exact S.I. Unit	Rounded S.I. Unit
Length/Distance	10 in	1 in = 25 mm	250 mm	250 mm
	25 in	1 in = 25 mm	625 mm	625 mm
	2 ft	1 in = 25 mm	600 mm	600 mm
	17 ft	1 ft = 0.3 m	5.1 m	5 m
	3 mi	1 mi = 1.6 km	4.8 km	5 km
	10 mi	1 mi = 1.6 km	16 km	16 km
Area	36 ft^2	1 ft^2 = 0.09 m^2	3.24 m^2	3 m^2
	300 ft^2	1 ft^2 = 0.09 m^2	27 m^2	30 m^2
	5 mi^2	1 mi^2 = 2.6 km^2	13 km^2	13 km^2
	14 mi^2	1 mi^2 = 2.6 km^2	36.4 km^2	35 km^2
Mass/Weight	16 oz	1 oz = 28 g	448 g	450 g
	20 oz	1 oz = 28 g	560 g	560 g
	3.75 lb	1 lb = 0.5 kg	1.875 kg	2 kg
	2,000 lb	1 lb = 0.5 kg	1 000 kg	1 000 kg
	1 T	1 T = 0.9 T	900 kg	900 kg
	2.5 T	1 T = 0.9 T	2.25 T	2 T
Volume	55 ft^3	1 ft^3 = 0.03 m^3	1.65 m^3	1.5 m^3
	2,000 ft^3	1 ft^3 = 0.03 m^3	60 m^3	60 m^3
	8 fl oz	1 fl oz = 30 mL	240 mL	240 mL
	20 fl oz	1 fl oz = 30 mL	600 mL	600 mL
	10 qt	1 qt = 1 L	10 L	10 L
	22 gal	1 gal = 4 L	88 L	90 L
	500 gal	1 gal = 4 L	2 000 L	2 000 L
Flow	100 gpm	1 gpm = 4 L/min	400 L/min	400 L/min
	500 gpm	1 gpm = 4 L/min	2 000 L/min	2 000 L/min
	16 ft^3/min	1 ft^3/min = 0.03 m^3/min	0.48 m^3/min	0.5 m^3/min
	200 ft^3/min	1 ft^3/min = 0.03 m^3/min	6 m^3/min	6 m^3/min
Flow per Area	50 gpm/ft^2	1 gpm/ft^2 = 40 L/(m^2.min)	2 000 L/(m^2.min)	2 000 L/(m^2.min)
	326 gpm/ft^2	1 gpm/ft^2 = 40 L/(m^2.min)	13 040 L/(m^2.min)	13 000 L/(m^2.min)
Pressure	100 psi	1 psi = 7 kPa	700 kPa	700 kPa
	175 psi	1 psi = 7 kPa	1225 kPa	1 200 kPa
	526 psf	1 psf = 0.05 kPa	26.3 kPa	25 kPa
	12,000 psf	1 psf = 0.05 kPa	600 kPa	600 kPa
	5 psi in Hg	1 psi = 3.4 kPa	17 kPa	17 kPa
	20 psi in Hg	1 psi = 3.4 kPa	68 kPa	70 kPa
Speed/Velocity	20 mph	1 mph = 1.6 km/h	32 km/h	30 km/h
	35 mph	1 mph = 1.6 km/h	56 km/h	55 km/h
	10 ft/sec	1 ft/sec = 0.3 m/s	3 m/s	3 m/s
	50 ft/sec	1 ft/sec = 0.3 m/s	15 m/s	15 m/s
Heat	1200 Btu	1 Btu = 1 kJ	1 200 kJ	1 200 kJ
Heat Flow	5 BTU/min	1 Btu/min = 18 W	90 W	90 W
	400 BTU/min	1 Btu/min = 18 W	7 200 W	7 200 W
Density	5 lb/ft^3	1 lb/ft^3 = 16 kg/m^3	80 kg/m^3	80 kg/m^3
	48 lb/ft^3	1 lb/ft^3 = 16 kg/m^3	768 kg/m^3	770 kg/m^3
Force	10 lbf	1 lbf = 0.5 N	5 N	5 N
	1,500 lbf	1 lbf = 0.5 N	750 N	750 N
Torque	100	1 lbf ft = 1.4 N.m	140 N.m	140 N.m
	500	1 lbf ft = 1.4 N.m	700 N.m	700 N.m
Dynamic Viscosity	20 lb/ft.s	1 lb/ft.s = 1.5 Pa.s	30 Pa.s	30 Pa.s
	35 lb/ft.s	1 lb/ft.s = 1.5 Pa.s	52.5 Pa.s	50 Pa.s
Surface Tension	6.5 lb/ft	1 lb/ft = 15 N/m	97.5 N/m	100 N/m
	10 lb/ft	1 lb/ft = 15 N/m	150 N/m	150 N/m

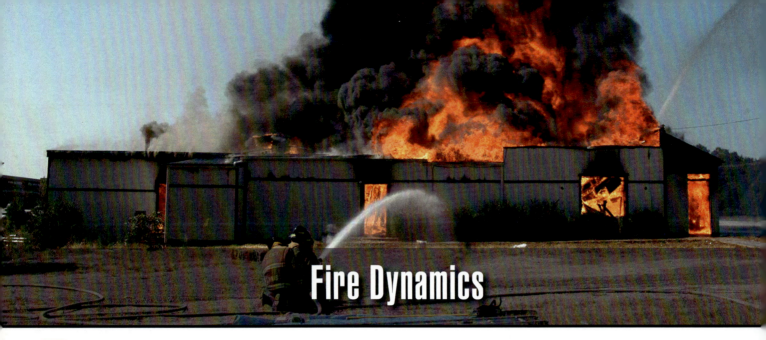

Fire Dynamics

Chapter Contents

- **Science of Fire 11**
  - Physical Science Terminology 12
  - Fire Triangle and Tetrahedron 15
  - Ignition 15
  - Combustion 17
  - Pressure Differences 22
- **Thermal Energy (Heat) 22**
  - Difference between Heat Release Rate and Temperature 22
  - Sources of Thermal Energy 23
  - Heat Transfer 26
- **Fuel 31**
  - Gases 32
  - Liquids 33
  - Solids 35
- **Oxygen 37**
- **Self-Sustained Chemical Reaction 39**
- **Compartment Fire Development 41**
  - Stages of Fire Development 41
  - Incipient Stage 43
  - Growth Stage 44
  - Fully Developed Stage 55
  - Decay Stage 58
- **Structure Fire Development 59**
  - Flow Path 59
  - Ventilation and Wind Considerations 61
  - Smoke Explosions 63
  - Effects of Fire Fighting Operations 63
- **Reaction of Building Construction to Fire .. 63**
  - Construction Type and Elapsed Time of Structural Integrity 64
  - Fuel Load of Structural Members and Contents 65
  - Building Compartmentation 69
  - Effects of Building Construction Features 70
  - Construction, Renovation, and Demolition Hazards ... 74
- **Chapter Review 75**
- **Discussion Questions 76**
- **Chapter Notes 76**

chapter 1

Key Terms

Asphyxiation 19	Flashover 50	Pressure...................................22
Autoignition.............................16	Flow Path 46	Products of Combustion 18
Autoignition Temperature 17	Free Radical............................39	Pyrolysis 16
Backdraft 50	Fuel..12	Radiation..................................27
Buoyant....................................22	Fuel-Limited............................41	Reducing Agent 31
Carbon-Based Fuels 18	Fuel Load 65	Rollover....................................51
Carbon Dioxide (CO_2)............21	Heat 11	Self-Heating............................22
Carbon Monoxide (CO) 19	Heat of Combustion................13	Smoke Explosion 63
Ceiling Jet...............................43	Heat Flux.................................22	Solubility..................................35
Chemical Flame Inhibition 40	Heat Release Rate..................22	Specific Gravity.......................35
Combustion 11	Hydrocarbon Fuel 18	Spontaneous Ignition 22
Combustion Zone 45	Hydrogen Cyanide (HCN)........20	Surface-To-Mass Ratio...........35
Compartmentation...................69	Incomplete Combustion..........18	Temperature 11
Conduction 27	Isolated Flames......................46	Thermal Conductivity 27
Convection...............................27	Joule (J) 14	Thermal Energy.......................13
Endothermic Reaction.............14	Kinetic Energy.........................13	Thermal Equilibrium 27
Energy......................................12	Lower Explosive (Flammable)	Thermal Layering (of Gases)...46
Entrain......................................18	Limit (LEL) 38	Upper Explosive (Flammable)
Entrainment.............................44	Matter 12	Limit (UEL)...........................38
Exothermic Reaction 14	Miscible...................................35	Vapor Density..........................33
Exposure Fire..........................29	Neutral Plane..........................46	Vapor Pressure........................35
Fire...11	Open Burning 42	Vaporization.............................16
Fire Point.................................35	Oxidation 12	Ventilation-Limited..................41
Fire Tetrahedron......................15	Oxidizer...................................12	Watt (W) 32
Fire Triangle............................15	Piloted Ignition 16	
Flammable (Explosive)	Polar Solvents 35	
Range 38	Potential Energy......................13	
Flash Point..............................35	Power 32	

Chapter 1 • Fire Dynamics 9

Fire Dynamics

FESHE Learning Outcomes

After reading this chapter, students will be able to:

1. Discuss fire behavior as it relates to strategies and tactics.

Chapter 1
Fire Dynamics

Fire dynamics describes the meeting point between fire science, materials science, fluid dynamics of gases, and heat transfer. Understanding the basic physics of these sciences can give firefighters the knowledge needed to forecast fire growth at a scene and predict the likely consequences of various tactical options available for controlling a fire. All of the following provide firefighters with pieces of the total picture about a fire's likely behavior during fireground operations:

- Fire science
- The combustion process
- Fire behavior and its relationship to various materials and environments
- Classifications of fires and their corresponding extinguishing agents
- Recognition of fire behavior indicators, fire development patterns, and the potential for rapid fire development
- Various ventilation and suppression tactics used as tools for controlling fires

Science of Fire

Firefighters should have a scientific understanding of **combustion**, **fire**, **heat**, and **temperature**. Fire can take various forms, but all fires involve a heat-producing chemical reaction between some type of **fuel** and an **oxidizer**, most commonly oxygen in the air. Oxidizers are not combustible but will support or enhance combustion. **Table 1.1** lists some common oxidizers.

Table 1.1 Common Oxidizers

Substance	Common Use
Calcium Hypochlorite (granular solid)	Chlorination of water in swimming pools
Chlorine (gas)	Water purification
Ammonium Nitrate (granular solid)	Fertilizer
Hydrogen Peroxide (liquid)	Industrial bleaching (pulp and paper and chemical manufacturing)
Methyl Ethyl Ketone Peroxide	Catalyst in plastics manufacturing

Courtesy of Ed Hartin.

Combustion — A chemical process of oxidation that occurs at a rate fast enough to produce heat and usually light in the form of either a glow or flame. (Reproduced with permission from NFPA 921-2011, *Guide for Fire and Explosion Investigations*, Copyright 2011, National Fire Protection Association).

Fire — A rapid oxidation process, which is a gas phase chemical reaction resulting in the evolution of light and heat in varying intensities.

Heat — Form of energy associated with the motion of atoms or molecules in solids or liquids that is transferred from one body to another as a result of a temperature difference between the bodies, such as from the sun to the earth. To signify its intensity, it is measured in degrees of temperature.

Temperature — Measure of the average kinetic energy of the particles in a sample of matter, expressed in terms of units or degrees designated on a standard scale.

Fuel — A material that will maintain combustion under specified environmental conditions (Reproduced with permission from NFPA 921-2011, *Guide for Fire and Explosion Investigations*, Copyright 2011, National Fire Protection Association).

Oxidizer — Any material that readily yields oxygen or other oxidizing gas, or that readily reacts to promote or initiate combustion of combustible materials. (Reproduced with permission from NFPA 400-2010, *Hazardous Materials Code*, Copyright 2010, National Fire Protection Association)

Matter — Anything that occupies space and has mass.

Energy — Capacity to perform work; occurs when a force is applied to an object over a distance, or when a substance undergoes a chemical, biological, or physical transformation.

Oxidation — Chemical process that occurs when a substance combines with an oxidizer such as oxygen in the air; a common example is the formation of rust on metal.

Physical Science Terminology

Physical science is the study of **matter** and **energy** and includes chemistry and physics. This theoretical foundation must be translated into a practical knowledge of fire dynamics. To remain safe, you need to be able to identify the fire dynamics present in a given situation and anticipate what the next stages of the fire will be along with how fire fighting operations may impact the fire's behavior **(Figure 1.1)**.

Figure 1.1 The conditions found at a fire scene offer indications of a fire's current behavior and potential behavior. *Courtesy of UL, LLC.*

The world around you is made up of matter in the form of physical materials that occupy space and have mass. While matter can undergo many types of physical and chemical changes, this chapter will concentrate on those changes related to fire.

A physical change occurs when a substance remains chemically the same but changes in size, shape, or appearance. Examples of physical change are water freezing (liquid to solid) and boiling (liquid to gas).

A chemical reaction occurs when a substance changes from one type of matter into another, such as two or more substances combining to form compounds. **Oxidation** is a chemical reaction involving the combination of an oxidizer, such as oxygen in the air, with other materials. Oxidation can be slow, such as the combination of oxygen with iron to form rust, or rapid, as in combustion of methane (natural gas) **(Figure 1.2)**.

Energy is the capacity to perform work. Work occurs when a force is applied to an object over a distance or when a substance undergoes a chemical, biological, or physical change. In the case of heat, work means increasing a substance's temperature.

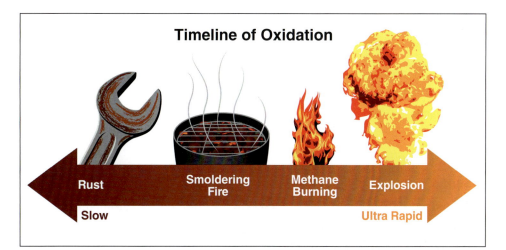

Figure 1.2 The timeline of oxidation illustrates the speed differences among types of oxidation.

Figure 1.3 Potential energy is stored energy, while kinetic energy is actively being released. *Courtesy of Dan Madrzykowski, NIST.*

Forms of energy are classified as either potential or kinetic (**Figure 1.3**). **Potential energy** represents the amount of energy that an object can release at some point in the future. Fuels have a certain amount of potential energy before they are ignited, based on their chemical composition. This potential energy available for release in the combustion process is known as the **heat of combustion**. The rate at which a fuel releases energy over time depends on many variables including:

- Chemical composition
- Arrangement
- Density of the fuel
- Availability of oxygen for combustion

Kinetic energy is the energy that a moving object possesses. While a fuel such as wood is not "moving" as you might define it, when heat is introduced, the molecules within the fuel begin to vibrate. As the heat (**thermal energy**) increases, these molecules vibrate more and more rapidly. The fuel's kinetic energy is the result of these vibrations in the molecules.

> **Potential Energy** — Stored energy possessed by an object that can be released in the future to perform work once released.
>
> **Heat of Combustion** — Total amount of thermal energy (heat) that could be generated by the combustion (oxidation) reaction if a fuel were completely burned. The heat of combustion is typically measured in kilojoules per gram (kJ/g) or megajoules per kilogram (MJ/kg).
>
> **Kinetic Energy** — Energy possessed by a moving object because of its motion.
>
> **Thermal Energy** — Kinetic energy associated with the random motions of the molecules of a material or object; often used interchangeably with the terms *heat* and *heat energy*.

Chapter 1 • Fire Dynamics

Joule (J) — Unit of work or energy in the International System of Units (SI); the energy (or work) when a unit force (1 newton) moves a body through a unit distance (1 meter). Joules are defined in terms of mechanical energy. In terms of thermal energy, joules refer to the amount of additional heat needed to raise the temperature of a substance, such as the 4.2 Joules needed to raise the temperature of 1 gram of water 1 degree Celsius. Takes the place of calorie for heat measurement (1 calorie = 4.19 J).

Exothermic Reaction — Chemical reaction between two or more materials that changes the materials and produces heat.

Endothermic Reaction — Chemical reaction in which a substance absorbs heat.

There are many types of energy including:
- Chemical
- Thermal
- Mechanical
- Electrical
- Light
- Nuclear
- Sound

All energy can change from one type to another. For example, mechanical energy from a machine can convert to thermal energy when friction between moving parts generates heat. In terms of fire behavior, the potential chemical energy of a fuel converts into heat and light during combustion.

Energy is measured in **joules (J)** in the International System of Units (SI). The quantity of heat required to change the temperature of 1 gram of water by 1 degree Celsius is 4.2 joules. In the customary system, the unit of measurement for heat is the British thermal unit (Btu). A British thermal unit is the amount of heat required to raise the temperature of 1 pound of water by 1 degree Fahrenheit. While not used in scientific and engineering texts, the Btu is still frequently used in the fire service. When comparing joules and Btu, 1 055 J = 1 Btu.

Chemical and physical changes almost always involve an exchange of energy. A fuel's potential energy releases during combustion and converts to kinetic energy. Reactions that emit energy as they occur are **exothermic reactions**. Fire is an exothermic chemical reaction that releases energy in the form of heat and sometimes light.

Reactions that absorb energy as they occur are **endothermic reactions** (**Figure 1.4**). For example, converting water from a liquid to a gas (steam) requires the input of energy resulting in an endothermic reaction. Converting water to steam is a tactic for controlling and extinguishing some types of fires.

Figure 1.4 Exothermic reactions release energy while endothermic reactions absorb energy.

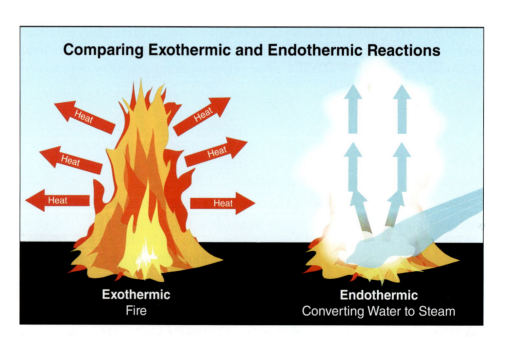

Fire Triangle and Tetrahedron

The **fire triangle** and **fire tetrahedron** models are used to explain the elements of fire and how fires can be extinguished **(Figure 1.5)**. The oldest and simplest model, the fire triangle, shows three elements necessary for combustion to occur: fuel, oxygen, and heat. Remove any one of these elements and the fire will be extinguished. The fire triangle was used prior to the general adaptation of the fire tetrahedron, which includes a chemical chain reaction.

An uninhibited chemical chain reaction must also be present for a fire to occur. The fire tetrahedron model includes the chemical chain reaction to explain flaming or gas-phase combustion (fire is an example of gas-phase combustion).

> **Fire Triangle** — Plane geometric model of an equilateral triangle that is used to explain the conditions/elements necessary for combustion. The sides of the triangle represent heat, oxygen, and fuel.
>
> **Fire Tetrahedron** — Model of the four elements/conditions required to have a fire. The four sides of the tetrahedron represent fuel, heat, oxygen, and self-sustaining chemical chain reaction.

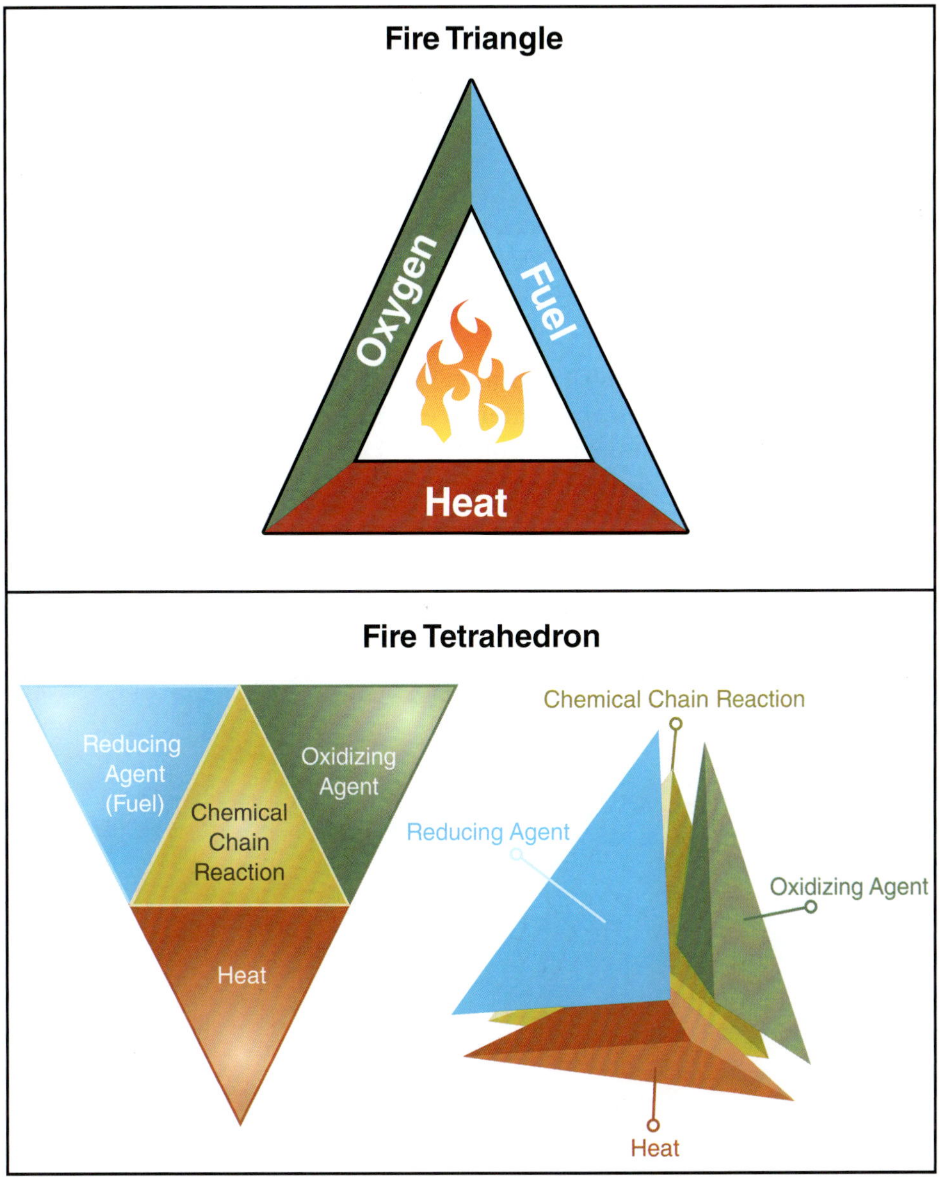

Figure 1.5 The fire triangle illustrates the three components needed for a fire, while the fire tetrahedron demonstrates the four components needed for a self-sustaining fire.

Ignition

Fuels must be in a gaseous state to burn; therefore, solids and liquids must become gaseous in order for ignition to occur. When heat is transferred to a liquid or solid, the substance's temperature increases and the substance starts

Pyrolysis — The chemical decomposition of a solid material by heating. Pyrolysis precedes combustion of a solid fuel.

Vaporization — Physical process that changes a liquid into a gaseous state; the rate of vaporization depends on the substance involved, heat, pressure, and exposed surface area.

Piloted Ignition — Moment when a mixture of fuel and oxygen encounters an external heat (ignition) source with sufficient heat or thermal energy to start the combustion reaction.

Autoignition — Initiation of combustion by heat but without a spark or flame (Reproduced with permission from NFPA 921-2011, *Guide for Fire and Explosion Investigations*, Copyright 2011, National Fire Protection Association).

to convert to a gaseous state (off-gassing). In solids, off-gassing is a chemical change known as **pyrolysis**; in liquids, a physical change called **vaporization** **(Figure 1.6).**

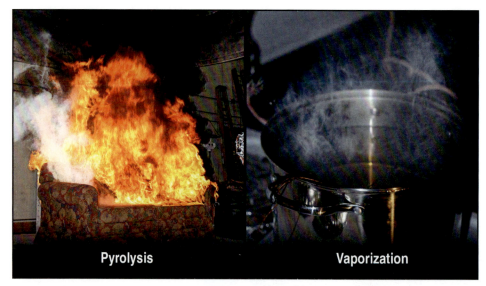

Figure 1.6 Pyrolysis occurs when a solid fuel is converted into a gaseous fuel. Vaporization is the conversion of a liquid to a vapor by the heat energy of combustion.

Piloted ignition is the most common form of ignition and occurs when a mixture of fuel and oxygen encounter an external heat source with sufficient heat or thermal energy to start the combustion reaction. **Autoignition** occurs without any external flame or spark to ignite the fuel gases or vapors. The fuel's surface is heated to the point at which the combustion reaction occurs **(Figure 1.7).**

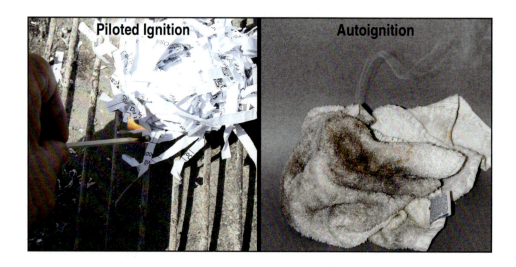

Figure 1.7 Piloted ignition involves the introduction of an external ignition source, while autoignition occurs under special conditions without the heat of a spark or other source.

Once the fuel is ignited, the energy released from combustion transfers to the remaining solid fuel resulting in the production and ignition of additional fuel vapors or gases. This exchange of energy from the burning gases to the fuel results in a sustained combustion reaction.

Autoignition temperature (AIT) is the minimum temperature at which a fuel in the air must be heated in order to start self-sustained combustion. The autoignition temperature of a substance is always higher than its piloted ignition temperature.

Combustion

Fire and combustion are similar conditions. Both words are commonly used to mean the same thing. Combustion, however, is a chemical reaction while flaming combustion is only one possible form of combustion. Combustion can occur without visible flames. There are two modes of combustion: nonflaming and flaming **(Figure 1.8)**.

> **Autoignition Temperature (AIT)** — The lowest temperature at which a combustible material ignites in air without a spark or flame (Reproduced with permission from NFPA 921-2011, *Guide for Fire and Explosion Investigations*, Copyright 2011, National Fire Protection Association).

Figure 1.8 Flaming combustion displays visible flames above the burning fuel. Nonflaming combustion features lower temperatures and smoldering conditions.

Nonflaming Combustion

Nonflaming combustion occurs more slowly and at a lower temperature, producing a smoldering glow in the material's surface. The burning may be localized on or near the fuel's surface where it is in contact with oxygen. Examples of nonflaming or smoldering combustion include burning charcoal or smoldering wood or fabric. The fire triangle illustrates the elements/conditions required for this mode of combustion.

Flaming Combustion

Flaming combustion is commonly referred to as fire. It produces a visible flame above the material's surface. Flaming combustion occurs when a gaseous fuel mixes with oxygen in the correct ratio and heats to ignition temperature. Flaming combustion requires liquid or solid fuels to be converted to the gas phase through the addition of heat (vaporization or pyrolysis, respectively). When heated, both liquid and solid fuels will emit vapors that mix with oxygen, producing flames above the material's surface if the gases ignite. The fire tetrahedron accurately reflects the conditions required for flaming combustion.

Each element of the tetrahedron must be in the proper proportion and in close physical proximity for flaming combustion to occur. Removing any element of the tetrahedron interrupts the chemical chain reaction and stops flaming combustion. However, the fire may continue to smolder depending on the characteristics of the fuel.

Ignition is where the combustion process begins. A heat source pyrolizes a fuel, creating fuel gases. Those gases mix with oxygen and ignite, creating a fire. The fire can be compared to a pump. Fresh oxygen is "pumped in" and mixes with fuel gases. Then as it burns, the fire "pumps out" combustion products that have larger amounts of mass and a higher level of energy than the inlet air. In the case of open burning, the "pump" does not have a well-defined inlet or outlet, as the air is being **entrained** (drawn in) from all around the burning fuel. **Figure 1.9** illustrates the intake flow to the fire and the exhaust flow from the fire.

Entrain — To draw in and transport solid particles or gases by the flow of a fluid.

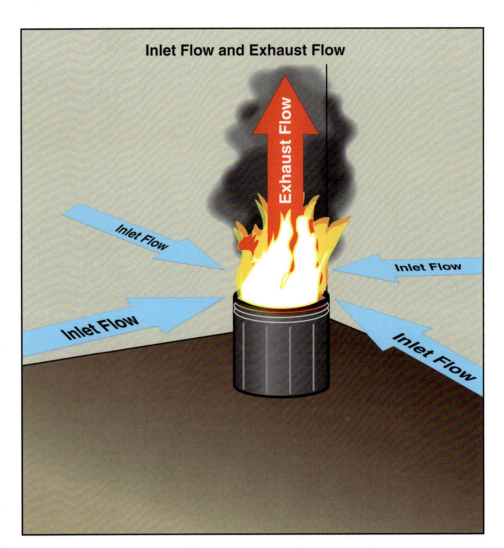

Figure 1.9 This illustration demonstrates the concepts of inlet flow and exhaust flow.

Products of Combustion — Materials produced and released during burning.

Incomplete Combustion — Result of inefficient combustion of a fuel; the less efficient the combustion, the more products of combustion are produced rather than burned during the combustion process.

Carbon-Based Fuels — Fuels in which the energy of combustion derives principally from carbon; includes materials such as wood, cotton, coal, or petroleum.

Hydrocarbon Fuel — Petroleum-based organic compound that contains only hydrogen and carbon; may also be used to describe those materials in a fuel load which were created using hydrocarbons such as plastics or synthetic fabrics.

The fire also generates heat. As the heat transfers to the gaseous combustion products, they expand and begin to rise and move away from the fire due to buoyancy. In other words, the density of the hot combustion products is less than the surrounding air, and the combustion products "float" on the dense cool air surrounding the fuel, creating the layers of smoke and fuel gases that fill a compartment during a fire.

Products of Combustion

As a fuel burns, its chemical composition changes, which produces new substances. These **products of combustion** are often simply described as heat and smoke. While the heat from a fire is a danger to anyone directly exposed to it, exposure to toxic gases found in smoke and/or lack of oxygen cause most fire deaths. Smoke is an aerosol comprised of gases, vapor, and solid particulates.

Smoke is the product of **incomplete combustion**. Simply stated, combustion is incomplete when any of the fuel is left after combustion has occurred. Smoke and ash are examples of left over fuel from incomplete combustion.

By comparison, under ideal conditions, the entire fuel would undergo a chemical conversion from its current form into an equal amount of new materials. For example, complete combustion of methane in air results in the production of heat, light, water vapor, and carbon dioxide.

However, combustion is incomplete in a structure fire, meaning that some of the fuel does not burn, but instead gets entrained with hot gases and rises aloft. This unburned fuel is smoke, and it has the potential to burn **(Figure 1.10)**.

Figure 1.10 Smoke is composed of a wide range of toxic and flammable gases and particulates.

Most structure fires involve multiple types of fuels (**carbon-based fuels** [wood, cotton], **hydrocarbon fuels** [plastics, synthetic fabrics], and other types), and the fires tend to have a limited air supply. When the air supply is limited, the level of incomplete combustion is higher, which produces more smoke. These factors result in complex chemical reactions that generate a wide range of products of combustion including toxic and flammable gases, vapors, and particulates that comprise smoke.

Gases such as carbon monoxide (CO) are generally colorless, while vapor and particulates give smoke its varied colors. Most components of smoke are toxic and dangerous to human life. The materials that make up smoke vary from fuel to fuel, but generally all smoke is toxic. The toxic effects of smoke inhalation are the result of the interrelated effect of all the toxic products present.

Keep in mind that the combustion process consumes oxygen from the air, effectively removing it from the environment. As part of the chemical reaction, the consumed oxygen combines with carbon in the smoke to form combustion products like CO or carbon dioxide (CO_2).

Low oxygen concentrations alone can result in hypoxia or death. The toxic gases in combination with a low oxygen concentration can reduce the time that a victim could survive. **Table 1.2, p. 20** lists some of the more common products of combustion and their toxic effects. Concentrations of the products of combustion and/or low oxygen concentrations can cause **asphyxiation** (fatal level of oxygen deficiency in the blood).

Carbon monoxide (CO) is a toxic and flammable product of the incomplete combustion of organic (carbon-containing) materials. Carbon monoxide is a colorless, odorless gas present at almost every fire. It is released when an organic material burns in an atmosphere with a limited supply of oxygen. CO exposure is frequently identified as the cause of death in civilian fatalities.

Asphyxiation — Fatal condition caused by severe oxygen deficiency and an excess of carbon monoxide and/or other gases in the blood.

Carbon Monoxide (CO) — Colorless, odorless, dangerous gas (both toxic and flammable) formed by the incomplete combustion of carbon. It combines with hemoglobin more than 200 times faster than oxygen does, decreasing the blood's ability to carry oxygen.

Hydrogen Cyanide (HCN) — Colorless, toxic, and flammable liquid until it reaches 79°F (26°C). Above that temperature, it becomes a gas with a faint odor similar to bitter almonds; produced by the combustion of nitrogen-bearing substances.

Table 1.2
Common Products of Combustion and Their Toxic Effects

Carbon Monoxide	Colorless, odorless gas. Inhalation of carbon monoxide causes headache, dizziness, weakness, confusion, nausea, unconsciousness, and death. Exposure to as little as 0.2 percent carbon monoxide can result in unconsciousness within 30 minutes. Inhalation of high concentration can result in immediate collapse and unconsciousness.
Formaldehyde	Colorless gas with a pungent irritating odor that is highly irritating to the nose. 50-100 ppm can cause severe irritation to the respiratory track and serious injury. Exposure to high concentrations can cause injury to the skin. Formaldehyde is a suspected carcinogen.
Hydrogen Cyanide	Colorless, toxic, and flammable liquid below 79°F (26°C) produced by the combustion of nitrogen-bearing substances. It is a chemical asphyxiant that acts to prevent the body from using oxygen. It is commonly encountered in smoke in concentrations lower than carbon monoxide.
Nitrogen Dioxide	Reddish-brown gas or yellowish-brown liquid, which is highly toxic and corrosive.
Particulates	Small particles that can be inhaled and deposited in the mouth, trachea, or the lungs. Exposure to particulates can cause eye irritation, respiratory distress (in addition to health hazards specifically related to the particular substances involved).
Sulfur Dioxide	Colorless gas with a choking or suffocating odor. Sulfur dioxide is toxic and corrosive, and can irritate the eyes and mucous membranes.

Source: *Computer Aided Management of Emergency Operations (CAMEO) and Toxicological Profile for Polycyclic Aromatic Hydrocarbons.*

Table 1.3
Health Effects of CO

CO-Carbon Monoxide

Short-Term Exposure

As Concentration Increases ↓
- Headache
- Dizziness
- Vomiting
- Nausea
- Unconsciousness
- Death

Long-Term Exposure

Cardiovascular Disease
Possible mental health problems

CO acts as a chemical asphyxiant. CO poisoning is a sometimes lethal condition in which carbon monoxide molecules attach to hemoglobin, decreasing the blood's ability to carry oxygen. CO combines with hemoglobin about 200 times more effectively than oxygen does. CO does not act on the body, but excludes oxygen from the blood, leading to hypoxia of the brain and tissues. Death will follow if the process is not reversed. **Table 1.3** illustrates the effects of CO on humans.

Hydrogen cyanide (HCN), a toxic and flammable substance produced in the combustion of materials containing nitrogen, is also commonly found in smoke, although at lower concentrations than CO. Incomplete combustion of substances that contain nitrogen and carbon produce HCN.

The following materials produce HCN
- Natural fibers such as wool, cotton and silk
- Resins such as carbon fiber or fiberglass
- Synthetic polymers such as nylon or polyester
- Synthetic rubber such as neoprene, silicone and latex

These materials are found in:
- Upholstered furniture
- Bedding
- Insulation
- Carpets

- Clothing
- Other common building materials and household items

HCN is a significant byproduct of the combustion of polyurethane foam used in many household furnishings. HCN is also released during off-gassing as an object is heated. It may also be found in vehicle fires, where new insulation materials give off high amounts of gases and cause fires to last longer.

HCN is 35 times more toxic than CO. HCN acts as a chemical asphyxiant but with a different mechanism of action than CO. HCN prevents the body from using oxygen at the cellular level. HCN can be inhaled, ingested, or absorbed into the body, where it then targets the heart and brain. Inhaled HCN enters the bloodstream and prevents the blood cells from using oxygen properly, killing the cells. The effects of HCN depend on the concentration, length, and type of exposure. Large amounts, high concentrations, and lengthy exposures are more likely to cause severe effects, including permanent heart and brain damage or death **Table 1.4** illustrates the effects of HCN on the human body.

**Table 1.4
Health Effects of HCN Exposure**

HCN Cyanide
Low Concentration
Eye Irritation
Headache
Confusion
Nausea
Vomiting
Coma (in some cases)
Fatality (in some cases)
High Concentration
Immediate central nervous system, cardiovascular, and respiratory distress leading to death within minutes.

CAUTION
Wear full PPE and SCBA anytime there is a chance of exposure to smoke, heat, or toxic gases.

Carbon dioxide (CO_2) is a product of complete combustion of organic materials. It is not toxic in the same manner as CO or HCN, but it displaces existing oxygen which creates an oxygen deficient atmosphere. CO_2 also acts as a respiratory stimulant, increasing respiratory rate.

NOTE: Fire gases also contain many other gases than the three highlighted in this section. These additional gases have their own effects and exposure times. The exposure time is based on the combination of gases or the lethal effective dose.

Irritants in smoke are substances that cause breathing discomfort and inflammation of the eyes, respiratory tract, and skin. Smoke can contain a wide range of irritating substances depending on the fuels involved. More than 20 irritants in smoke have been identified including hydrogen chloride, formaldehyde, and acrolein.

Smoke also contains significant amounts of unburned fuels in the form of solid and liquid particulates and gases. Smoke must be treated with the same respect as any other flammable gas because it may burn or explode. Particulates can interfere with vision and breathing.

Carbon Dioxide (CO_2) — Colorless, odorless, heavier than air gas that neither supports combustion nor burns; used in portable fire extinguishers as an extinguishing agent to extinguish Class B or C fires by smothering or displacing the oxygen. CO_2 is a waste product of aerobic metabolism.

WARNING!
Smoke is fuel and is potentially flammable. Smoke can be oxygen deficient and contain chemicals, which may be acutely toxic, and/or carcinogens which may cause cancer.

Pressure — Force per unit area exerted by a liquid or gas measured in pounds per square inch (psi) or kilopascals (kPa).

Buoyant — The tendency or capacity of a liquid or gas to remain afloat or rise.

Heat Release Rate — Total amount of heat released per unit time. The heat release rate is typically measured in kilowatts (kW) or Megawatts (MW) of output.

Heat Flux — The measure of the rate of heat transfer to or from a surface, typically expressed in kilowatts per square meter (kW/m^2).

Self-Heating — The result of exothermic reactions, occurring spontaneously in some materials under certain conditions, whereby heat is generated at a rate sufficient to raise the temperature of the material (Reproduced with permission from NFPA 921-2011, *Guide for Fire and Explosion Investigations*, Copyright 2011, National Fire Protection Association).

Spontaneous Ignition — Initiation of combustion of a material by an internal chemical or biological reaction that has produced sufficient heat to ignite the material (Reproduced with permission from NFPA 921-2011, *Guide for Fire and Explosion Investigations*, Copyright 2011, National Fire Protection Association).

Pressure Differences

Pressure is the force per unit of area applied perpendicular to a surface. For example, atmospheric pressure (1 atmosphere [app. 101 kPa]) at standard temperature (68° F [20° C]) indicates the amount of pressure the atmosphere applies to the surface of the earth. At standard temperature and atmospheric pressure, gases remain calm and move very little. Differences in pressure above or below standard pressure create movement in gases. Gases always move from areas of higher pressure to areas of lower pressure. Even small differences in pressure, such as the 0.1 kPa or less differences created in most compartment fires, create this movement.

Heat from a fire increases the pressure of the surrounding gases. This increased pressure will seek to expand and equalize with areas of lower pressure. Heated gases will rise, remain aloft (**buoyant**) and generally travel up and out. At the same time, cooler, fresh air will generally travel inward toward the fire. This exchange of air creates a convective flow. As the pressure difference between high and low pressure areas increases, the speed with which gases will move from high to low also increases. It is critical for firefighters to understand how small changes to the gas pressure within a structure can dramatically affect fire behavior (**Figure 1.11**).

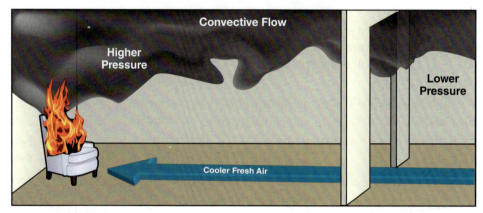

Figure 1.11 Heated gases will travel upward and outward from a fire while cooler, fresher air is drawn in toward the fire creating a convective flow.

Thermal Energy (Heat)

A working knowledge of fire dynamics requires an understanding of temperature, energy, and power or **heat release rate**. Firefighters often use these terms interchangeably because the differences between the terms are not always understood.

Difference between Heat Release Rate and Temperature

Heat is the thermal kinetic energy needed to release the potential chemical energy in a fuel. As heat begins to vibrate the molecules in a fuel, the fuel begins a physical change from a solid or liquid to a gas. The fuel emits flammable vapors which can ignite and release thermal energy. This new source of thermal energy begins to heat other, uninvolved fuels converting their energy and spreading the fire.

Temperature is the measurement of heat. More specifically, temperature is the measurement of the average kinetic energy in the particles of a sample of matter. A block of wood at room temperature has stable molecules and is in no danger of ignition. When thermal energy transfers to the wood, the wood is heated, and the temperature of the wood rises because the molecules have begun to vibrate and move more freely and rapidly.

Temperature can be measured using several different scales. The most common are the Celsius scale, used in the International System of Units (SI) (metric system), and the Fahrenheit scale, used in the customary system. The freezing and boiling points of water provide a simple way to compare these two scales **(Figure 1.12)**.

A dangerous misconception is that temperature is an accurate predictor or measurement of heat transfer. It is not. For example, one candle burns at the same temperature as ten candles. However, the heat release rate (kW) of the ten candles is ten times greater than one candle at the same temperature. The increased heat release rate results in an increased heat transfer rate to an object. This energy flow to a unit area (**heat flux**) is measured in kilowatts per square meter. Translated to an interior fire environment, the temperature in the structure may be within tolerances for personal protective equipment however, the heat flux to the PPE from the fire indicates the real measurement of how long the PPE will protect you. In other words, the temperature tells you it is safe to go in, but the heat transfer rate – not the temperature – tells you how long you can stay in.

Sources of Thermal Energy

Chemical, electrical, and mechanical energy are common sources of heat that result in the ignition of a fuel. They can all transfer heat, cause the temperature of a substance to increase, and are most frequently the ignition sources of structure fires.

Chemical Energy

Chemical energy is the most common source of heat in combustion reactions. The potential for oxidation exists when any combustible fuel is in contact with oxygen. The oxidation process almost always results in the production of thermal energy **(Figure 1.13)**.

Self-heating, a form of oxidation, is a chemical reaction that increases the temperature of a material without the addition of external heat. Self-heating can lead to **spontaneous ignition**, which is ignition without the addition of external heat.

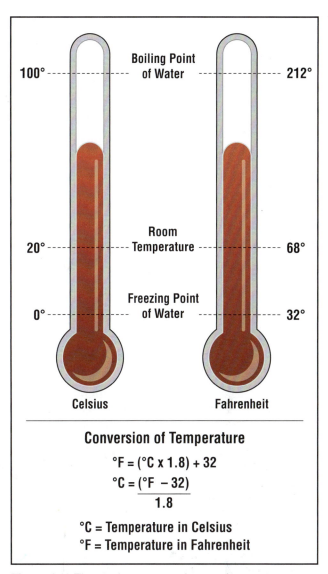

Figure 1.12 The two common scales used to measure temperature are the Celsius scale (International System of Units [SI or metric system]) and the Fahrenheit scale used in the Customary System.

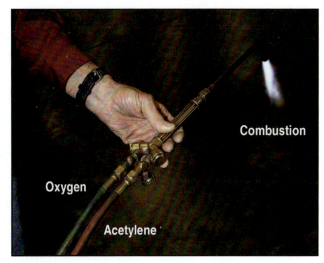

Figure 1.13 The flame of a cutting torch illustrates the generation of heat from a chemical reaction.

Oxidation normally produces thermal energy slowly. The energy dissipates almost as fast as it is generated. An external heat source such as sunshine can initiate or accelerate the process. For self-heating to progress to spontaneous ignition, the following factors are required:

- The insulation properties of the material immediately surrounding the fuel must be such that the heat cannot dissipate as fast as it is generated.
- The rate of heat production must be great enough to raise the temperature of the material to its autoignition temperature.
- The available air supply in and around the heated material must be adequate to support combustion.

Rags soaked in linseed oil, rolled into a ball, and thrown into a corner have the potential for spontaneous ignition. The natural oxidation of this vegetable oil and the cloth will generate heat if some method of heat transfer such as air movement around the rags does not dissipate the heat. The cloth could eventually increase in temperature enough to cause ignition.

The rate of most chemical reactions increases as the temperature of the reacting materials increases. The oxidation reaction that causes heat generation accelerates as the fuel generates and absorbs more heat. When the heat generated exceeds the heat being lost, the material may reach its autoignition temperature and ignite spontaneously. **Table 1.5** lists some common materials that are subject to self-heating.

Table 1.5
Spontaneous Heating Materials and Locations

Type of Material	Possible Locations
Charcoal	Convenience stores
	Hardware stores
	Industrial plants
	Restaurants
	Residences
Linseed oil-soaked rags	Woodworking shops
	Lumber yards
	Furniture repair shops
	Picture frame shops
	Residential/Commercial
	Construction/Remodeling sites
Hay and manure	Farms
	Feed stores
	Arenas
	Feedlots

Electrical Energy

Electrical energy can generate temperatures high enough to ignite any combustible materials near the heated area. Electrical heating can occur in several ways, including the following **(Figure 1.14)**:

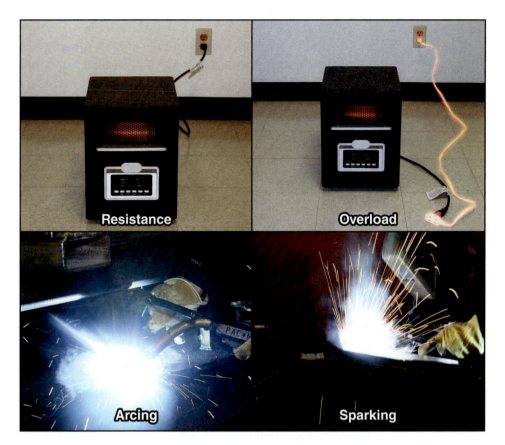

Figure 1.14 Examples of resistance heating, overcurrent or overload heating, arcing, and sparking.

- **Resistance heating** — Electric current flowing through a conductor produces heat. Some electrical appliances, such as incandescent lamps, ranges, ovens, or portable heaters, are designed to make use of resistance heating. Other electrical equipment is designed to limit resistance heating under normal operating conditions.

- **Overcurrent or overload** — When the current flowing through a conductor exceeds its design limits, the conductor may overheat and present an ignition hazard. Overcurrent or overload is unintended resistance heating.

- **Arcing** — In general, an arc is a high-temperature luminous electric discharge across a gap or through a medium such as charred insulation. Arcs may be generated when there is a gap in a conductor such as a cut or frayed wire or when there is high voltage, static electricity, or lightning.

- **Sparking** — When an electric arc occurs, luminous (glowing) particles can form and splatter away from the point of arcing.

Mechanical Energy

Friction or compression generates mechanical energy **(Figure 1.15, p. 26)**. The movement of two surfaces against each other creates heat of friction that generates heat and/or sparks. Heat is generated when a gas is compressed. Diesel engines use this principle to ignite fuel vapors without spark plugs. This principle is also the reason that SCBA cylinders feel warm to the touch after they are filled. When a compressed gas expands, the gas absorbs heat. This absorption accounts for the way the cylinder cools when a CO_2 extinguisher is discharged.

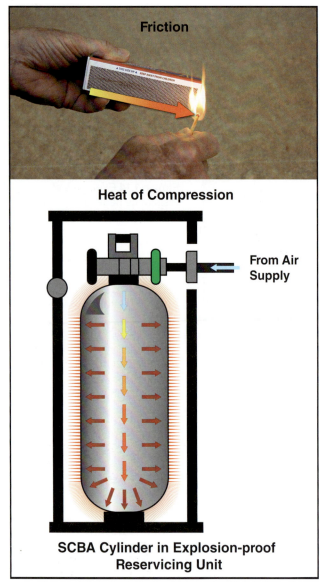

Figure 1.15 The friction of the match head rubbing across the box's striker generates the heat needed to ignite the match.

Heat Transfer

The transfer of heat from one point or object to another is part of the study of thermodynamics. Heat transfer from the initial fuel package (burning object) to other fuels in and beyond the area of fire origin affects the growth of any fire and is part of the study of fire dynamics. Heat transfers from warmer objects to cooler objects because heated materials will naturally return to a state of **thermal equilibrium** in which all areas of an object are a uniform temperature. Objects at the same temperature do not transfer heat.

The rate at which heat transfers is related to the temperature differential of the bodies and the **thermal conductivity** of the materials involved. The greater the temperature differences between the bodies, the greater the transfer rate. A material with higher thermal conductivity will transfer heat more quickly than other materials. Heat transfers from one body to another by three mechanisms: **conduction**, **convection**, and **radiation**.

Conduction

Conduction is the transfer of heat through and between solids. Conduction occurs when a material is heated as a result of direct contact with a heat source **(Figure 1.16)**. Conduction results from increased molecular motion and collisions between a substance's molecules, resulting in the transfer of energy through the substance. The more closely packed the molecules of a substance are, the more readily it will conduct heat. For example, if a fire heats a metal pipe on one side of a wall, heat conducted through the pipe can ignite wooden framing components in the wall or nearby combustibles on the other side of the wall.

Heat transfer due to conduction is dependent upon three factors:

Figure 1.16 Conduction occurs when heat is transferred between solid objects, in this case, between the door and the firefighter's hand.

- Area being heated
- Temperature difference between the heat source and the material being heated
- Thermal conductivity of the heated material

Table 1.6 shows the thermal conductivity of various common materials at the same ambient temperature (68°F [20°C]). For example, copper will conduct heat more than seven times faster than steel. Likewise, steel is nearly forty times as thermally conductive as concrete. Air is the least able to conduct heat of most substances, so it is a very good insulator.

Thermal Equilibrium — The point at which two regions that are in thermal contact no longer transfer heat between them because they have reached the same temperature.

Thermal Conductivity — The propensity of a material to conduct heat within its volume. Measured in energy transfer over distance per degree of temperature.

Table 1.6
Thermal Conductivity of Common Substances

Substance	Temperature	Thermal Conductivity (W/mK)
Copper	68°F (20°C)	386.00
Steel	68°F (20°C)	36.00 – 54.00
Concrete	68°F (20°C)	0.8 – 1.28
Gypsum Wall Board	68°F (20°C)	0.5
Wood (pine)	68°F (20°C)	0.13
Air	68°F (20°C)	0.03

Insulating materials slow the conduction of heat from one solid to another. Good insulators are materials that do not conduct heat well because their physical makeup disrupts the point-to-point transfer of heat or thermal energy. The best commercial insulators used in building construction are those made of fine particles or fibers with void spaces between them filled with a gas such as air. Gases do not conduct heat very well because their molecules are relatively far apart.

Conduction — Physical flow or transfer of heat energy from one body to another, through direct contact or an intervening medium, from the point where the heat is produced to another location, or from a region of high temperature to a region of low temperature.

Convection — Transfer of heat by the movement of heated fluids or gases, usually in an upward direction.

Radiation — Transmission or transfer of heat energy from one body to another body at a lower temperature through intervening space by electromagnetic waves.

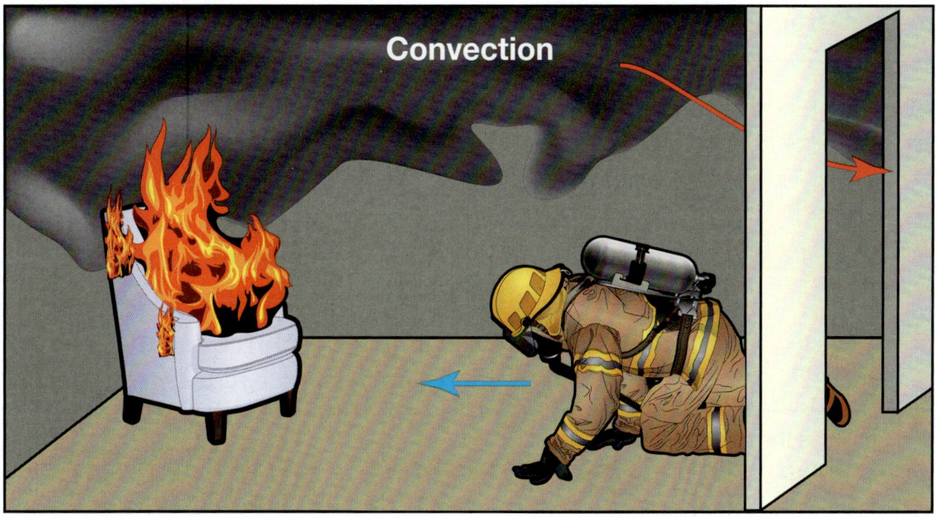

Convection
Convection is the transfer of thermal energy by the circulation or movement of a fluid (liquid or gas) **(Figure 1.17)**. In the fire environment, convection

Figure 1.17 Convection is the transfer of heat by the circulation of liquids or gases.

usually involves transfer of heat through the movement of hot smoke and fire gases. As with all heat transfer, the heat flows from the hot fire gases to the cooler structural surfaces, building contents, and air. Convection may occur in any direction. Vertical movement is due to the buoyancy of smoke and fire gases. Lateral movement is usually the result of pressure differences (movement from high to low pressure).

Heat transfer due to convection is dependent upon three factors:

- Area being heated
- Temperature difference between the hot fluid or gas and the material being heated
- Turbulence and velocity of moving gases

Radiation

Radiation is the transmission of energy as electromagnetic waves, such as light waves, radio waves, or X-rays, without an intervening medium **(Figure 1.18)**.

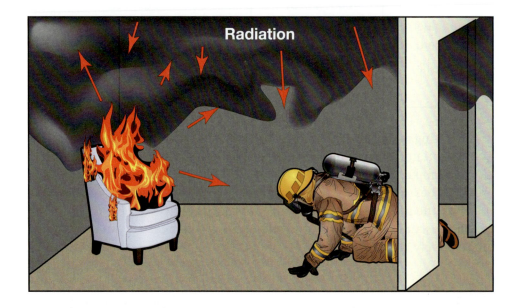

Figure 1.18 Radiation is the transfer of heat by electromagnetic waves without another medium to transfer the heat energy.

Radiant heat can become the dominant mode of heat transfer as the fire grows in size and can have a significant effect on the ignition of objects located some distance from the fire. Radiant heat transfer is also a significant factor in fire development and spread in compartments.

Numerous factors influence radiant heat transfer, including:

- **Nature of the exposed surfaces** — Dark-colored materials emit and absorb heat more effectively than light-colored materials; smooth or highly-polished surfaces reflect more radiant heat than rough surfaces.
- **Distance between the heat source and the exposed surfaces** — Increasing distance reduces the effect of radiant heat **(Figure 1.19)**.

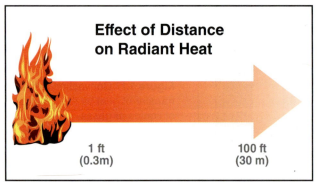

Figure 1.19 The effects of radiant heat diminish as the distance between the origin point and an exposure increases.

28 Chapter 1 • Fire Dynamics

- **Temperature of the heat source** — Unlike other methods of heat transfer that depend on the temperature of both the heat source and exposed surface, radiant heat transfer primarily depends on the temperature of the heat source. As the temperature and heat release rate of the heat source increases, the radiant energy also increases **(Figure 1.20)**.

Figure 1.20 As the heat release rate or temperature of the source increases, the thermal radiation given off will also increase. The fire on the right is giving off more thermal radiation than the fire on the left.

As an electromagnetic wave, radiated heat energy travels in a straight line at the speed of light. The heat of the sun is the best example of radiated heat transfer. The energy travels at the speed of light from the sun through space (a vacuum) until it strikes and warms the surface of the earth.

Radiation is a common cause of **exposure fires**. As a fire grows, it radiates more energy which other objects absorb as heat. In large fires, it is possible for the radiated heat to ignite buildings or other fuel packages a considerable distance away. Radiated heat travels through vacuums and air spaces that would normally disrupt conduction or convection. However, materials that reflect, absorb, or scatter radiated energy will disrupt the heat transmission. While flames have high temperature resulting in significant radiant energy emission, hot smoke or flames in the upper layer can also radiate significant heat.

Exposure Fire — A fire ignited in fuel packages or buildings that are remote from the initial fuel package or building of origin.

The Importance of Understanding Temperature and Heat Transfer Rate

Heat flux (kW/m²) from radiated heat emitted from flames or hot surfaces such as the walls and ceiling may cause PPE failure even when the temperature of the gases within a compartment are within acceptable limits. Traditionally, firefighters have focused on the gas temperature, stated in degrees on the Fahrenheit or Celsius scale, within a compartment that is on fire as being the best indicator of the thermal hazard. However, National Institute of Standards and Technology (NIST) laboratory tests show that these temperature measurements may not accurately account for radiated heat. SCBA facepieces, especially, are susceptible to radiated heat flux (Putorti, 2013).

Personal protective equipment (PPE) is designed to insulate the wearer from a specified amount of heat long enough to extinguish the fire or exit the compartment under a limited set of conditions. PPE will not protect you indefinitely. While temperature measurements are a useful tool, relying upon personal, situational awareness "in the moment" is still essential for monitoring PPE's performance during operations.

Interaction among the Methods of Heat Transfer

The methods of heat transfer rarely occur individually during a fire. The fire radiates heat, causes convection of heat through hot fuel gases, and conducts heat through burning materials or metals that are involved in the fire.

Convected heat and radiated heat that reaches walls and ceilings. heats those surfaces which, in turn, begin to conduct heat to whatever extent possible based upon the material's thermal conductivity. One side of the object is warm and slowly warms through the object until the opposite side is of equal temperature with the heated side.

A heated surface will then, in turn, begin to radiate heat which could lead to ignition, combustion, convection, and so on. This cycle continues until interrupted.

A good example of this interaction is how your PPE absorbs heat during interior operations. Convected and radiated heat will begin to heat the exterior of your PPE. The longer you are in the heated environment, the more heat that surface will absorb. The PPE has low thermal conductivity, so it will conduct heat slowly. However, eventually the interior surface of the PPE will heat to the same level as the exterior. Wherever the gear is compressed against skin or underclothing, heat will be conducted faster. **Table 1.7** shows various responses of human skin and PPE as they are heated.

Where the PPE is not in contact, it will radiate heat to the insulating air layer between your body and the interior surface of the gear. This transferred heat can cause heat stress and will eventually cause PPE to fail. The heat absorption and build-up in PPE is a direct result of all of the heat transfer methods acting at the same time.

Table 1.7 Response to Temperature of an Object Being Heated	
Temperature °F (°C)	**Response**
98.6 °F (37 °C)	Normal human oral/body temperature
111 °F (44 °C)	Human skin begins to feel pain
118 °F (48 °C)	Human skins receives a first degree burn injury
131°F (55 °C)	Human skins receives a second degree burn injury
140 °F (62 °C)	A phase where burned human tissue becomes numb
162 °F (72 °C)	Human skin is instantly destroyed
212 °F (100 °C)	Water boils and produces steam
284 °F (140 °C)	Glass transition temperature of polycarbonate
446 °F (230 °C)	Melting temperature of polycarbonate
482 °F (250 °C)	Charring of natural cotton begins
>572 °F (>300 °C)	Charring of modern protective clothing fabrics begins
>1,112 °F (>600 °C)	Temperatures inside a post-flashover room fire

Source: UL, LLC.

Fuel

Fuel is the oxidized or burned material or substance in the combustion process. A fuel may be found in any of three physical states of matter: gas, liquid, or solid.

The fuel in a combustion reaction is known as the **reducing agent**. Fuels may be inorganic or organic. Inorganic fuels, such as hydrogen or magnesium, do not contain carbon. Most common fuels are organic, containing carbon and other elements. Organic fuels can be divided into hydrocarbon-based fuels, such as:

- Gasoline
- Fuel oil
- Plastics
- Cellulose-based materials (wood and paper)

A fuel's chemical content influences both its heat of combustion and heat release rate. The fuel's heat of combustion is the total amount of thermal energy released when a specific amount of that fuel burns. In other words, different materials release more or less heat than others based on their chemical makeup. Many plastics, flammable liquids, and flammable gases contain more potential thermal energy than wood **(Table 1.8)**.

> **Reducing Agent** — Fuel that is being oxidized or burned during combustion.

Table 1.8
Representative Peak Heat Release Rates (HRR) During Unconfined Burning

Fuel Material	Peak HRR in kilowatts	Common Locations for Material
Small wastebasket	4-50	Homes, businesses, shops
Cotton mattress	40-970	Homes, furniture stores, motels
Cotton easy chair	290-370	Homes, furniture stores, office buildings
Small pool of gasoline	400	Traffic crash, fuel stations
Dry Christmas tree	3000-5000	Homes, trash facilities, dumpsters, recycling sites
Polyurethane mattress	810-2630	Homes, furniture stores, motels, dormitories, jails
Polyurethane easy chair	1350-1990	Homes, furniture stores, motels
Polyurethane sofa	3120	Homes, furniture stores, motels, dormitories, office buildings

Adapted from NFPA® 921, 2014 edition

Synthetic materials are common in modern construction and furnishings. These materials are synthesized from petroleum products, and as a result, they have higher heats of combustion and may generate higher heat release rates than wood on a per-mass basis.

Power is the rate at which energy transfers. Another way to describe power is the rate at which energy converts from one form to another. The standard international (SI) unit for power is the **watt** (W). One watt is 1 joule per second (J/s).

In terms of fire behavior, power is the heat release rate during combustion. When a fuel is heated, work is being performed (energy is being transferred). The speed with which this work occurs, heat release rate, is the amount of generated power. Heat release rate is the energy released per unit of time as a fuel burns and is usually expressed in kilowatts (kW) or megawatts (MW). Heat release rate depends on the type, quantity, and orientation of the fuel. Heat release rate directly relates to oxygen consumption because the combustion process requires a continuous supply of oxygen to continue. Typically, the more oxygen is available, the higher the heat release rate. Similarly, the heat release rate decreases if all available oxygen is consumed and not replenished. **Figure 1.21** shows an example of the heat release rate produced by various sizes of fires.

> **Power** — Amount of energy delivered over a given period of time.
>
> **Watt (W)** — The SI unit of power or rate of work equal to 1 joule per second (J/s).

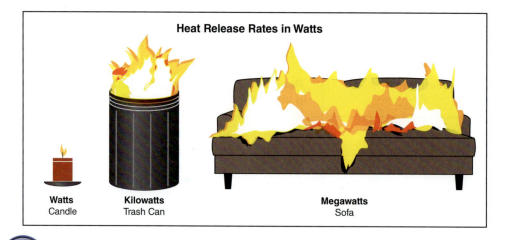

Figure 1.21 Examples of heat release rate conditions that may be measured in watts, kilowatts, or megawatts.

Prefixes for Units of Measure: Kilo and Mega
The standard international system of units (SI) specifies a set of prefixes that precede units of measure to indicate a multiple or fraction of that unit. Two common prefixes encountered when discussing energy (joules) and heat release rate (watts) are *kilo* and *mega*. The prefix *kilo* indicates a multiple of thousands and *mega* indicates a multiple of millions. For example, a kilowatt is 1 thousand watts and a megawatt is 1 million watts.

Gases

For flaming combustion to occur, fuels must be in the gaseous state. As previously described, thermal energy is required to change solids and liquids into the gaseous state. *Vapor* is the common term used to describe the gaseous state of a fuel that would normally exist as a liquid or a solid at standard temperature and pressure.

Gaseous fuels such as methane (natural gas), hydrogen, and acetylene, can be the most dangerous of all fuel types because they are already in the physical state required for ignition. When wood burns inefficiently, the combustion products may contain methane, acetylene and other fuel gases. **Table 1.9** contains characteristics of common flammable gases.

Table 1.9
Characteristics of Common Flammable Gases

Material	Vapor Density	Ignition Temperature
Methane (Natural Gas)	0.55	(1004°F) 540°C
Propane (Liquefied Petroleum Gas)	1.52	(842°F) 450°C
Carbon Monoxide	0.96	(1,128°F) 620°C

Source: *Computer Aided Management of Emergency Operations* (CAMEO)

Vapor density describes the density of gases in relation to air. Air has a vapor density of 1. Gases with a vapor density of less than 1, such as methane, will rise while those having a vapor density of greater than 1, such as propane, will sink **(Figure 1.22)**. Vapor densities are based upon the assumption that the density is measured at standard temperature and pressure. Heated gases expand and become less dense; when cooled they contract and become more dense.

> **Vapor Density** — Weight of pure vapor or gas compared to the weight of an equal volume of dry air at the same temperature and pressure. A vapor density less than 1 indicates a vapor lighter than air; a vapor density greater than 1 indicates a vapor heavier than air.

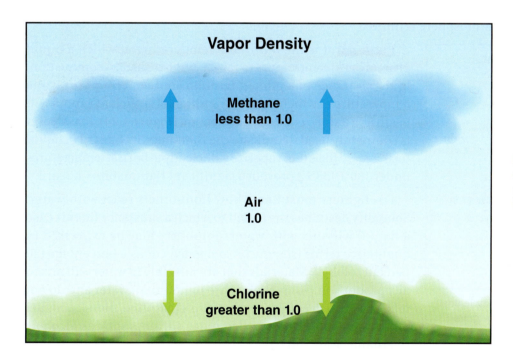

Figure 1.22 The vapor density of a gas provides an indication of where a gas will collect at an incident.

Liquids

Liquids have mass and volume but no definite shape except for a flat surface or the shape of their container. Unlike gases, liquids will not expand to fill all of a container. When released on the ground, liquids will flow downhill and pool in low areas. Just as gas density is compared to air, liquid density

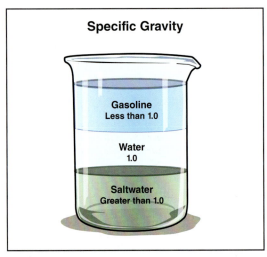

Figure 1.23 The specific gravity of a liquid indicates whether the liquid will float on the surface of water or sink.

is compared to water. **Specific gravity** is the ratio of the mass of a given volume of a liquid compared to the mass of an equal volume of water at the same temperature. Water is assigned a specific gravity of 1. Liquids with a specific gravity less than 1, such as gasoline and most other flammable liquids, are lighter than water and will float on its surface. Liquids with a specific gravity greater than 1, such as corn syrup, are heavier than water and will sink **(Figure 1.23)**.

To burn, liquids must vaporize. Vaporization is the transformation of a liquid to vapor or a gaseous state. Unlike solids, liquids retain their state of matter partly due to standard atmospheric pressure. For vaporization to occur, the escaping vapors must be at a greater pressure than atmospheric pressure. The pressure that vapors escaping from a liquid exert is known as **vapor pressure**. Vapor pressure indicates how easily a substance will evaporate into air. Flammable liquids with a high vapor pressure present a special hazard to firefighters.

The vapor pressure of the substance and the amount of thermal energy applied to it determines the rate of vaporization. For example, a puddle of water eventually evaporates because of slow heat transfer from the sun. When the same amount of water is heated on a stove, however, it vaporizes much more rapidly because there is more thermal energy applied. The volatility or ease with which a liquid gives off vapor influences how easily it can ignite. The size of a liquid's surface area also influences the extent to which the liquid will give off vapor. In many open containers, the surface area of liquid exposed to the atmosphere is limited.

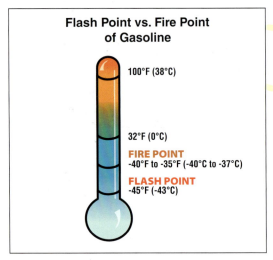

Figure 1.24 The flashpoint of a liquid indicates the temperature at which the liquid will ignite temporarily, while the fire point indicates the temperature at which the liquid, once ignited, will continue to burn.

Flash point is the minimum temperature at which a liquid gives off sufficient vapors to ignite, but not sustain combustion, in the presence of a piloted ignition source. **Fire point** is the temperature at which a piloted ignition of sufficient vapors will begin a sustained combustion reaction **(Figure 1.24)**. Flash point is commonly used to indicate the flammability hazard of liquid fuels. Liquid fuels that vaporize sufficiently to burn at temperatures under 100°F (38°C) present a significant flammability hazard.

Firefighters must know how liquid fuels react with water. Solubility describes the extent to which a substance (in this case a liquid) will mix with water. **Solubility** may be expressed in qualitative terms (*slightly* or *completely*) or as a percentage (20 percent soluble). Materials that are **miscible** in water will mix in any proportion. Some liquids are lighter than water and do not mix with it, such as hydrocarbon fuels (gasoline, diesel, and fuel oil). Flammable liquids called **polar solvents** such as alcohols (methanol, ethanol) will mix readily with water.

Liquids that are less dense (lighter) than water are more difficult to extinguish using water as the sole extinguishing agent. Because the liquid fuel is less dense and will not mix with water, adding water to the liquid fuel may disperse the burning liquid instead of extinguishing it, which could potentially spread the fire to other areas. Firefighters should use the appropriate foam or chemical agent to extinguish liquid fuels that are not water-soluble.

Water-soluble liquids will mix with some water-based extinguishing agents, such as many types of fire fighting foam. The extinguishing agent will mix with the burning liquid and become much less effective at extinguishing the fire. To avoid this mixture, firefighters should use alcohol-resistant fire fighting foams specifically designed for polar solvents. **Table 1.10** lists the characteristics of common flammable and combustible liquids.

Table 1.10
Characteristics of Common Flammable and Combustible Liquids

Material	Water Soluble	Specific Gravity	Flash Point	Autoignition Temperature
Gasoline	No	0.72	(-36°F) -38°C	(853°F) 486°C
Diesel	No	<1.00	(125°F) 52°C	(410°F) 210°C
Ethanol	Yes	0.78	(55°F) 13°C	(689°F) 365°C
Methanol	Yes	0.79	(52°F) 11°C	(867°F) 464°C

Source: *Computer Aided Management of Emergency Operations* (CAMEO)

Solids

Solids have definite size and shape. Different solids react differently when exposed to heat. Some solids such as wax and metals will change their state and melt, while others such as wood and plastics will not. When solid fuels are heated, they begin to pyrolize (off-gas) and release fuel gases and vapors The solid fuels begin to decompose and emit combustible vapors. If there is enough fuel and heat, the process of pyrolysis generates sufficient flammable vapors to ignite in the presence of sufficient oxygen or another oxidizer.

When wood first heats, it begins to pyrolize and decompose into its volatile components and carbon. These vapors are usually white in color. Pyrolysis of wood begins at temperatures below 400°F (204°C), which is lower than the temperature required for ignition of the released vapors. Home construction still uses wood based products for the framing, subfloor and roof decking. However the insulation, exterior and interior finish materials, and the contents are likely to be or contain synthetic materials, like polyvinyl chloride, polyethylene, polystyrene, polypropylene, and polyurethane. Today, flexible polyurethane foam is one of the most common materials used in upholstered furniture. **Table 1.11, p. 36** shows how the pyrolysis process of wood and polyurethane foam is different.

Solid fuels have a definite shape and size which significantly affects how easily they ignite. The primary consideration is the surface area of the fuel in proportion to its mass, called the **surface-to-mass ratio**. One of the best examples is that of a large tree:

1. To produce lumber, the tree must be felled and cut into a log. The surface area of this log is low compared to its mass; therefore, the surface-to-mass ratio is low.

2. The log is then cut into planks. This reduces the mass of the individual planks compared to the log. The resulting surface area increases, thus increasing the surface-to-mass ratio.

Specific Gravity — Mass (weight) of a substance compared to the weight of an equal volume of water at a given temperature. A specific gravity less than 1 indicates a substance lighter than water; a specific gravity greater than 1 indicates a substance heavier than water.

Vapor Pressure — The pressure at which a vapor is in equilibrium with its liquid phase at a given temperature; liquids that have a greater tendency to evaporate have higher vapor pressures at a given temperature.

Flash Point — Minimum temperature at which a liquid gives off enough vapors to form an ignitable mixture with air near the surface of the liquid.

Fire Point — Temperature at which a liquid fuel produces sufficient vapors to support combustion once the fuel ignites. The fire point is usually a few degrees above the flash point.

Solubility — Degree to which a solid, liquid, or gas dissolves in a solvent (usually water).

Miscible — Materials that are capable of being mixed in all proportions.

Polar Solvents — Flammable liquids that have an attraction to water, much like a positive magnetic pole attracts a negative pole; examples include alcohols, esters, ketones, amines, and lacquers.

Surface-To-Mass Ratio — Ratio of the surface area of the fuel to the mass of the fuel.

Table 1.11
Pyrolysis of Wood and Polyurethane Foam

Wood	Polyurethane Foam (PUF)
Stage 1 **Temperature:** Less than 392° F (200° C) **Physical and Chemical Changes:** Moisture is released as the wood begins to dry; combustible and noncombustible materials are released to the atmosphere although there is insufficient heat to ignite them.	**Stage 1** **Temperature:** Less than 392° F (200° C) **Physical and Chemical Changes:** As flexible polyurethane foam (PUF) thermally degrades (pyrolyzes), it transforms into combustible gases and liquid.
Stage 2 **Temperature:** 392° F – 536° F (200° C) – (280° C) **Physical and Chemical Changes:** The majority of the moisture has been released; charring has begun; the primary compound being released is carbon monoxide (CO); ignition has yet to occur.	**Stage 2** **Temperature:** 392° F – 536° F (200° C) – (280° C) **Physical and Chemical Changes:** As the liquid polyols continue to be heated, they will also vaporize into combustible gases, as well. Ignition of these gases may occur in this stage.
Stage 3 **Temperature:** 536° F – 932° F (280° C) – (500° C) **Physical and Chemical Changes:** Rapid pyrolysis takes place; combustible compounds are released and ignition can occur; charcoal is formed by the burning process.	**Stage 3** **Temperature:** 536° F – 932° F (280° C) – (500° C) **Physical and Chemical Changes:** Pyrolysis continues at an increased rate. Ignition of PUF occurs 698° F (370° C). Auto-ignition of PUF can occur at temperatures in the range of 797° F to 833° F (425° to 445° C). No char layer is formed.
Stage 4 **Temperature:** Greater than 932° F (500° C) **Physical and Chemical Changes:** Free burning exists as the wood material is converted to flammable gases.	

Sources: Wood data adapted from NFPA *Fire Protection Handbook®*, 19th edition, Volume II, pages 8-35 and 36. Polyurethane foam data from UL-FSRI, ASTM 1929 test (NIST NCSTAR 2); SFPE *Handbook of Fire Protection Engineering*, 4th edition; and *The Ignition Handbook* (V. Babrauskas).

3. The chips and sawdust produced as the planks are cut into boards have an even higher surface-to-mass ratio.

4. If the boards are milled or sanded, the shavings or sawdust have the highest surface-to-mass ratio of any of the examples.

As this ratio increases, the fuel particles become more finely divided like shavings or sawdust. Therefore, the particles' ability to ignite increases tremendously. As the surface area increases, more of the material is exposed to the heat and generates combustible pyrolysis products more quickly **(Figure 1.25)**.

The proximity and orientation of a solid fuel relative to the source of heat also affects the way the fuel burns **(Figure 1.26)**. For example, if you ignite one corner of a sheet of ⅛-inch (3 mm) plywood paneling that is lying horizontally (flat), the fire will consume the fuel at a relatively slow rate. The same type of paneling in a vertical position (standing on edge) burns much more rapidly because the heated vapors rise over more surface area and transfer more heat to the paneling.

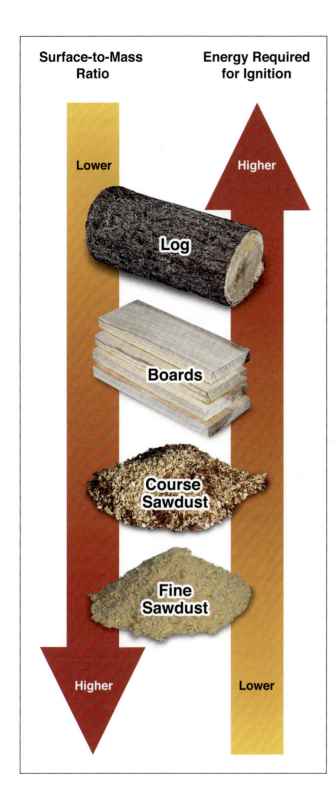
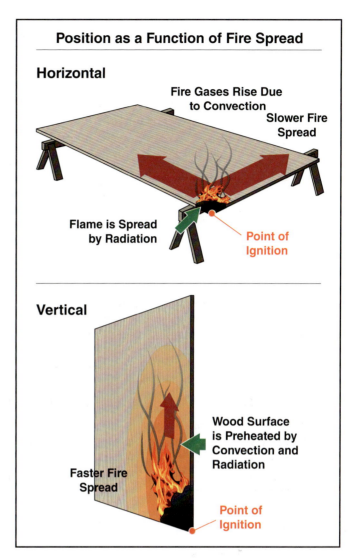

Figure 1.25 (left) As the surface-to-mass ratio of a fuel becomes higher (increases), the energy required for ignition is lower (reduced).

Figure 1.26 (above) This illustration demonstrates how the position (orientation) of a fuel impacts fire spread.

Oxygen

Oxygen in the air is the primary oxidizing agent in most fires. Normally, air consists of about 21 percent oxygen. The energy release in fire is directly proportional to the amount of oxygen available for combustion. When a fire ignites in an open area where air is plentiful, the fire will release energy based on the given surface area. In contrast, when a fire ignites within a compartment with limited air supply the fire can only react with oxygen from the compartment's air and any additional oxygen supplied through openings. Thus, in most compartment fires, the energy released is proportional to the limited amount of oxygen available, not the amount of fuel available to burn.

At normal ambient temperatures (68°F [20°C]), materials can ignite and burn at oxygen concentrations as low as 15 percent. When oxygen concentration is limited, the flaming combustion will diminish, causing combustion to continue in the nonflaming mode. Nonflaming or smoldering combustion can continue at extremely low oxygen concentrations even when the surrounding environment's temperature is relatively low. However, at high ambient temperatures, flaming combustion may continue at considerably lower oxygen concentrations.

Effects of Oxygen Concentration

Oxygen concentration in the atmosphere has a significant effect on both fire behavior and our ability to survive. Typically, an atmosphere having less than 19.5 percent oxygen is considered oxygen deficient and presents a hazard to persons not wearing respiratory protection, such as SCBA, to provide fresh air. When the oxygen concentration in the atmosphere exceeds 23.5 percent, the atmosphere is considered oxygen enriched and presents an increased fire risk.

Flammable (Explosive) Range — Range between the upper flammable limit and lower flammable limit in which a substance can ignite.

Lower Explosive (Flammable) Limit (LEL) — Lower limit at which a flammable gas or vapor will ignite and support combustion; below this limit the gas or vapor is too *lean* or *thin* to burn (lacks the proper quantity of fuel). *Also known as* Lower Flammable Limit (LFL).

Upper Explosive (Flammable) Limit (UEL) — Upper limit at which a flammable gas or vapor will ignite; above this limit the gas or vapor is too *rich* to burn (lacks the proper quantity of oxygen). *Also known as* Upper Flammable Limit (UFL).

When the oxygen concentration is higher than normal, materials exhibit different burning characteristics. Materials that burn at normal oxygen levels will burn more intensely and may ignite more readily in oxygen-enriched atmospheres. Some petroleum-based materials will autoignite in oxygen-enriched atmospheres.

Many materials that do not burn at normal oxygen levels will burn in oxygen-enriched atmospheres. One such material is Nomex® fire-resistant fabric, which is used in many types of protective clothing. At normal oxygen levels, Nomex® does not burn. When placed in an oxygen-enriched atmosphere of approximately 31 percent oxygen, Nomex® ignites and burns vigorously.

Fires in oxygen-enriched atmospheres are more difficult to extinguish and present a potential safety hazard. Firefighters may find these conditions in hospitals and other healthcare facilities, some industrial occupancies, and even private homes where occupants use breathing equipment containing pure oxygen.

For combustion to occur after a fuel converts into a gaseous state, the fuel must be mixed with air (an oxidizer) in the proper ratio. The range of concentrations of the fuel vapor and air is called the **flammable (explosive) range**. The fuel's flammable range is reported using the percent by volume of gas or vapor in air for the **lower explosive (flammable) limit (LEL)** and for the **upper explosive (flammable) limit (UEL)**. The LEL is the minimum concentration of fuel vapor and air that supports combustion. Concentrations below the LEL are said to be *too lean* to burn. The UEL is the concentration above which combustion cannot take place. Concentrations above the UEL are said to be *too rich* to burn. Within the flammable range, there is an ideal concentration at which there is exactly the correct amount of fuel and oxygen required for combustion **(Figure 1.27)**.

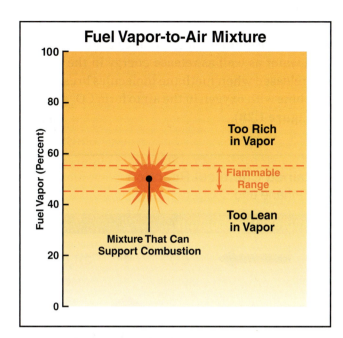

Figure 1.27 The flammable range is a relatively narrow band of conditions at which a mixture of fuel vapors and air will burn.

Table 1.12 Flammable Ranges of Common Flammable Gases and Liquids (Vapor)	
Substance	Flammable Range
Methane	5%–15%
Propane	2.1%–9.5%
Carbon Monoxide	12%–75%
Gasoline	1.4%–7.4%
Diesel	1.3%–6%
Ethanol	3.3%–19%
Methanol	6%–35.5%

Source: *Computer-Aided Management of Emergency Operations* (CAMEO)

Table 1.12 presents the flammable ranges for some common materials. Chemical handbooks and documents such as the National Fire Protection Association (NFPA) *Fire Protection Guide to Hazardous Materials* present the flammable limits for combustible gases. The Guide and other sources normally reports the limits at standard temperature and atmospheric pressures. Variations in temperature and pressure can cause the flammable range to vary considerably.

Self-Sustained Chemical Reaction

The self-sustained chemical reaction involved in flaming combustion is complex. As flaming combustion occurs, the molecules of a fuel gas and oxygen (O_2) break apart to form **free radicals** (electrically charged, highly reactive parts of molecules). Free radicals combine with oxygen or with the elements released from the fuel gas to form new substances (molecules) and even more free radicals. The process also increases the speed of the oxidation reaction.

Free Radical — Electrically charged, highly reactive parts of molecules released during combustion reactions.

The combustion of a simple fuel such as methane and oxygen provides a good example. Complete oxidation of methane releases the elements needed to create carbon dioxide and water as well as release energy in the form of heat and light. The elements released when methane molecules break down (carbon and hydrogen) recombine with oxygen in the air to form CO_2 and H_2O (carbon dioxide and water) **(Figure 1.28)**.

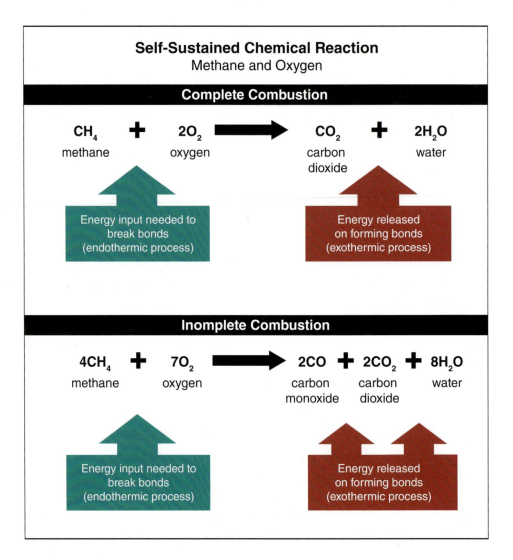

Figure 1.28 Illustrating the concepts of complete and incomplete combustion of methane.

At various points in the combustion of methane, this process results in production of carbon monoxide and formaldehyde, which are both flammable and toxic. When more chemically complex fuels burn, their combustion creates different types of free radicals and intermediate combustion products, many of which are also flammable and toxic.

Flaming combustion is one example of a chemical chain reaction. Sufficient heat will cause fuel and oxygen to form free radicals and initiate the self-sustained chemical reaction. The fire will continue to burn until it consumes the fuel or oxygen or an extinguishing agent, applied in sufficient quantity, interferes with the ongoing reaction. **Chemical flame inhibition** occurs when an extinguishing agent, such as dry chemical or Halon-replacement agent, interferes with this chemical reaction, forms a stable product, and terminates the combustion reaction.

Chemical Flame Inhibition — Extinguishment of a fire by interruption of the chemical chain reaction.

Compartment Fire Development

Typically, when we think about a fire, we tend to limit our perspective to the burning fuel itself. However, in the sections that follow, we will see that the compartment surrounding that burning fuel has a significant impact on the available ventilation, access to additional fuel, and heat losses or gains.

Compartment fire development depends upon whether the fire is **fuel-limited** or **ventilation-limited**. When sufficient oxygen is available for flaming combustion, the fire is said to be fuel-limited. Under fuel-limited conditions, the fuel's characteristics such as heat release rate and configuration control fire development. As long as the fire can reach more ignitable fuel, it will continue to burn.

Conversely, ventilation-limited fires have access to all of the fuel needed to maintain combustion. However, the fire does not have access to enough oxygen to continue to burn or to spread to all available fuels.

All compartment fires begin in the incipient stage as fuel-limited fires. Once the fire reaches the growth stage, the fire will either remain fuel-limited if there is enough oxygen to support continued growth, or the fire will consume all available oxygen and become ventilation-limited. A fuel-limited fire will usually progress through the stages of fire development in order. Ventilation-limited fires tend to enter an early state of decay at the end of the growth stage because there is no longer enough available oxygen for the fire to become fully developed.

This section will define the stages of fire development and then describe the progression of a fire in a compartment. The examples in the information boxes describe fire behavior in a room with one exterior window, an exterior doorway, and typical modern furnishings found in a residential living room.

Stages of Fire Development

Fires develop through four stages: incipient, growth, fully developed, and decay. These stages can occur with any fire; however, there are three key factors that control how the fire develops: the fuel properties, the ventilation available, and heat conservation. Depending on these factors, the fire development stages exhibit different characteristics or may occur in a different sequence.

The four stages of fire development can be generally defined as follows:

- **Incipient Stage** — The incipient stage starts with ignition when the three elements of the fire triangle come together and the combustion process begins. At this point, the fire is small and confined to a small portion of the fuel first ignited.

- **Growth Stage** — As the fire transitions from incipient to growth stage, more of the initial fuel package becomes involved and the production of heat and smoke increases. If there are other fuels close to the initial fuel package, radiant heat from the fire may begin to pyrolize nearby fuels which could spread the fire to new fuel packages. The fire may continue to grow to become fully developed or may enter an early state of decay depending upon available oxygen.

- **Fully Developed Stage** — The fully developed stage occurs when all combustible materials in the compartment are burning at their peak heat release

---

Fuel-Limited — Fire with adequate oxygen in which the heat release rate and growth rate are determined by the characteristics of the fuel, such as quantity and geometry. Also known as Fuel-controlled (Reproduced with permission from NFPA 921-2011, *Guide for Fire and Explosion Investigations*, Copyright 2011, National Fire Protection Association).

Ventilation-Limited — Fire with limited ventilation in which the heat release rate or growth is limited by the amount of oxygen available to the fire. Also known as Ventilation-controlled (Reproduced with permission from NFPA 921-2011, *Guide for Fire and Explosion Investigations*, Copyright 2011, National Fire Protection Association).

rate based on the available oxygen. The fire is consuming the maximum amount of oxygen that it can. If the fire is limited to one fuel package, the fully developed stage occurs when the entire fuel package is on fire and the fire has reached its peak heat release rate.

- **Decay Stage** — As the fire consumes the available fuel or oxygen and the heat release rate begins to decline, the fire enters the decay stage. Fuel-limited fires may self-extinguish in this phase or reduce to smoldering fires. Ventilation-limited fires may also self-extinguish. However, if oxygen becomes available during the decay stage before complete extinguishment, these fires are likely to reenter the growth stage and rapidly become fully developed.

Open burning or a *free burn* condition provides the most basic fire growth curve, as shown in **Figure 1.29a and b**. Open burning is representative of a

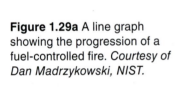

Open Burning — Description of a fire burning in the open with no restrictions to its oxygen supply.

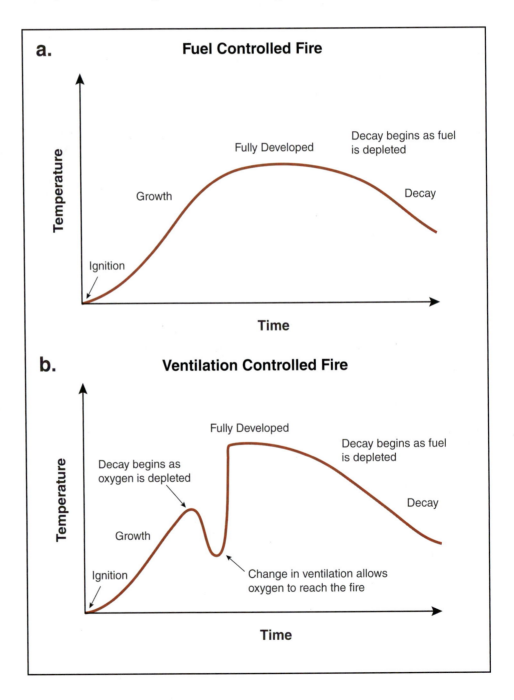

Figure 1.29a A line graph showing the progression of a fuel-controlled fire. *Courtesy of Dan Madrzykowski, NIST.*

Figure 1.29b A line graph showing the progression of a ventilation-controlled fire. *Courtesy of Dan Madrzykowski, NIST.*

fuel-limited fire, such as a campfire, a pile of wood pallets, or a sofa in a large, open, empty warehouse. This fire is considered fuel controlled because a single item burning either outside or in a large, well-ventilated space means there is sufficient oxygen available to burn the fuel until it can no longer sustain combustion. As heat and fire gases are produced, they move away from the fuel and disperse throughout the environment remote from the burning fuel. The only limit or control on the heat release rate of a fire burning out in the open is the fuel itself.

Incipient Stage

The incipient stage is where a fire begins (**Figure 1.30**). Once ignition occurs and the combustion process begins, development in the incipient stage depends largely upon the characteristics and configuration of the fuel involved (fuel-limited fire). Air in the compartment provides adequate oxygen to continue fire development. The following describe what occurs when a compartment fire enters the incipient stage:

- Radiant heat warms the adjacent fuel and continues the process of pyrolysis. A thin plume of hot gases and flame rises from the fire and mixes with the cooler air in the compartment.

- The hot gases in the plume rise until they encounter the ceiling and then begin to spread horizontally. This flow of fire gases is called the **ceiling jet**.

- Hot gases in contact with the surfaces of the compartment and its contents transfer heat to other materials.

Figure 1.30 An example of an incipient fire on a couch. *Courtesy of Dan Madrzykowski, NIST.*

Ceiling Jet — Horizontal movement of a layer of hot gases and combustion by-products from the center point of the plume, when a horizontal surface such as a ceiling redirects the vertical development of the rising plume.

In this early stage of fire development, the fire has not yet influenced the environment within the compartment to a significant extent. The temperature, while increasing, is only slightly above ambient in areas that the fire, plume, and ceiling jet directly affect. During the incipient stage, occupants can safely escape from the compartment, and a portable extinguisher or small hoseline can safely extinguish the fire.

The transition from incipient to growth stage can occur quickly (in some cases in seconds), depending on the type and configuration of fuel involved. A visual indicator that a fire is leaving the incipient stage is flame height. When flames reach 2.5 feet (750 mm) high, radiated heat begins to transfer more heat than convection. The fire will then enter the growth stage.

CAUTION
Transition from the incipient to the growth stage can occur in a matter of seconds depending upon the type and configuration of fuel.

Chapter 1 • Fire Dynamics 43

Example Compartment: Incipient Stage

To begin our example of a compartment fire's development, let's assume that a fire started in the cushions of a chair in the corner of a room. The window and the door are closed. Within the room, there is enough oxygen for the incipient fire to entrain ("pump in") air and create ("pump out") fuel gases and smoke. The fire begins to spread from the cushions to the rest of the chair. The polyurethane foam of the cushions burns quickly, creating black, fuel-rich smoke which begins to form a plume above the chair **(Figure 1.31)**.

Figure 1.31 The graph in the lower left of the photo shows the progression of the fire in its incipient stage. *Courtesy of Dan Madrzykowski, NIST.*

Growth Stage

Within the growth stage, a variety of fire behaviors can occur, depending upon the number of ventilation sources. The fire may consume all of its available oxygen and enter a ventilation-limited state of decay or ventilation may provide enough oxygen for rapid fire development and/or growth to full development. Rapid fire development usually occurs during the growth stage. Understanding fire dynamics is largely an understanding of everything that can happen during the growth stage.

NOTE: Keep in mind that if the fire enters ventilation-limited decay that does not indicate that the fire is in its final stage of development.

As the fire transitions from incipient to growth stage, it begins to influence more of the compartment's environment and has grown large enough for the compartment configuration and amount of ventilation to influence it. The first effect is the amount of air that is entrained into the fire.

Unconfined fires draw air from all sides and the **entrainment** (drawing in) of air cools the plume of hot gases, reducing flame length and vertical extension **(Figure 1.32)**. In a compartment fire, the location of the fuel package in relation to the compartment walls affects the amount of air that is entrained and thus the amount of cooling that takes place. The following tenets describe entrainment based on the positioning of fuel packages:

> **Entrainment** — The drawing in and transporting of solid particles or gases by the flow of a fluid.

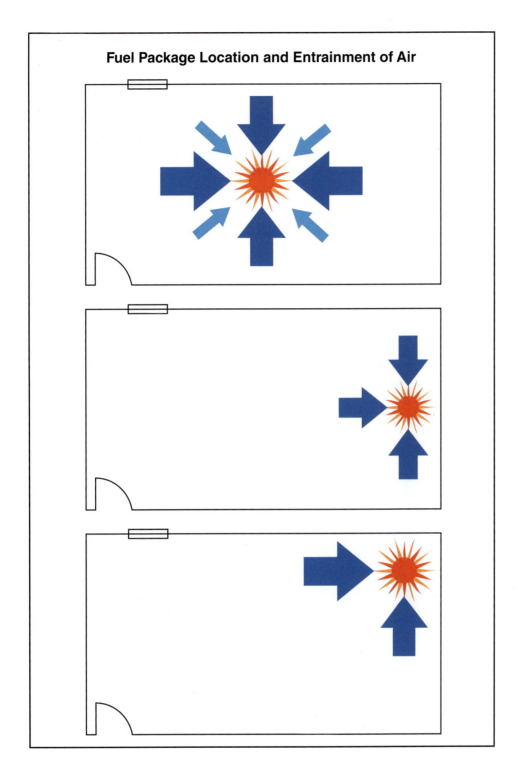

Figure 1.32 The location of a fire in a compartment influences the entrainment of air into the fire.

- Fires in fuel packages in the middle of the room can entrain air from all sides.
- Fires in fuel packages near walls can only entrain air from three sides.
- Fires in fuel packages in corners can only entrain air from two sides.

Therefore, when the fuel package is not in the middle of the room, the **combustion zone** (the area where sufficient air is available to feed the fire) expands vertically and a higher plume results. A higher plume increases the temperatures in the developing hot gas layer at ceiling level and increases

Combustion Zone — Area surrounding a heat source in which there is sufficient air available to feed a fire.

Chapter 1 • Fire Dynamics **45**

Thermal Layering — Outcome of combustion in a confined space in which gases tend to form into layers, according to temperature, gas density, and pressure with the hottest gases found at the ceiling and the coolest gases at floor level.

the speed of fire development. In addition, heated surfaces around the fire radiate heat back toward the burning fuel which further increases the speed of fire development.

A fire is said to be in the growth stage until the fire's heat release rate has reached its peak, either because of a lack of fuel or a lack of oxygen. In other words, when a fire cannot grow without the introduction of a new fuel source or a new oxygen source, it has left the growth stage and become fully developed. Two common routes to full development are as follows:

- Fires that consume all available oxygen and transition to a state of ventilation-limited decay.
- Fires that have enough oxygen and move to the growth phase and possibly through rapid fire development.

Example Compartment: Growth Stage – Transition from Incipient Stage

As the fire spreads to the chair, the entire chair becomes involved. Air entrains from only two sides, so the fire releases more energy than it would in the middle of the room and thus accelerates the fire's growth. The side table next to the chair begins to heat and pyrolize. The ceiling above the chair begins to blacken as the plume grows taller and transfers heat to the ceiling and surrounding walls. A ceiling jet begins to form **(Figure 1.33)**.

Figure 1.33 The graph in this image shows the transition of a fire from the incipient stage into the growth stage. *Courtesy of Dan Madrzykowski, NIST.*

Flow Path — The space between at least one intake and one exhaust outlet. The difference in pressure determines the direction of the flow of gases through this space. Heat and smoke in a high pressure area will flow toward areas of lower pressure.

Isolated Flames — Flames in the hot gas layer that indicate the gas layer is within its flammable range and has begun to ignite; often observed immediately before a flashover.

Neutral Plane — Level at a compartment opening where there is an equal difference in pressure exerted by expansion and buoyancy of hot smoke flowing out of the opening and the inward pressure of cooler, ambient temperature air flowing in through the opening.

Thermal Layering

Once the ceiling jet reaches the walls of the fire compartment, the hot gas layer begins to develop. **Thermal layering** is the tendency of gases to form into layers according to temperature, gas density, and pressure. Provided that there is no mechanical mixing from a fan or a hose stream, the hottest gases will form the highest layer, while the cooler gases will form the lower layers **(Figure 1.34)**. In addition to the effects of heat transfer through radiation and

Figure 1.34 (above) The image on the left shows the visible conditions within a compartment that is on fire. The infra-red image on the right shows the thermal layering of gases from the top (hottest) of the compartment to the bottom (coolest). *Courtesy of Dan Madrzykowski, NIST.*

Figure 1.35 (right) Illustrating the location of the neutral plane in a compartment fire. Hot gases are exiting through the upper part of the doorway while cooler air enters through the lower part of the doorway. *Courtesy of Dan Madrzykowski, NIST.*

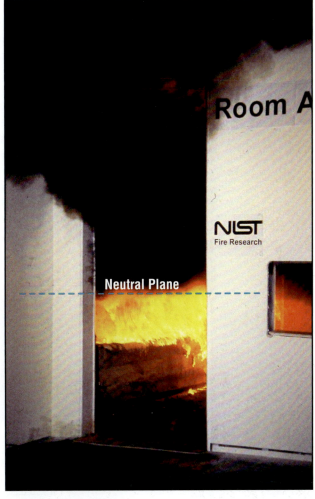

convection described earlier, radiation from the hot gas layer also acts to heat the interior surfaces of the compartment and its contents. Changes in ventilation and **flow path** can significantly alter thermal layering. The flow path is defined as the space between the air intake and the exhaust outlet. Multiple openings (intakes and exhausts) create multiple flow paths.

The products of combustion from the fire begin to affect the environment within the compartment. As the fire continues to grow, the hot gas layer within the fire compartment gains mass and energy. As the mass and energy of the hot gas layer increases, so does the pressure. Higher pressure causes the hot gas layer to spread downward within the compartment and laterally through any openings such as doors or windows. If there are no openings for lateral movement, the higher pressure gases have no lateral path to follow to an area of lower pressure. As a result, the hot gases will begin to fill the compartment starting at the ceiling and filling down.

Isolated or *intermittent* **flames** may move through the hot gas layer. Combustion of these hot gases indicates that portions of the hot gas layer are within their flammable range, and that there is sufficient heat to cause ignition. As these hot gases circulate to the outer edges of the plume or the lower edges of the hot gas layer, they find sufficient oxygen to ignite. This phenomenon frequently occurs before more substantial involvement of flammable products of combustion in the hot gas layer. The appearance of isolated flames is sometimes an immediate indicator of flashover.

The interface between the hot gas layers and cooler layer of air is commonly referred to as the **neutral plane** because the net pressure is zero, or neutral, where the layers meet. The neutral plane exists at openings where hot gases exit and cooler air enters the compartment. At these openings, hot gases at higher than ambient pressure exit through the top of the opening above the neutral plane. Lower pressure air from outside the compartment entrains into the opening below the neutral plane **(Figure 1.35)**.

> **Example Compartment: Growth Stage – Thermal Layering**
>
> The plume coming from the chair has now reached the ceiling and become a ceiling jet. Hot fire gases and fuel-rich smoke begin to spread horizontally across the ceiling. Both the door and the window are closed, so the smoke has no way to leave the compartment. The compartment begins to fill with hot gases and smoke. The dividing line between the dwindling air in the compartment and the increasing amount of smoke in the compartment steadily lowers toward the floor. The end table is completely involved with flames now. The walls and ceiling have also heated and are radiating heat back into the room. The coffee table has begun to pyrolize and the flat-screen television has begun to melt. The hydrocarbon materials in the compartment burn quickly and inefficiently, creating fuel-rich, black smoke. The heat release rate is high, though the environment near the floor where there is still air might be tenable for firefighters **(Figure 1.36)**.
>
>
>
> **Figure 1.36** This image shows the growth stage of a compartment fire as thermal layering begins to occur within the compartment. *Courtesy of Dan Madrzykowski, NIST.*

Transition to Ventilation-Limited Decay

Most residential fires that develop beyond the incipient stage become ventilation-limited. Even when doors and windows are open, insufficient air entrainment may prohibit the fire from developing based on the available fuel. When windows are intact and doors are closed, the fire may move into a ventilation-limited state of decay even more quickly. While a closed compartment reduces the heat release rate, fuel may continue to pyrolize, creating fuel-rich smoke.

As the interface height of the hot gas layer descends toward the floor, the greater volume of smoke begins to interrupt the entrainment of fresh air and oxygen to the seat of the fire and into the plume. This interruption causes the fire within the compartment to burn less efficiently. As the efficiency of combustion decreases (incomplete combustion), the heat release rate decreases and the amount of unburned fuel within the hot gas layer increases.

The fire is now in a state of ventilation-limited decay because:

- There is not enough oxygen to maintain combustion.
- The heat release rate has decreased to the point that fuel gases will not ignite.

Although the heat release rate decreases when a fire is ventilation-limited, the temperature in the room may remain high. Because there is not enough oxygen to maintain combustion, the fire has a lower heat release rate, but that does not mean that the environment is tenable. The compartment fills with fuel-rich gases that only need more oxygen to ignite because of the higher temperatures in the compartment.

Even if temperatures decrease, pyrolysis can continue. Under these conditions, a large volume of flammable products of combustion can accumulate within the compartment. These gases are fuel that can ignite, given a new source of oxygen.

If no other source of oxygen exists, the compartment will fill with black smoke and slowly cooling fuel gases. The compartment will show no visible flames. The characteristics of the fuel and fuel load in today's typical fires will cause fires to quickly become ventilation-limited.

In order for a ventilation-limited fire to grow, it needs a new supply of oxygen. Ventilation introduces outside air to the fire as this new source of oxygen. If windows or doors fail, the sudden introduction of fresh air creates a rapid increase in the heat release rate and growth of the fire. This rapid increase can also occur when firefighters open a door or window to enter the compartment for extinguishment, which creates a new flow path **(Figure 1.37)**.

The pressure outside the compartment is lower than the pressure inside the compartment **(Figure 1.38)**. Because of these pressure differences, any ventilation to the outside – opening an interior or exterior door, or breaking or opening a window – provides a flow path along which the hot gases can now move from the high pressure area inside to the low pressure area outside.

Figure 1.37 By breaking the window, the firefighter introduces new oxygen into the compartment increasing the fire's heat release rate and growth. *Courtesy of Dan Madrzykowski, NIST.*

WARNING!
Even coordinated tactical ventilation increases the combustion rate in ventilation-limited fires.

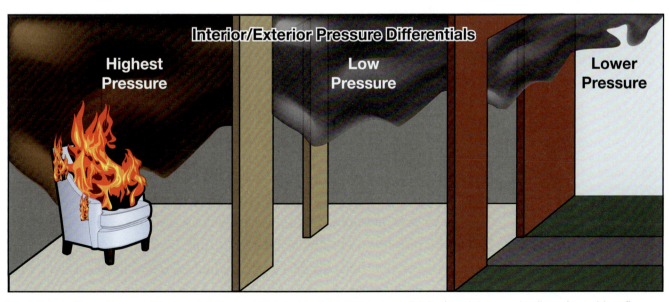

Figure 1.38 The pressure differences between the interior of a compartment (higher) and the exterior (lower) provide a flow path that hot gases may move along.

Example Compartment: Growth Stage – Transition to Ventilation-Limited Decay

The window in the compartment has not failed and the door is still closed. The hot gas layer in the room has lowered to about 2 feet (600 mm) off the floor. The small amount of oxygen left below the gas layer is no longer sufficient to sustain flaming combustion **(Figure 1.39)**. No flames are visible, but the remainder of the furnishings in the room slowly continue to pyrolize and add fuel gases to the compartment. The walls and ceiling still radiate heat. Though the heat release rate is low, use of a thermal imager from outside the compartment shows temperatures hot enough to ignite flammable gases. From outside the window, only smoke is visible, and there are pulses of smoke in the cracks around the door.

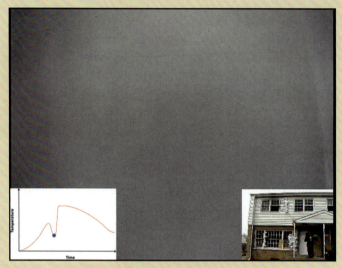

Figure 1.39 This image shows a fire's transition to ventilation-limited decay. *Courtesy of Dan Madrzykowski, NIST.*

Rapid Fire Development

Rapid fire development refers to the rapid transition from the growth stage or early decay stage to a ventilation-limited, fully developed stage **(Figure 1.40)**. Among these events are **flashover** and **backdraft**.

NOTE: Smoke explosions are also incidents of rapid fire development, but they involve more than just one compartment of a structure. Smoke explosions will be described later in this chapter.

Rapid fire development has been responsible for numerous firefighter deaths and injuries. To protect yourself and your crew, you must be able to:

- Recognize the indicators of rapid fire development
- Know the conditions created by each of these situations
- Determine the best action to take before they occur

In this section, rapid fire development conditions are described along with their indicators.

Flashover. Rapid transition from the growth stage to the fully developed stage is known as flashover. When flashover occurs, the combustible materials and fuel gases in the compartment ignite almost simultaneously; the result is

Flashover — Rapid transition from the growth stage to the fully developed stage.

Backdraft — Instantaneous explosion or rapid burning of superheated gases that occurs when oxygen is introduced into an oxygen-depleted confined space. The stalled combustion resumes with explosive force; may occur because of inadequate or improper ventilation procedures.

full-room fire involvement. Flashover typically occurs during the fire's growth stage, but may occur during the fully developed stage as the result of a change in ventilation.

Flashover conditions are defined in various ways; however, during flashover, the environment of the room changes from a two-layer condition (hot on top, cooler on the bottom) to a single, well mixed hot gas condition from floor to ceiling. The environment is untenable, even for fully protected firefighters. As flashover occurs, the gas temperatures in the room reach 1,100 °F (593°C) or higher.

A significant indicator of flashover is **rollover**. Rollover describes a condition where the unburned fire gases that have accumulated at the top of a compartment ignite and flames propagate through the hot gas layer or across the ceiling.

Rollover may occur during the growth stage as the hot gas layer forms at the ceiling of the compartment. Flames may appear in the layer when the combustible gases reach their ignition temperature. While the flames add to the total heat generated in the compartment, this condition is not flashover. Rollover will generally precede flashover, but it may not always result in flashover. Rollover contributes to flashover conditions because the burning gases at the upper levels of the room generate tremendous amounts of radiant heat which transfers to other fuels in the room. The new fuels begin pyrolysis and release the additional gases necessary for flashover.

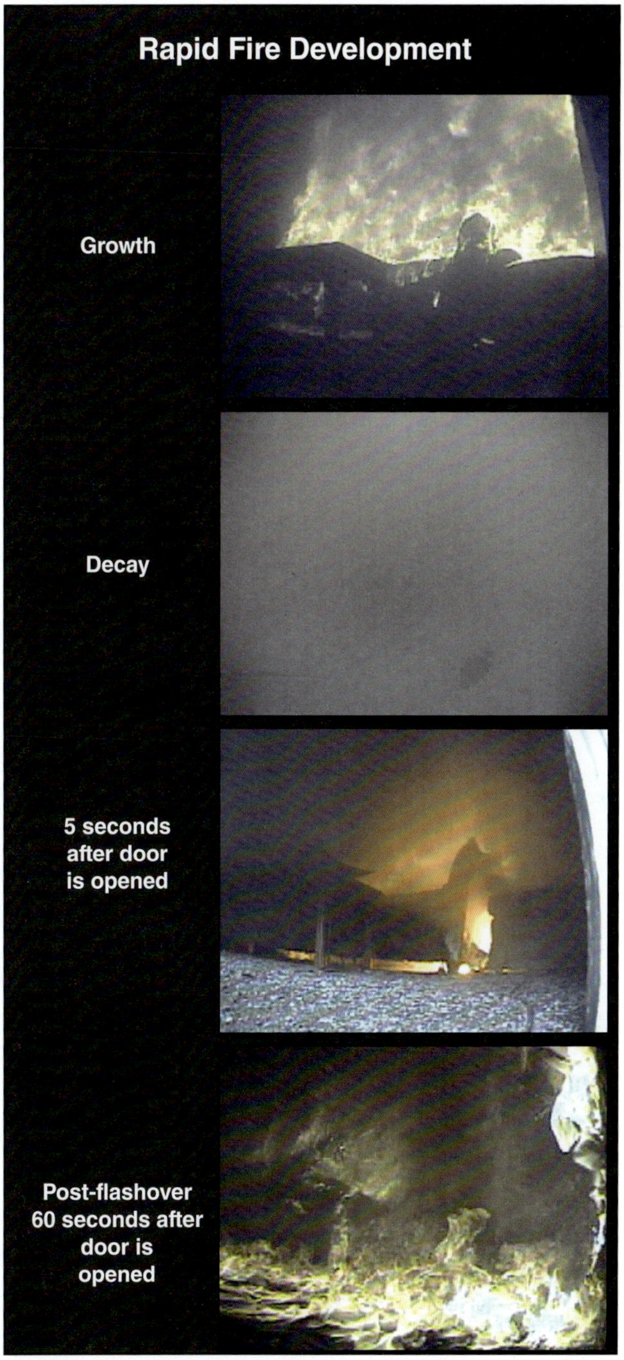

The transition period between preflashover fire conditions (growth stage/ventilation-limited decay) to postflashover (fully developed stage) can occur rapidly. Radiation from the compartment's upper layer heats the compartment's contents until they reach their ignition temperature simultaneously. When the upper layer ignites, the amount of radiation increases to levels that rapidly ignite contents in the room, even if they are remote from the fire. During flashover, the volume of burning gases can increase from approximately one-fourth to one-half of the room's upper volume to fill the room's entire volume and extend out of any openings from the room. When flashover occurs, burning gases push out of compartment openings (such as a door to another room) at a substantial velocity.

There are four common elements of flashover:

- **Transition in fire development** — Flashover represents a transition from the growth stage to the fully developed stage.
- **Rapidity** — Although it is not an instantaneous event, flashover happens rapidly, often in a matter of seconds, to spread fire completely throughout the compartment.

Figure 1.40 This series of images show the evolution of a fire from the growth stage through decay and then shows rapid fire development that occurs once a door is opened. *Courtesy of Dan Madrzykowski, NIST.*

Rollover — Condition in which the unburned fire gases that have accumulated at the top of a compartment ignite and flames propagate through the hot gas layer or across the ceiling.

- **Compartment** — There must be an enclosed space such as a single room or enclosure.

- **Pyrolysis of all exposed fuel surfaces** — Fire gases from all of the combustible surfaces in the enclosed space ignite, provided that there is sufficient oxygen to support flaming combustion.

Two interrelated factors determine whether a fire within a compartment will progress to flashover. First, there must be sufficient fuel and the heat release rate must be sufficient for flashover conditions to develop. For example, ignition of discarded paper in a small metal wastebasket may not have sufficient heat to develop flashover conditions in a large room lined with gypsum drywall. On the other hand, ignition of a sofa with polyurethane foam cushions placed in the same room will likely result in flashover provided the fire has sufficient oxygen.

The second factor is ventilation. Regardless of the type, quantity, or configuration of fuel, heat release depends on oxygen. A developing fire must have sufficient oxygen to reach flashover, an amount that a sealed room may not provide. The available air supply limits the heat release. If there is insufficient natural ventilation, the fire may enter the growth stage but not reach the heat release rate or gaseous fuel production to transition through flashover to a fully involved fire.

NOTE: The autoignition temperature of CO, the most abundant fuel gas created in most fires, is approximately 1,100° F (593°C).

Survival rates for firefighters are extremely low in a flashover. At the floor level, a heat flux of approximately 20 kW/m² is also typical of rollover conditions at the start of the flashover. Once flames begin to affect a surface, the heat flux could range from 60 to 200 kW/m². For frame of reference on heat flux, consider that NIST testing conducted in 2013 (Purtoti, 2013) has shown that SCBA face pieces begin to fail after 5 minutes of exposure to a heat flux of 15 kW/m². You must be aware of the following flashover indicators to protect yourself:

- **Building indicators** — Interior configuration, fuel load, thermal properties, and ventilation

- **Smoke indicators** — Rapidly increasing volume, turbulence, darkening color, optical density, and lowering of the hot gas layer and/or neutral plane

- **Heat indicators** — Rapidly increasing temperature in the compartment, pyrolysis of contents or fuel packages located some distance away from the fire, or hot surfaces

- **Flame indicators** — Isolated flames or rollover in the hot gas layers or near the ceiling

Levels of the neutral plane observed from the exterior of the structure are also good indicators of fire behavior within the structure as follows **(Figure 1.41)**:

- **High neutral plane** — May indicate that the fire is in the early stages of development. Remember that high ceilings can hide a fire that has reached a later development stage. A high neutral plane can also indicate a fire above your level.

- **Mid-level neutral plane** — Could indicate that the compartment has not yet ventilated or that flashover is approaching.

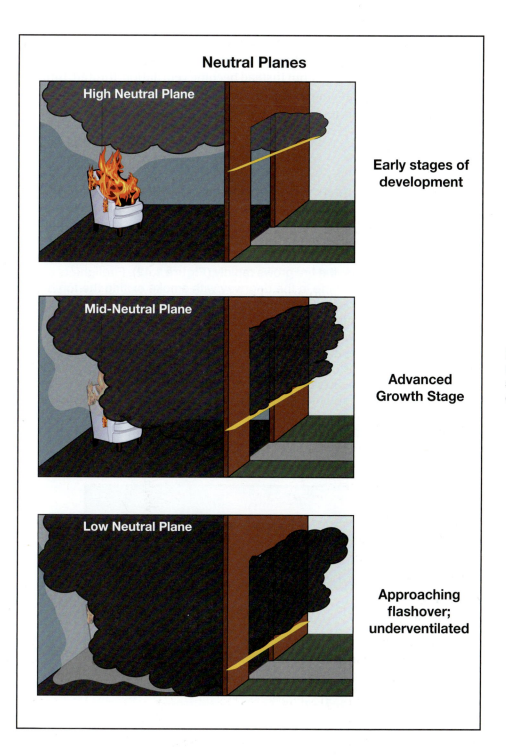

Figure 1.41 Observing the neutral plane from outside a structure can provide indications of the behavior of the fire within.

- **Very low-level neutral plane** — May indicate that the fire is reaching backdraft conditions. This occurrence could also mean that a fire is below you (basement fire or lower story).

When a fire is in ventilation-limited decay, the introduction of new oxygen can trigger flashover quickly. Flashover may occur whenever sufficient oxygen and ventilation are available for fire growth. However, in an uncontrolled situation, it may be difficult to identify what stage a fire is in, so firefighters should assume that flashover may occur at any time that the conditions are right.

Flashover may not occur in every compartment fire, such as in large-area compartments or compartments with high ceilings. Fire development may

Chapter 1 • Fire Dynamics   53

take an alternative path in a compartment that quickly becomes ventilation-limited, before the thermal energy can build within the compartment. The fire may not progress to flashover but instead become ventilation-limited, limiting heat release rate and causing the fire to enter the decay stage while continuing the process of pyrolysis and increasing the fuel content of the smoke.

Example Compartment: Growth Stage – Flashover

Let's back up a moment and change the conditions of our example. When the hot gas layer reaches about halfway down the compartment, a firefighter opens the door to ventilate the room. When the door was closed, the fire was on its way to ventilation-limited decay. With the introduction of fresh air from outside, the fire grows rapidly **(Figure 1.42)**. Firefighters can observe the neutral plane in the doorway with smoke exiting the top half of the door. The amount of smoke increases, and the neutral plane lowers in the doorway as more of the surfaces in the room pyrolize. Firefighters observe flames moving through the top of the hot gas layer. These flames radiate a large amount of energy to the compartment contents, causing the contents to pyrolize and release more fuel while rapidly heating to their ignition temperature. Suddenly, the hot gases and incoming oxygen reach the correct mixture, and the isolated flames become a room full of flames. All of the hot gases ignite at once. The heat release rate rises dramatically, igniting all of the flammable fuels within the compartment. Flames extend up and around the door frame of the open door.

Figure 1.42 In this image, the fire grows rapidly as fresh oxygen is introduced into the compartment. *Courtesy of Dan Madrzykowski, NIST.*

Backdraft. A ventilation-limited compartment fire can produce a large volume of flammable smoke and other gases due to incomplete combustion. While the heat release rate from a ventilation-limited fire decreases, elevated temperatures may still be present within the compartment. An increase in ventilation such as opening a door or window can result in an explosively rapid combustion of the flammable gases, called a backdraft. Backdraft occurs in a space containing a high concentration of heated flammable gases that lack sufficient oxygen for flaming combustion.

When potential backdraft conditions exist in a compartment, the introduction of a new source of oxygen will return the fire to a fully involved state rapidly (often explosively). A backdraft can occur with the creation of a horizontal or vertical opening. All that is required is the mixing of hot, fuel-rich smoke with air. Backdraft conditions can develop within a room, a void space, or an entire building. Anytime a compartment or space contains hot combustion products, firefighters must consider potential for backdraft before creating any openings into the compartment. Backdraft indicators include:

- **Building indicators** — Interior configuration, fuel load, thermal properties, amount of trapped fuel gases, and ventilation
- **Smoke indicators** — Pulsing smoke movement around small openings in the building; smoke-stained windows
- **Air flow indicators** — High velocity air intake
- **Heat indicators** — High heat, crackling or breaking sounds
- **Flame indicators** — Little or no visible flame

The effects of a backdraft can vary considerably depending on a number of factors, including:

- Volume of smoke
- Degree of confinement
- Temperature of the environment
- Pressure
- Speed with which fuel and air mix

Do not assume that a backdraft will always occur immediately after an opening is made into the building or involved compartment. You must watch the smoke for indicators of potential rapid fire development including the air currents changing direction, or smoke rushing in or out. To some degree, the violence of a backdraft depends upon the extent to which the fuel/air mixture is confined in the compartment. The more confined, the more violent the backdraft will be.

> **Example Compartment: Growth Stage – Backdraft**
> Let's again change the conditions in our compartment. The compartment is ventilation-limited, and all of the petroleum products in the room (chair cushions, plastic in the flat-screen TV) have produced large amounts of fuel-rich, black smoke. The bottom of the hot gas layer is barely off the floor. There is nothing visible but dense, black smoke through the window. At this moment, the window fails. The high volume of confined smoke rushes to the new area of low pressure and billows out the window. Air from the outside rushes in at the same speed. The hot gases and air mix so quickly that they reach their explosive limit almost immediately. Flames propagate through the hot gases seemingly all at once, and an explosion of fire erupts out of the open window.

Fully Developed Stage

The fully developed stage occurs when the heat release rate of the fire has reached its peak, because of lack of either fuel or oxygen. There are two main types of fully developed fires: ventilation-limited and fuel-limited fires. The factor limiting the peak heat release rate is used to identify which type of fully developed fire exists.

Firefighters often misinterpret the term "fully developed" to mean that the fire can no longer grow. A more accurate description would be that the fire has grown as much as it can. New sources of fuel introduced after full development will allow fuel-limited fires to grow. Likewise, new sources of oxygen introduced after full development will allow ventilation-limited fires to grow.

Fuel-Limited Conditions

The available fuel limits the peak heat release in a fuel-limited, fully developed fire. The most effective method of increasing the heat release rate is to provide more fuel. A campfire located in a fire ring is a good example of fuel-limited conditions. The fire reaches its peak when all the fuel becomes involved. The fire ring separates the burning fuel from other potential fuel resulting in a fuel-limited, fully developed fire. Adding additional fuel or firewood would increase the energy release of the fire to a new peak heat release rate.

Technically speaking, most compartment fires, even those that are ventilated and have untenable interior environments, are ventilation-limited. Adding ventilation points to a compartment fire that is already ventilated will add oxygen that will allow the fire to grow. Fuel-limited full development usually occurs when fires are not contained within compartments such as wildland fires, vehicular fires, or fires burning in collapsed structures.

> **Example Compartment: Fully Developed – Fuel-Limited Conditions**
> Let's assume that the compartment flashed over shortly after a firefighter opened the door for ventilation. As a result of flashover, all of the available fuels in the compartment burned. Let's further assume that the compartment burned until one of the walls and the adjoining part of the ceiling collapsed. The fire then had access directly to the outside air without the ventilation limitations of the doorway. The fire now has access to an unlimited, unimpeded supply of oxygen and is consuming oxygen at the highest capacity that it can. Any available fuel burns, and flames, smoke, heated fire gases, and embers exit the compartment through the opening. At this point, firefighters have no choice but to protect surrounding exposures (neighboring buildings or vegetation) and contain the fire from the exterior.

Ventilation-Limited Conditions

In contrast, a fully developed, ventilation-limited fire lacks the oxygen available to grow because the number and size of openings in the compartment limit the entrainment of air. The fire reaches a peak when it consumes all the available oxygen from the air intake, typically with incomplete combustion. Additional fuel is available and gaseous fuel is leaving the compartment in the smoke; however, the fire cannot release any more energy. Allowing additional air into the compartment via an additional opening or enlarging the existing opening will provide more oxygen, resulting in a higher peak heat release rate.

> **WARNING!**
> Additional ventilation will cause an already ventilated fire to grow.

Ventilation-limited, fully developed fires present a hazardous situation to firefighters. The potential for a window failure to provide fresh oxygen and increase the peak heat release rate can endanger both firefighters and potential victims. To reduce the risk of the unpredictable window failure, firefighters must transition the fire from ventilation-limited to fuel-limited. With the high heat of combustion found in modern furnishings, the only mechanism to transition the fire is to extinguish some of the burning fuel. It is not possible to make enough openings in a compartment to transition a fire from ventilation-limited to fuel-limited conditions.

> **WARNING!**
> Additional ventilation alone will not transition a ventilation-limited fire to a fuel-limited fire.

Example Compartment: Fully Developed – Ventilation-Limited Conditions

Again, let's take a step back in the development of our fire. Let's assume this time that the window is already ventilated, and firefighters are planning to enter the compartment through the doorway for an interior attack. The amount of smoke exiting the window is significant and larger than earlier in the fire's development, but the amount of smoke is no longer increasing, an indicator that the fire is consuming as much oxygen as it can through the window opening **(Figure 1.43)**.

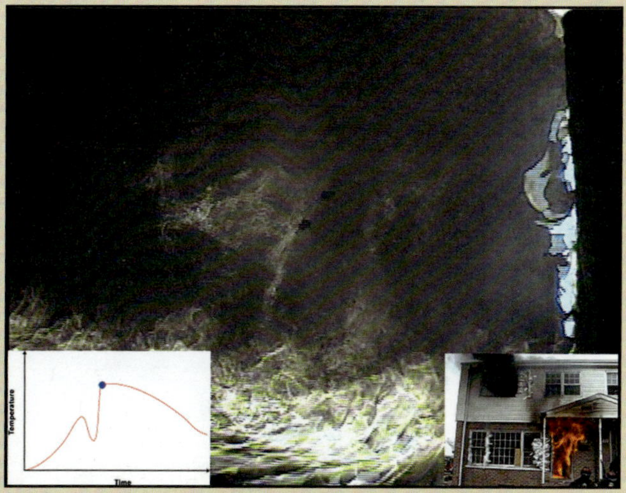

The firefighters have two choices for controlling this type of fire. First, they could enter through the door low and under the smoke to attack the seat of the fire directly. They know, however, that opening the door will introduce a new flow path for air that could allow the fire to grow. So they choose their second option. One team cools the hot gas layer using a straight stream from outside the window, which lowers the temperature in the compartment. Once this tactic reduces the heat release rate, another team takes a charged handline through the door. The exterior team disrupts the fully developed conditions of the fire, minimizing the impact of a new ventilation source at the doorway. The interior attack crew encounters an environment with a lower heat release rate and more tenable conditions than they would have in option one. They are able to accomplish full extinguishment of the fire with greater ease.

Figure 1.43 In this image, the fire reaches the fully developed stage. *Courtesy of Dan Madrzykowski, NIST.*

Decay Stage

A fire is said to be in the decay stage when it runs out of either fuel or oxygen. Either fuel or oxygen is an integral part of the fire triangle introduced earlier in the chapter. Without all three components of the triangle, the fire will decay and extinguish.

In fuel-limited fires, the decay stage is usually the fire's final stage, leading to the fire's self-extinguishment when it runs out of available fuel. Ventilation-limited fires can also self-extinguish due to lack of oxygen. Both of these situations can result in the termination of the combustion reaction. However, just like throwing another log on top of a smoldering campfire, introducing new oxygen to a ventilation-limited fire can cause it to reenter the growth stage.

Fuel-Limited Decay

After a fuel-limited fire reaches the fully developed stage the fire will decay as the fuel is consumed. As the fire consumes the available fuel and the heat release rate begins to decline, the fire enters the decay stage. The heat release rate will decrease, but the temperature of surrounding objects may remain high for some time due to absorbed heat.

Compartment fires rarely enter a state of fuel-limited decay unless the compartment burns all the way to the ground. If the compartment fails and the fire opens to the atmosphere, then the amount of fuel available would limit the fire's ability to grow.

Figure 1.44 After water application, the fire's heat has been reduced so that pyrolysis of the contents and structural members will no longer occur. *Courtesy of Ron Moore, McKinney (TX) Fire Department.*

Example Compartment: Fuel-Limited Decay

After the wall and ceiling fail and the fire continues to burn, the compartment now collapses into a pile of slowly cooling fuel and embers. Some flames still show, but there is no longer enough fuel to sustain flaming combustion. There are embers and small amounts of smoke, but the fire is basically over.

Assuming there were firefighters at the scene, the fire could have reached the decay stage when the firefighters applied enough water to the fuels that the fire's heat would no longer pyrolize the contents or structural members. The water would have also greatly reduced the heat release rate to tenable or even negligible levels **(Figure 1.44)**.

Ventilation-Limited Decay

When a fire enters a ventilation-limited state of decay, this stage is not necessarily the last stage of the fire's development. As stated earlier, the fire awaits a new supply of oxygen to return to the growth stage. This statement is true even if compartment ventilation has already occurred.

To ensure that the decay stage of a ventilation-limited fire is the fire's final stage, a controlled transition from ventilation-limited to fuel-limited must take place. To provide this control, firefighters must cool the hot fire gases before any further ventilation occurs or immediately following any forcible entry. This tactic will lessen the likelihood of the gases igniting when supplied fresh oxygen.

> **Example Compartment: Ventilation-Limited Decay**
>
> Let's back up to the point where the compartment was in ventilation-limited decay. Remember, the temperatures in the compartment indicate high enough levels of heat to ignite fuel gases, but there is not enough oxygen to support combustion.
>
> Firefighters on the scene wait until they have charged and ready hoselines in place before opening the door to access the fire. The Incident Commander assigns one firefighter to door control, two more to fire attack, and two more as a rapid intervention crew. The door control firefighter opens the door. The heat release rate increases for a moment, but the attack firefighters open their nozzle utilizing a straight stream to cool the gases. The smoke changes in color (it whitens) and after a few moments a high neutral plane appears in the doorway. Thermal imagers show a reduction in temperature in the compartment. The hoseline team slowly advances through the doorway and applies water to the seat of the fire **(Figure 1.45)**.
>
> The fire crew's coordination has controlled the transition from a ventilation-limited environment to a fuel-limited environment. As a result, the compartment is still standing and in a state of fuel-limited decay. It is ventilated and free of smoke, releases very little heat, and presents a tenable temperature.

If the compartment has no ventilation openings, the heat release rate will eventually decrease to the point that the heat in the compartment naturally transfers through the compartment itself to the outside. This process takes time and is rarely a viable fire fighting strategy because firefighters must ensure that no ventilation occurs until the compartment transfers all of the heat.

Structure Fire Development

Structures are essentially composed of individual compartments connected by hallways, stairways, or openings such as doorways. If a fire starts within one of the compartments, how it grows or decays is based on the model growth curves presented in previous sections. However, in a structure, the fire has the potential to involve more than one compartment or could spread beyond the contents of a compartment and involve the structural members of the building itself. Fighting a fire in a structure, as opposed to a stand-alone single compartment, is challenging because firefighters will need to size up the building, find the fire, and then find a way to attack the fire.

Flow Path

In a structure fire, the method by which the fire receives the needed oxygen to sustain the combustion reaction occurs through one or many flow paths.

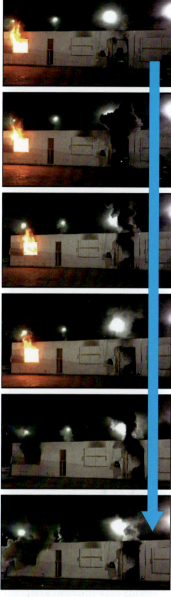

Figure 1.45 This sequence of images shows the effects of water as it cools the heated fire gases within the structure. *Photos courtesy of Dan Madrzykowski, UL-FSRI.*

The flow path is composed of two regions: the ambient air flow in and the hot exhaust flow out **(Figure 1.46)**. The flow is always unidirectional due to pressure differences where the ambient air flows toward the seat of the fire and reacts with the fuel. The products of combustion flow away from the fire toward the low pressure outlet.

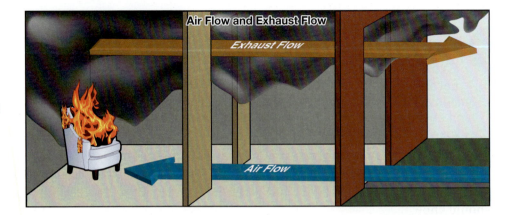

Figure 1.46 Illustrating the concepts of air flow and exhaust flow.

In a structure fire, the floor plan and openings within the structure determine the available flow path. For example, hot gases from a fire in a bedroom will travel out of the doorway and into the hallway if the door is open. If other doors in the structure are also open, the adjoining rooms also become possible parts of the flow path. The pressure in these other rooms is lower than the pressure in the fire room; therefore, hot fire gases and smoke will travel toward those areas unless the direction of flow is altered, for example, through tactical ventilation or door control **(Figure 1.47)**. Air in those rooms will entrain toward the fire as the structure fills with fuel gases and the fire grows and spreads.

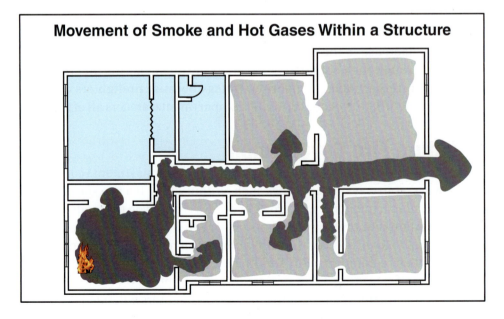

Figure 1.47 This image illustrates how the floor plan and openings within a structure determine the available flow path of hot gases.

A flow path's effectiveness to transport ambient air to the seat of the fire is based on the following:

- Size of the ventilation opening
- Length of the path traveled

- Number of obstructions
- Elevation differences between the base of the fire and the opening

When firefighters advance a hoseline or ventilate windows to make entry into a building, they establish new flow paths between the fire compartment and exterior vents of the building. These new flow paths may allow air and thus oxygen to reach the fire, increasing the heat release rate. In addition, hot, fuel-rich fire gases may flow toward a vent opened to the building's exterior because the hot gases are at a higher pressure than the lower pressure air outside the vent. When the hot gases mix with the outside oxygen, they may be hot enough to autoignite. Since any ventilation creates new flow paths for oxygen and hot gases, firefighters must use tactics that control the oxygen available to the fire and the fire's generated heat to prevent unwanted fire spread.

When hot gases follow the flow path from areas of high to low pressure, they convect heat to a larger portion of the structure. They also carry the products of combustion into new areas of the structure. Since these gases are also fuel, fire can propagate through them, out of the fire room. As a result, firefighters should know the location of the flow path for these gases in the structure and coordinate their ventilation and interior activities accordingly.

Firefighters working in the exhaust portion of the flow path will feel the increase in temperature as the velocity and/or turbulence increases, causing increased convective heat transfer. Convective heat transfer is a similar phenomenon to wind chill, except energy is transferred from a hot fluid (gas) to a solid surface (your PPE) rather than from a hot surface (your skin) to a cooler fluid (air). If ventilation is not well coordinated, this heat transfer – such as that associated with flashover or backdraft – can be unsurvivable even when wearing PPE. If you must perform operations in the flow path, recognize that these operations are risky and potentially life threatening. The time that firefighters are operating in the flow path should be strictly limited. They should not be in the area any longer than necessary.

A structure fire that extends beyond the room of origin may have two compartments involved, each in different stages of development. The room of origin may be in a fully developed, ventilation-limited stage while the adjacent compartment may be in the growth stage and nearing flashover. Understanding the model growth curves and what conditions to expect based on fire dynamics will aid firefighters in finding or establishing a tenable environment for their interior operations. Tactics employed for fire suppression, ventilation, and search and rescue will directly relate to the fire dynamics occurring on a given incident.

Ventilation and Wind Considerations

Beginning an attack on a ventilation-limited structure fire with ventilation alone will progressively increase the fire's heat release rate and spread as additional vents are made. Once the fire has filled the structure's compartments with hot, unburned, gaseous fuel, using ventilation as the only tactic will not enable you to get ahead of the fire and limit fire growth and spread **(Figure 1.48)**.

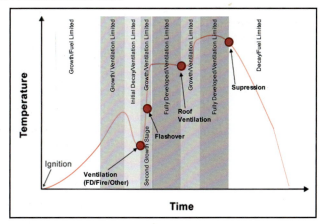

Figure 1.48 A model showing the evolution of a compartment fire. *Courtesy of UL, LLC.*

Unplanned Ventilation

Unplanned ventilation occurs when a structural member fails – usually because of exposure to heat – and introduces a new source of oxygen to the fire. This new oxygen source could result from the failure of a:

- Window **(Figure 1.49)**
- Roof
- Doorway
- Wall

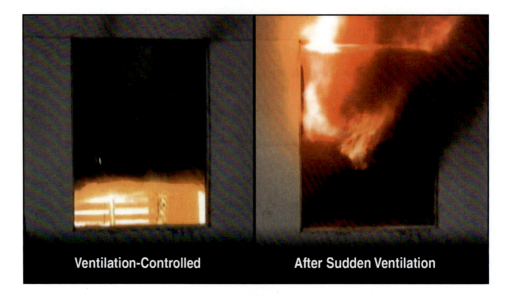

Figure 1.49 The sudden ventilation of a compartment fire can lead to rapid fire development. *Courtesy of Dan Madrzykowski, NIST.*

The source of new oxygen does not have to originate from outside the building. When floors fail above basement fires, the interior air in the structure becomes a new oxygen source.

Unplanned ventilation is often the result of:

- Occupant action
- Fire effects on the building (such as window glazing)
- Actions other than planned, systematic, and coordinated tactical ventilation

Unplanned ventilation, by definition, is unexpected. When it occurs, situational awareness is essential to ensure your safety and that of other crew members.

Wind Conditions

The wind can increase the pressure inside the structure, drive smoke and flames into unburned portions of the structure and onto advancing firefighters, and/or upset tactical ventilation efforts. You must be aware of the wind direction and velocity and use it to your advantage to assist in tactical ventilation.

WARNING! Wind-driven conditions can occur in any type of structure. Wind speeds as low as 10 mph (16 km/h) can create wind-driven fire conditions.

Wind conditions can also create differences in pressure that can cause windows to fail. The exterior pressure on the upwind side of a structure will be higher than the pressure on the downwind side of the structure. As a result, the ambient air on the outside of the structure is constantly trying to move through the structure along the path from high to low pressure. If heat exposure weakens windows in this path, wind pressure could cause them to fail which introduces a new flow path for oxygen and hot fire gases.

> **CAUTION**
> A strong wind can overpower the natural convective effect of a fire and drive the smoke and hot gases back into the building.

Smoke Explosions

A **smoke explosion** occurs when a mixture of unburned fuel gases and oxygen comes in contact with an ignition source. When smoke travels away from the fire it can accumulate in other areas and mix with air. When the fuel and oxygen are within the flammable range and contact an ignition source, the result will be explosive, rapid combustion. Smoke explosions are violent because they involve premixed fuel and oxygen.

> **Smoke Explosion** — Form of fire gas ignition; the ignition of accumulated flammable products of combustion and air that are within their flammable range.

Effects of Fire Fighting Operations

Limiting or interrupting one or more of the essential elements in the combustion process depicted in the fire tetrahedron controls and extinguishes fire. Firefighters can influence fire dynamics in a number of ways:

- **Temperature reduction** — Using water or a foam agent to cool fire gases and hot surfaces for the purposes of extinguishment.
- **Fuel removal** — Eliminating sources of fuel in the path of the fire's spread that could provide a new source of fuel; typically a tactic in wildland fires or liquid and gas fires.
- **Oxygen exclusion/flow path control** — Using door control and tactical ventilation techniques to control the amount of air available to the fire.
- **Chemical flame inhibition** — Using extinguishing agents other than water and foam, such as some dry chemicals, halogenated agents (Halons), and Halon-replacement "clean" agents, to inhibit or interrupt the combustion reaction and stop flame production.

NOTE: Tactics for these methods and further descriptions of their benefits are provided in Chapter 14, Fire Control.

Reaction of Building Construction to Fire

As buildings burn, the fire creates a variety of dangerous conditions. You must be aware of these conditions to remain safe during an emergency incident. An already serious situation can worsen if firefighters fail to recognize the potential of the situation and take the wrong actions.

Two primary types of dangerous building conditions are:

- Conditions that contribute to the spread and intensity of the fire

- Conditions that make the building susceptible to collapse

These two conditions are related; conditions that contribute to the spread and intensity of the fire will increase the likelihood of structural collapse. The following sections describe some of these conditions.

Construction Type and Elapsed Time of Structural Integrity

Most building codes rate the various construction types according to how long each construction type maintains its structural integrity over a certain period of time. **Table 1.13** shows some examples of the expected fire resistance

Table 1.13
Fire Resistance Ratings for Type I through Type V Construction (In Hours)

	Type I		Type II			Type III		Type IV	Type V	
	442	332	222	111	000	211	200	2HH	111	000
Exterior Bearing Walls										
Supporting more than one floor, columns, or other bearing walls	4	3	2	1	0	2	2	2	1	0
Supporting one floor only	4	3	2	1	0	2	2	2	1	0
Supporting a roof only	4	3	1	1	0	2	2	2	1	0
Interior Bearing Walls										
Supporting more than one floor, columns, or other bearing walls	4	3	2	1	0	1	0	2	1	0
Supporting one floor only	3	2	2	1	0	1	0	1	1	0
Supporting roofs only	3	2	1	1	0	1	0	1	1	
Columns										
Supporting more than one floor, columns, or other bearing walls	4	3	2	1	0	1	0	H	1	0
Supporting one floor only	3	2	2	1	0	1	0	H	1	0
Supporting roofs only	3	2	1	1	0	1	0	H	1	0
Beams, Girders, Trusses, and Arches										
Supporting more than one floor, columns, or other bearing walls	4	3	2	1	0	1	0	H	1	0
Supporting one floor only	2	2	2	1	0	1	0	H	1	0
Supporting roofs only	2	2	1	1	0	1	0	H	1	0
Floor-Ceiling Assemblies	2	2	2	1	0	1	0	H	1	0
Roof-Ceiling Assemblies	2	1½	1	1	0	1	0	H	1	0
Interior Nonbearing Walls	0	0	0	0	0	0	0	0	0	0
Exterior Nonbearing Walls	0	0	0	0	0	0	0	0	0	0

H: Heavy timber members
Adapted from Table 4.1.1 of NFPA 220.

Reprinted with permission from NFPA 220-2015, *Standard on Types of Building Construction*, Copyright © 2014, National Fire Protection Association, Quincy, MA. This reprinted material is not the complete and official position of the NFPA on the referenced subject, which is represented only by the standard in its entirety.

of the five types of construction. How long a building will maintain structural integrity is not an exact science. Observations made at a fire scene must be used to reevaluate estimations based upon building construction type. When you respond to a fire, there are many unknown factors that are not reflected in fire resistance estimates, such as the following:

- The duration of the fire up to the time of arrival
- The building's contents
- Way(s) the building contents affect the heat release rate
- The heat release rate and intensity of the fire
- Renovations to the interior that may have compromised fire resistance

Information gathered at the scene provides the best indicators of structural integrity. You can then compare this immediate knowledge with any known information about the structure, such as its construction type, to form opinions about safety on the fireground.

What This Means To You

You and your company respond to a residential house fire in a one-story, modern home with an open floor plan. Upon arrival, you can see the fire through the front windows burning in the living room. All of the windows are intact, so the fire has not yet ventilated the structure. The house is beginning to fill with smoke.

Your company officer instructs you to ready ladders, hoses, and equipment while he conducts a 360-degree size-up of the structure. While he is away, you notice tan or brown smoke coming from the eaves under the roof. When the company officer returns, he orders you and another firefighter to move to the back of the house and perform vertical ventilation on the roof to release smoke from the home so that a second team can enter through the front door. You point out the smoke coming from under the eaves of the roof. The company officer immediately changes his orders. He warns the entire crew that the fire may be in the attic or void space of the home and that no one is allowed on the roof or in the structure until water has been applied to the fire from the exterior.

During size-up, the company officer had not seen the smoke from the eaves because the smoke appeared in his absence. Since there was no smoke coming from the eaves at the rear of the house, he assumed that the roof might not yet be involved, and that he could ventilate the structure quickly with a roof cut. Once he knew that the fire had reached the attic space, he no longer had confidence in predicting the structural integrity of the roof and changed his orders. Observations of changing conditions at the scene caused him to revise his tactical priorities.

Fuel Load of Structural Members and Contents

The total quantity of combustible contents of a building, space, or fire area is referred to as the **fuel load** (some documents may use the term fire load). All combustible materials in the building's construction comprise the fuel load, such as:

- Wood framing

Fuel Load — The total quantity of combustible contents of a building, space, or fire area, including interior finish and trim, expressed in heat units of the equivalent weight in wood.

- Floors
- Ceilings
- Furnishings
- Combustible materials within the building

The more materials that are combustible, the more fuel is available to pyrolize and burn. Your knowledge of building construction and occupancy types will be essential to determining fuel loads. At a scene, you will only be able to estimate the fuel load based upon your knowledge and experience. For example, a concrete block structure with a steel roof assembly containing stored steel pipe will have a much smaller fuel load than a wood-frame structure used for storing flammable liquids. In buildings where the construction materials are flammable, the materials themselves add to the structure's fuel load. For example, in wood-frame buildings, the structure itself is a source of fuel **(Figure 1.50)**.

Figure 1.50 The wood frame of this house served as fuel to the fire. *Courtesy of Mike Wieder.*

The orientation of the fuels as well as their surface-to-mass ratio will also influence the rate and intensity of fire spread. The contents of a structure are often the most readily available fuel source, significantly influencing fire development in a compartment fire. When contents release a large amount of heat rapidly, both the intensity of the fire and speed of development will increase. For example, synthetic furnishings, such as polyurethane foam, begin to pyrolize rapidly under fire conditions even when the contents are located some distance from the fire's origin. The chemical makeup of the foam and its high surface-to-mass ratio speed the process of fire development **(Figure 1.51)**.

The proximity and continuity of contents and structural fuels also influence fire development. Fuels located in the upper level of adjacent compartments will pyrolize more quickly because of heat radiating from the hot gas layer. Continuous fuels such as combustible interior finishes will rapidly spread the fire through compartments.

Figure 1.51 The polyurethane foam in the chair pyrolizes rapidly under fire conditions. *Courtesy of Dan Madrzykowski, NIST.*

Similarly, the fire's location within the building will influence fire development. Fires originating on upper levels generally extend downward much more slowly following the fuel path or as a result of structural collapse. When the fire originates in a low level of the building, such as in the basement or on the first floor, convected heat currents will cause vertical extension through:

- Atriums
- Stairways
- Vertical shafts
- Concealed spaces

Additionally, if the structural elements of the building become involved in the fire, not only does the structure itself provide a new source of fuel, but the fire may be burning in hidden cavities throughout the building. These hidden spaces make finding and extinguishing the fire more difficult and increase the potential risk of building collapse.

In commercial, industrial, and storage facilities with large fuel loads, the fire can overwhelm the capabilities of a fire suppression system and make it difficult for firefighters to gain access during fire suppression operations **(Figure 1.52)**. Performing and updating preincident surveys is the most effective means of establishing awareness of these hazards.

Figure 1.52 These rolls of printing paper are a large fuel load.

Chapter 1 • Fire Dynamics

Assuming that there is available oxygen, the higher the fuel load, the more likely the fire will behave in the following ways:

- If structural members are part of the fuel load, structural integrity of the building will deteriorate faster.
- The longer the fire burns, the more fire spread accelerates.
- The fire may have a higher heat release rate.
- The structure may self-ventilate, introducing even more oxygen to the fuel-limited fire and accelerating fire development and involvement of combustible structural members.

If fires are ventilation-limited, higher fuel loads indicate a greater amount of unburned fuel that could reignite with the introduction of a new oxygen source. There may also be a greater amount of unburned fuel gases in the air because fuel packages pyrolized but did not begin combustion before the building became oxygen-limited. Such buildings are subject to backdrafts and flashovers if firefighters do not coordinate ventilation.

Furnishings and Finishes

In addition to structural members, combustible interior finishes and furnishings, can be a significant factor that influences fire spread and are a major factor in the loss of lives in fires. The interior finishes include the window, wall, and floor coverings such as drapes, wallpaper, and carpet. Furnishings may include:

- Tables
- Sofas
- Desks
- Beds
- Other items found in occupancies

Combustible Exterior Wall Coverings

Flammable material that contributes to the structure's fuel load often covers exterior walls. Exterior wall coverings may add carbon fuels (wooden siding) or petroleum fuels (vinyl siding) to the fuel load. The wall coverings may be installed atop exterior insulation which, in turn, is another fuel source. When exterior coverings become exposed to heat and catch fire, they can spread the fire to other areas of the structure or to adjacent exposures such as vegetation or neighboring buildings **(Figure 1.53)**.

Figure 1.53 The vinyl siding on the exterior of this building contributed greatly to the fire. *Courtesy of Ron Jeffers.*

Combustible Roof Materials

The combustibility of a roof's surface is a basic concern to the fire safety of an entire community. Some of the earliest fire regulations imposed in North America related to combustible roof coverings because they were blamed for several conflagrations caused by flaming embers flying from roof to roof.

Wood shakes, even when treated with fire retardant, can significantly contribute to fire spread. This is a problem in wildland/urban interface fires where wood shake roofs have contributed to large fires. Firefighters must use exposure protection tactics to protect combustible roofs on structures adjacent to a fire building or wildland fire.

Fire-resistant metal roof decking may be covered with combustible layers of foam insulation and felt paper covered with asphalt waterproofing. A fire below the metal deck can melt and ignite combustible materials, causing a second fire above the roof.

Building Compartmentation

The arrangement of compartments in a building directly affects fire development, severity, possible duration, and intensity. Building **compartmentation** is the layout of the various open spaces in a structure and includes:

- Number of stories above or below ground
- Floor plan
- Openings between floors
- Continuous voids or concealed spaces
- Barriers to fire spread

Each of these elements may contribute to either fire spread or containment. For instance, an open floor plan space may contain furnishings that provide fuel sources on all sides of a point of ignition. Conversely, a compartmentalized configuration may have fire rated barriers, such as walls, ceilings, and doors, separating fuel sources and limiting fire development to an individual compartment **(Figure 1.54)**.

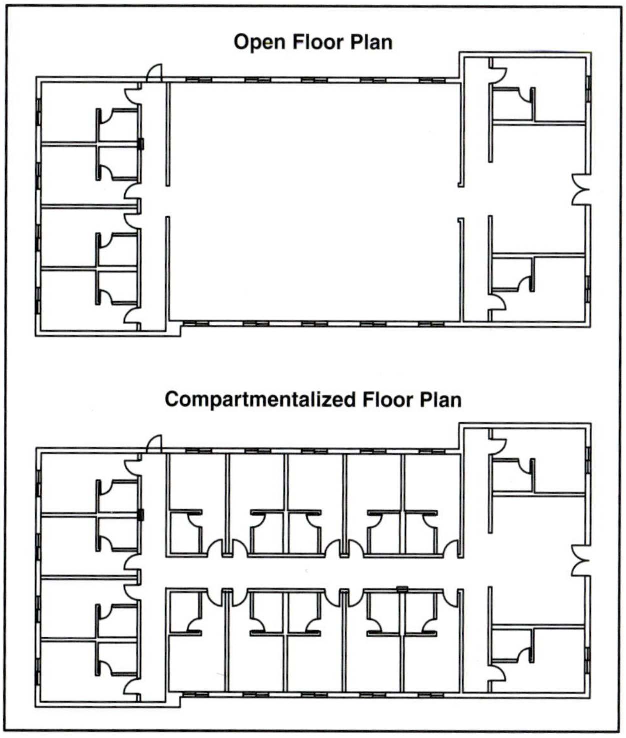

Compartmentation — The way that the arrangement of compartments creates or does not create a series of barriers designed to keep flames, smoke, and heat from spreading from one room or floor to another.

Figure 1.54 Illustrating the differences between open and compartmentalized floor plans.

Any open space with no complete fire barrier dividing it is considered a compartment. Two rooms that a doorway connects are considered two compartments only if the door between them is closed. When the door is open, a fire in either room can access the oxygen from the adjoining room. The rooms will affect one another more slowly than if the intervening wall weren't there at all, but a closed door will slow the effect of a fire in one room on the adjoining room even further. Firefighters should use doors to their advantage during interior operations, closing doors whenever appropriate to control available oxygen sources.

Given enough available fuel, fire will follow oxygen through a building along any available flow path. Firefighters can take advantage of compartmentation to control the flow path to create more predictable fire behavior during operations.

Effects of Building Construction Features

Building features such as lightweight materials or open floor plans have direct effects on how fire will spread in the structure. If you do not take building construction features into account, then fire fighting activities may worsen an emergency – possibly catastrophically – rather than help to extinguish the fire. Remember, a structure fire is the place where fire dynamics and building construction interact. The sections that follow highlight some of the construction features that firefighters should consider when fighting structure fires.

Modern vs. Legacy Construction

Over the past 50 years, building construction in North America has changed drastically. In single-family residential structures, the square footage of houses increased over 150 percent between 1973 and 2008. At the same time lot sizes have shrunk, reducing firefighter access and increasing potential exposure risks (Kerber, "Analysis ..." 2012).

Residential interior layouts and construction materials have also changed. Older structures (prior to approximately 1990) had the following features:

- Smaller compartments
- More compartments within the same square footage as modern homes
- Windows that could be opened for ventilation
- Air pockets in empty wall cavities; this construction technique used the air as insulation

Modern single-family structures may feature:

- Open floor plans
- High ceilings
- Atriums
- Lightweight manufactured structural components
- Sealed windows
- Wall cavities
- Synthetic insulation

Construction materials and interior finishes consisting of synthetic materials and light composite wood components add to the fuel load of the structure and contribute to the creation of toxic gases during a fire. Because of energy-efficient designs, the structures also tend to contain fires for a longer period of time, thus creating fuel-rich, ventilation-limited environments. These problems are magnified in large-area residential structures.

Commercial, institutional, educational, and multifamily residential structures also rely on energy conservation measures that increase the intensity of a fire and make the use of tactical ventilation difficult. Open plan commercial structures, such as warehouse-style stores, have high fuel loads in the contents and no physical barriers to prevent the spread of fire and smoke in the space **(Figure 1.55)**.

Figure 1.55 An example of a commercial structure with an open floor plan and a high fuel load.

The use of plastics and other synthetic materials has also dramatically increased the fuel load in all types of occupancies. These synthetic materials produce large quantities of toxic and combustible gases. Fires in these fuel types can escalate rapidly, reach high temperatures, and consume the structure's available oxygen quickly.

Knowledge of the building involved is a great asset when firefighters make decisions concerning tactical ventilation. This information can come from preincident plans, inspection reports, or observations of similar types of structures. Building characteristics to consider include the following:

- Occupancy classification
- Construction type
- Square footage and compartmentation
- Ceiling height
- Number of stories above and below ground level
- Number and size of exterior windows, doors, and other wall openings
- Number and location of staircases, elevator shafts, dumbwaiters, ducts, and roof openings

- External exposures
- Extent to which a building connects to adjoining structures
- Type and design of roof construction
- Type and location of fire protection systems
- Contents
- Heating, ventilation, and air conditioning (HVAC) system

Compartment Volume and Ceiling Height

A fire in a large compartment will normally develop more slowly than one in a small compartment. Slower fire development is due to the greater volume of air and the increased distance radiated heat must travel from the fire to the contents that must be heated. However, a large volume of air will support the development of a larger fire before the lack of ventilation becomes the limiting factor.

A high ceiling can also make determining the extent of fire development more difficult. In structures with high ceilings, a large volume of hot smoke and fire gases can accumulate at the ceiling level, while conditions at floor level remain relatively unchanged. Firefighters may mistake floor-level conditions for the actual state of fire development. If the large hot gas layer ignites, the situation becomes immediately hazardous.

Large, open spaces in buildings contribute to the spread of fire throughout. Such spaces may be found in **(Figure 1.56)**:

- Warehouses
- Churches
- Large atriums
- Large-area mercantile buildings
- Theaters

Figure 1.56 Large open spaces are common in warehouses, churches, and mercantile structures. *Church and mercantile photos courtesy of Ron Moore, McKinney (TX) FD.*

Large spaces may also exist between roofs and ceilings and under rain roofs. In these concealed spaces, fire can travel undetected, feeding on combustible, exposed wood rafters. When smoke appears through openings in the roof or around the eaves, the exact point of origin may be deceiving.

Thermal Properties of the Building

The thermal properties of the building can contribute to rapid fire development. The thermal properties can also make extinguishment more difficult and reignition possible. Thermal properties of a building include:

- **Insulation** — Contains heat within the building that causes a localized increase in the temperature and fire growth and may introduce an additional fuel source

- **Heat reflectivity** — Increases fire spread through the transfer of radiant heat from wall surfaces to adjacent fuel sources
- **Retention** — Maintains temperature by slowly absorbing and releasing large amounts of heat

Failure of Lightweight Trusses and Joists

The increased use of engineered or lightweight construction and trussed support systems pose a danger to firefighters. Unprotected engineered steel and wooden trusses can fail after five to ten minutes of exposure to fire (Kerber, et al, "Improving ..." 2012). These trusses can fail from exposure to heat alone without flame contact. For steel trusses, 1,000°F (538°C) is the critical temperature of steel – the temperature at which steel begins to weaken (SFPE 2016). Metal gusset plates in wooden trusses can fail quickly when exposed to heat. Although protective fire-retardant treatments can enhance the fire resistance of both steel and wooden trusses, most trusses lack this protection.

The traditional wood-joist roof uses solid wood joists that tend to lose their strength gradually when exposed to fire. This loss of strength causes a roof to become soft or "spongy" before failure, especially with a wood plank roof deck. Although a soft or sagging roof is an obvious indication of structural failure, it should not be considered the only sign of imminent collapse.

More modern homes may use engineered joists that burn more quickly and fail before the fire affects the roof decking, so the plywood or OSB used for roof sheathing may not show any signs of sagging during a fire. When the trusses fail first, entire pieces of the decking may fall into the fire **(Figure 1.57)**. Until they fall, there may be no indication that a firefighter is in danger of falling through from "sounding" the roof or even standing on the roof. Observations about the fire's location, its behavior and activity, and the location of generated smoke are all key to establishing the safety of the roof. Visual observations of the roof's exterior may not be enough. Exterior, interior, and roof crews should communicate their observations to one another throughout an incident to monitor the safety of the roof.

NOTE: Floor joists used to support floors in multilevel structures may react to fire much the same way as roof trusses. Floors above basement fires are especially prone to joist failures.

Figure 1.57 Parts of the roof of this modern home failed during a fire and fell into the fire within the building.

> **WARNING!**
> Entire pieces of decking may fall into the fire when lightweight trusses fail. There may be no indications from the exterior that the trusses no longer support the roof decking.

An arched or curved outline often indicates a bowstring truss roof. Before 1960, the bowstring truss roof design was one of the most common design types for large commercial and industrial structures. The bowstring truss roof was commonly used in facilities wherever large open floor spaces with limited interior supports were needed, such as:

- Automobile dealerships and repair facilities
- Bowling alleys
- Grocery stores
- Industrial complexes

The principles of bowstring truss construction are similar to other types of truss construction. Web members form a series of triangles that transfer tension from the bottom chord and compression from the top chord of the truss onto the load-bearing walls **(Figure 1.58)**. One difference with the bowstring truss is that the compressional forces within the top chord act to force the load-bearing walls outward as well as downward. Another difference is that the space between trusses is greater than the spaces in other types of trusses. Bowstring truss roof systems constructed before the late 1960s have a common code deficiency: the bottom chord members may have inadequate tensile strength to support code-prescribed roof loads.

Figure 1.58 An example of a bowstring truss.

Some construction features, such as parapets, obstruct the ability to easily identify bowstring truss systems. Preincident surveys should provide information on hidden construction features and should be consulted during size-up and operations. Aerial mapping or photographs can also help identify bowstring truss structures.

Construction, Renovation, and Demolition Hazards

The risk of fire rises sharply when a structure is under construction, being renovated or awaiting demolition. Contributing factors include the additional fuel loads and ignition sources such as open flames from cutting torches and sparks from grinding or welding operations.

Some local fire codes mandate that standpipe systems must remain in operation during the demolition of multistory buildings. Unfortunately, contractors do not always adhere to these requirements. The result is that inoperative standpipes and sprinkler systems have become a contributing factor in fires in buildings under demolition.

Figure 1.59 The exposed wood in this building would contribute to rapid fire spread.

Buildings under construction are subject to rapid fire spread when they are partially completed because many of the protective features such as gypsum wallboard and automatic fire suppression systems are not yet in place. Buildings under construction with exposed wooden framing are often thought of as the equivalent of a vertical lumberyard **(Figure 1.59)**.

The lack of doors or other barriers that would normally slow fire spread also contribute to rapid fire growth.

Abandoned buildings or structures undergoing renovation or demolition are also subject to faster-than-normal fire growth. Breached walls, open stairwells, missing doors, and deactivated fire suppression systems are all potential contributors. Abandoned structures may have missing or altered components, may be in a deteriorated state, and may be more susceptible to structural collapse. Arson is also a factor at construction or demolition sites because of easy access into the building.

Hazardous situations may arise during renovation because occupants and their belongings may remain in one part of the building while work continues in another **(Figure 1.60)**. Fire detection or alarm systems may be taken out of service or damaged during renovation. If good housekeeping is not maintained, accumulations of debris and construction materials can block exits. This debris may impede occupants trying to escape from the building in an emergency. Debris can also make firefighter entry more difficult. The contractors or owner/occupants performing the renovations do not always follow local building codes.

Figure 1.60 A number of hazardous situations that might contribute to fire spread could arise during the renovation of this hospital. *Courtesy of Ron Moore, McKinney (TX) FD.*

Chapter Review

1. What is the difference between a physical change and a chemical reaction?
2. What is the difference between the fire triangle and the fire tetrahedron?
3. What is the difference between piloted ignition and autoignition?
4. List the various products of combustion that may be found in a fire.
5. How does fire influence the pressure of the surrounding gases?
6. What is the difference between heat and temperature?
7. Contrast the three methods of heat transfer.
8. How are gaseous, liquid, and solid fuels different?
9. How does oxygen concentration relate to flammability?
10. How are free radicals produced in the chemical reaction that takes place during flaming combustion?
11. How do the four stages of fire development differ?

12. How can firefighters impact fire behavior during fire fighting operations?
13. How does the construction or configuration of a building impact fire development within it?

Discussion Questions

1. Why is it important for firefighters to understand the principles of fire science?
2. Provide an example of each of the methods of heat transfer: conduction, convection, and radiation.
3. Why is it important for firefighters to understand the properties of fuels?
4. How does understanding the stages of fire development help firefighters during structural fire fighting operations?
5. Make a list of five buildings in your jurisdiction. How will the building construction features of those structures impact fire development within them?

Chapter Notes

Kerber, Stephen, "Analysis of Changing Residential Fire Dynamics and Its Implications on Firefighter Operational Timeframes," Underwriters Laboratories, Inc. 2012.

Kerber, Stephen; Daniel Madrzykowski; James Dalton; Bob Backstrom, 2012. "Improving Fire Safety by Understanding the Fire Performance of Engineered Floor Systems and Providing the Fire Service with Information for Tactical Decision Making," Underwriters Laboratories, Inc. March, 2012.

Putorti, Anthony Jr.; Amy Mensch; Nelson Bryner; George Braga, "Thermal Performance of Self-Contained Breathing Apparatus Facepiece Lenses Exposed to Radiant Heat Flux," NIST Technical Note 1785 (NIST.TN 1785), February 2013.

SFPE Handbook of Fire Protection Engineering. Hurley, M.J. et al. editors, Springer, 2016

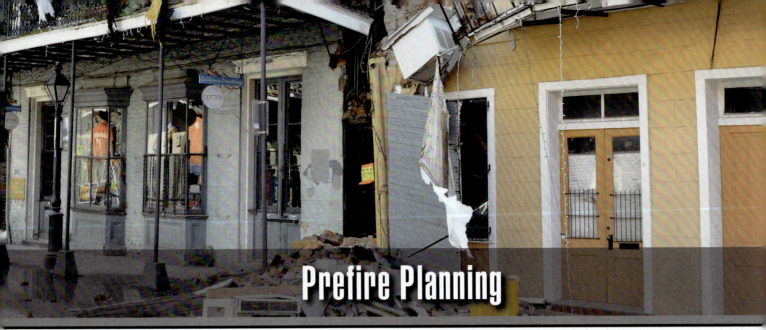

Prefire Planning

Photo courtesy of Chris Mickel, New Orleans (LA) F.D. Photo Unit.

Chapter Contents

Understanding Surveys 81	Water Supply 98
Updating Prefire Plan 82	Required Fire Flow Calculations 99
Area Familiarization 83	Resource Needs 100
Preparing for Preincident Surveys 84	**Understanding a Building's Construction** .102
Emergency Response Considerations 85	Lightweight/Engineered Construction 103
Survey Information Records 86	United States Construction 103
Building Interior 88	Canadian Construction 110
Life-Safety Information 90	Interior Building Arrangement 111
Ventilation Systems 93	Roofs Types and Hazards 125
Fire Protection Systems 95	Occupancy Types 128
Available Resources 97	**Chapter Review** 130

78 Chapter 2 • Prefire Planning

chapter 2

Key Terms

Attic 115	Oriented Strand Board 107
Bowstring Truss 127	Pitot Gauge 96
Collapse Zone 117	Plenum 116
Concealed Space 116	Plot Plan 85
Draft Curtains 94	Preincident Planning 81
Fire Stop 106	Preincident Survey 82
Glue-Laminated Beam 107	Protected Steel 104
Green Roof 129	Purlin 103
Load-Bearing Wall 104	Structural Collapse 117
Noncombustible 104	Thermoplastic 95
Nonload-Bearing Wall 104	

Chapter 2 • Prefire Planning

Prefire Planning

FESHE Learning Outcomes

After reading this chapter, students will be able to:

2. Explain the main components of prefire planning and identify steps needed for a prefire plan review.
3. Identify the basics of building construction and how they interrelate to prefire planning and strategy and tactics.

Chapter 2
Prefire Planning

This chapter focuses on preincident planning, a multistep process that includes:

- Understanding surveys
- Preparing for preincident surveys **(Figure 2.1)**
- Available resources
- Understanding a building's construction

Figure 2.1 Personnel will conduct preincident surveys at structures in their area as part of the preincident planning process.

Understanding Surveys

Incident Commanders (IC) should be trained and knowledgeable in **preincident planning** and how it affects emergency services delivery. The company responsible for the response area often collects information during preincident surveys. This information is vital to the preincident planning process, and in turn, influences the size-up and incident management activities of emergency services delivery **(Figure 2.2, p. 82)**. The information gathered during preincident surveys/area familiarization can alert fire and emergency personnel to the hazards found in an occupancy.

Preincident planning is the process of gathering and evaluating information, developing initial actions based on that information, and ensuring that the information remains current. To obtain this information, personnel may

> **Preincident Planning** — Act of preparing to manage an incident at a particular location or a particular type of incident before an incident occurs. *Also known as* Prefire Inspection, Prefire Planning, Preincident Inspection, Preincident Survey, *or* Preplanning.

Figure 2.2 Information gathered during a preincident survey will assist the Incident Commander in making informed decisions at a scene. *Courtesy of Ron Jeffers, Union City, NJ.*

> **Preincident Survey** — Assessment of a facility or location made before an emergency occurs, in order to prepare for an appropriate emergency response. *Also known as* Preplan.

be assigned to conduct **preincident surveys** of target hazards within their response areas.

ICs need to understand the difference between preincident surveys and fire and life-safety code enforcement inspections. Although similar, preincident surveys are not intended to locate code violations. If violations are discovered during a survey, the company officer may request/require that the owner/occupant correct the violation or report the problem to the inspection division.

Some departments require the same personnel to conduct both preincident surveys and code enforcement inspections during a single visit. However, preincident surveys and code enforcement inspections are conducted for entirely different purposes and should not be combined. Preincident surveys are conducted to assist fire fighting operations should there ever be an incident at the building or facility. Code enforcement inspections are performed to ensure that buildings are up to code and therefore less likely to be at risk for fires or other hazards. Because preincident surveys require close cooperation with owners/occupants, they should not be conducted alongside inspections since property management may not be as cooperative during inspections.

ICs should be familiar with their organization's process for conducting preincident plans and the required forms and formats that the organization uses. NFPA 1620, *Standard for Pre-Incident Planning*, would be a helpful reference.

Updating Prefire Plan

Fire departments should review their prefire plans to assure that the buildings and facilities in their communities have not undergone major renovations since the plans were created. This review should occur on a regular basis, such as annually or if it is known that a structure has gone through a recent renovation **(Figure 2.3)**. Personnel should take the existing prefire plan to the structure and compare the building to the plan to identify any changes. Any changes to the building should be noted in the prefire plan, and personnel should take new photographs and draw sketches that show the changes to the

Figure 2.3 A fire department should regularly update its prefire plans, especially when a structure has undergone major renovations that could alter the response to an incident.

structure. Once back at the station, personnel should update the prefire plan and images of the structure and then disseminate them to all units within the response area.

Area Familiarization

Prefire planning should include your primary response area, also referred to as your "first due" or "first in" area. Within that area, multiple types of occupancies, ages of structures and residents, types of building construction, and barriers to rapid response may exist. With each day, changes will occur within your response area that will alter existing hazards you will face or how you will respond to those hazards.

Because life safety is the first priority, you must be familiar with the population demographics within your response area. This approach includes the age, population density, and behavioral characteristics of the individuals and groups that populate the area. You should know the types and locations of high life hazards and become familiar with those places through preincident building surveys and preplans.

The structures where people live and work should be your second concern. Information about the structures that you need to consider includes:

Figure 2.4 Buildings constructed of combustible materials will collapse under severe fire conditions. *Courtesy of Chris Mickel, New Orleans (LA) F.D. Photo Unit.*

- Types of building construction, building materials used, and age of construction **(Figure 2.4)**
- Code, zoning, and fire protection requirements
- Condition and design of streets, roads, highways, and setbacks
- Temporary road and weather conditions
- Water supply concerns

Preparing for Preincident Surveys

Building the relationship between the company officer and the business owner/occupant is the first step to completing a successful preincident survey. Respect for the owner/occupants should be maintained throughout the survey **(Figure 2.5)**. When owner/occupants are shown respect, it is much more likely that accurate information will be obtained during the survey.

Figure 2.5 Company officers are more likely to get accurate information if they are respectful when meeting with business owners during preincident surveys.

Good preparation ensures that preincident survey results will be valuable and the process will not inconvenience the facility owners/occupants. Company officers should:

- Inform unit members in advance
- Discuss the survey process
- List factors that should be considered during the survey
- Assign duties if required

If one member of the unit must remain with the apparatus during the survey, communication between the company officer and personnel at the apparatus should be ensured in the event of an emergency dispatch. The company officer should also verify the necessary survey documents, tools, and equipment are immediately available.

If possible, the company officer should obtain a copy of the facility **plot plan** from the owners/occupants or the building code department. Copies of the last code enforcement inspection and preincident survey should be consulted, when possible, to provide a basis for identifying any changes or discrepancies during the survey. The company officer must contact the owners/occupants to explain the reason for the visit and establish a mutually acceptable time.

> **Plot Plan** — Architectural drawing showing the overall project layout of building areas, driveways, fences, fire hydrants, and landscape features for a given plot of land; view is from directly above.

Emergency Response Considerations

Preincident surveys provide emergency response personnel with vital occupancy information that they will need to successfully mitigate a fire or other emergency on the premises. During a survey, personnel should concentrate on answering questions about what fire fighting and rescue tactics could be successful in the occupancy, such as:

- Where and how will fires or other emergencies most likely occur?
- How are those emergencies likely to develop?
- What are the primary evacuation routes for the occupants, or if necessary, shelter in place?
- What is the layout of the work area?
- Will any of the occupants need assistance evacuating during an emergency or area of refuge?
- What will likely happen as a result of a fire or emergency?
- What will be needed in order to mitigate contingencies?
- How can building features such as fire walls and ventilation systems be used to confine a fire to one section of the building? **(Figure 2.6)** A good idea is to keep a knowledgeable building representative at the Command Post.
- Are any of the following hazards to firefighter safety on the premises:
  — Hazardous materials or processes **(Figure 2.7)**
  — High-voltage equipment
  — Unprotected openings
  — Metal-clad doors
  — Overhead power lines
  — Extreme elevation differences
- What will firefighters need to know about this occupancy in order to function safely under obscured visibility conditions?

In some jurisdictions, the sheer number of commercial, multifamily residential, and industrial occupancies

Figure 2.6 A fire wall in a warehouse building isolated the protected area from the destruction seen on the other side of the wall. *Courtesy of Ed Prendergast.*

Figure 2.7 Firefighters should know if they are responding to a facility that stores pesticides, which have the potential to explode and cause great harm to individuals at a scene. *Courtesy of Rich Mahaney.*

makes it virtually impossible for responsible companies to conduct preincident surveys in all of them. Therefore, companies must prioritize the occupancies to be surveyed.

The priorities are normally based on life-safety risk (including the risk to firefighters), property values at risk, and potential frequency and severity of fires or other emergencies occurring. Once these target hazards (occupancies with the highest priority) have been identified, the responsible companies can focus their efforts on those occupancies.

Survey Information Records

Survey information may be recorded on a standardized form, improving the chance that essential information is gathered. Field sketches should be made of the structure or facility showing its size, location, and components **(Figure 2.8)**. Photographs or videos may be made of the structure or facility. Owners/occupants may consider some processes and areas proprietary or private, and photography or videography may not be permitted in those locations. When collecting and storing data/graphics, it should be remembered that this information must be accessed rapidly and under less than ideal conditions. Survey information may be gathered in a variety of methods, including checklists, written essay-style or voice-recorded commentaries, sketches, photographs, or videos.

NOTE: Taking photographs and videos must include the express permission of the owners/occupants.

Common Survey Form Data

Survey forms are used to record information identified during the course of a survey. Common information recorded on survey forms may include for strategic and tactical considerations:

- Occupancy information
- Access
- Water supply to include estimated fire flow
- Location of utility shutoffs
- Hazards
- Roof construction and ventilation system and tactical challenges

Multistory and large buildings have unique information that should be recorded on survey forms. This information may include:

- Building system controls (such as fire alarms, elevator, and smoke control) **(Figure 2.9)**
- Number of stories

Figure 2.8 A firefighter draws a field sketch of a warehouse.

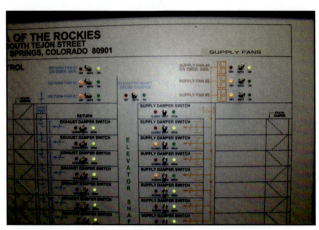

Figure 2.9 Personnel should record a building's smoke control system on a preincident survey form. *Courtesy of Colorado Springs (CO) Fire Department.*

- Occupant egress plan
- FDC and stand pipe location

Field Sketches

A field sketch is a rough drawing of a building that is prepared during the facility survey. This drawing should show general information about building dimensions and other related information around the exterior of the building, such as the locations of fire hydrants, streets, water tanks, and distances to nearby exposures. All of the basic information for survey drawings that accompany the survey report should be shown on field sketches, but not all of the details need to be included. Making field sketches on graph paper makes it easier to draw them to scale. Drawing to scale is not necessary, but it helps to keep the drawing proportional. This procedure will make it easier to transfer the information onto the survey drawings.

Photography, Videography, and Global Positioning

Photography can supplement the information contained in sketches. Digital cameras can be used to create, edit, store, and reproduce photographs quickly and economically. Graphic symbols can be added to the photographs, and images can be inserted into the preincident plan.

Videos can show relationships between buildings, manufacturing processes, and how the facility might appear as someone moves through it. Videos and photographs can be used as training aids for personnel who may respond to an emergency at the facility.

Global Positioning System (GPS) can be used to identify locations of key items of interest. They can also assist with mapping of a building and its surroundings to enhance preplanning. Personnel should work with their department to ensure that GPS information is included in the Global Information System (GIS). Satellite imaging is another tool to utilize.

Building Exterior

The facility exterior survey focuses on obtaining information to create a plot plan or compare observations to an existing plot plan. Buildings should be measured, and their dimensions recorded, including distances from each building to exposures. Note on the plot plan the locations of the following items:

- Fire hydrants and valves
- Sprinkler and standpipe connections **(Figure 2.10)**
- Power lines and utility controls (also consider outside generators)
- Obstructions to property or structure access or egress
- Underground storage tanks
- Types of roof coverings (solar panels and green roofs) and dead loads

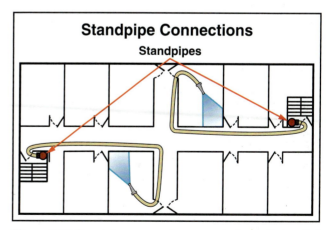

Figure 2.10 Standpipe connections are strategically located to reach all parts of the response area when used with the correct length hose.

Roof Collapse at a Fast-Food Restaurant

A medic company and an engine company responded to a call of a fire coming through the roof of a fast-food restaurant. Upon arrival, the captain of the engine company called for a "fast attack" with his crew entering the restaurant with a 1 ¾-inch hoseline to knock down the fire. A ladder company arrived minutes later. The two firefighters from the engine company made a forcible entry into the restaurant and advanced a hoseline. A firefighter from another engine company that had arrived on scene also entered the restaurant with a 1 ¾-inch hoseline.

Debris soon began falling inside the restaurant, and a fire in the middle of the kitchen could be seen. A district chief, who had arrived on scene, assumed command and performed a size-up. He saw heavy fire and ordered all companies to exit the restaurant and move to a defensive attack on the fire. The crews could also see the heavy fire conditions above them.

Moments later, the middle section of the roof — which was over the kitchen — collapsed. When the first two engine company firefighters did not exit the restaurant during the evacuation, the IC sent several firefighters into the structure to look for them. The two firefighters were rescued, but both suffered injuries as a result of the heavy HVAC unit falling from the roof.

The cause of the fire was determined to be arson, and an investigation of the incident stated that the heavy fire conditions in the overhead wood frame members coupled with the dead load of the HVAC unit contributed to the ceiling and roof members collapsing.

NOTE: A building's exterior does not provide a good vantage point for gathering building construction information since the type and location of interior supports may be disguised by brick, stone, or aluminum siding.

Site access should be noted on the survey plot plan, including:

- Access to parking lots, driveways, bridges, and gates **(Figure 2.11)**
- The proximity of access routes to possible exposures
- Private roadways and bridges that do not meet the weight requirements for emergency apparatus
- Fire lanes on solid-surface roads
- Fire lanes constructed with concrete modules that grass conceals and are indistinguishable from regular turf
- Narrow alleyways or other access routes
- Overhead obstructions in access routes that can create barriers to emergency apparatus

Figure 2.11 An apparatus may have difficulty gaining access around certain buildings because streets or pedestrian malls are blocked.

Building Interior

After surveying the facility's exterior, company personnel may move to either the building's top floor (or roof if it is accessible) or lowest floor (basement, subbasement, or ground floor) to begin the interior survey **(Figure 2.12 a and b)**. Unless the organizational policy dictates otherwise, the starting point for

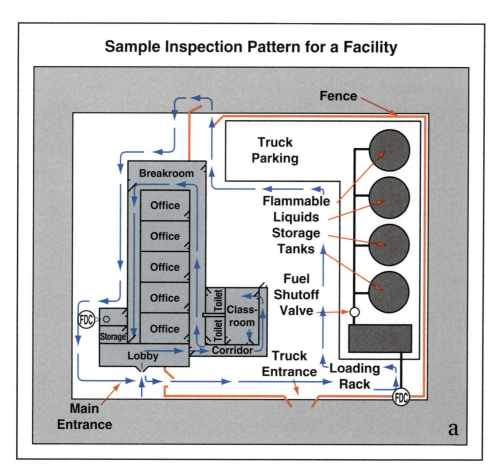

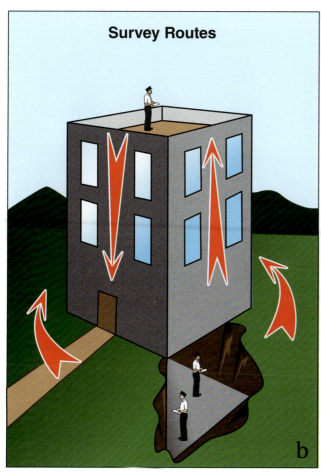

Figure 2.12 a and b Illustration (a) example of a survey pattern for a large facility; (b) example of a survey pattern of a multistory building.

the interior survey is a matter of personal preference; most people prefer to start on the top floor or roof.

Personnel then conduct the interior survey, systematically drawing floor plans of each floor to show the locations of permanent walls, partitions, fixtures, and heavy machinery. Furniture and similar items should not be included on floor plans because their locations are not fixed.

The locations of any of the following items should be noted on the floor plan drawings:

- Vertical shafts and horizontal openings
- Fire protection equipment such as standpipe or sprinkler control valves
- Fire control centers
- Safe haven areas where occupants may be sheltered in place
- Open pits and other process hazards

Life-Safety Information

Life-safety information is collected in two basic topic areas: protection and evacuation of occupants and protection of firefighters. Occupant protection information to be gathered and recorded during the interior survey includes:

- Locations and number of exits
- Locations of escalators and elevators
- Locations of windows and other openings suitable for rescue access **(Figure 2.13)**

Figure 2.13 The interior survey should note the location of windows, which could be used for rescuing trapped occupants and firefighters. *Courtesy of Ron Jeffers, Union City, NJ.*

- Special evacuation considerations for disabled occupants, old or young occupants, and large numbers of occupants **(Figure 2.14)**
- Locations of areas of safe refuge
- Flammable and toxic interior finishes or processes

Figure 2.14 Evacuating a building can be more difficult if the occupants are disabled, elderly, or very young, such as infants.

Survey personnel should also gather information about conditions within the building that may threaten firefighter safety. Some of the potential hazards to firefighters that should be noted include:

- Flammable and combustible liquids
- Toxic chemicals
- Biological hazards
- Explosives
- Reactive metals
- Radioactive materials
- Manufacturing processes that are inherently dangerous

Company personnel should also note building conditions that may present or contribute to hazardous situations. The company officer should also record the materials and items that are not part of the structure but that contribute to the structure's fuel load.

Building Conditions

The physical condition of the structure should also be noted. Conditions that may be hazardous to emergency responders during a fire include:

- Lightweight/engineered construction and structural components that may fail when exposed to fire or heavy loads
- Unsupported partitions or walls
- Stacked or high-piled storage
- Transformers and high-voltage electrical equipment vaults
- Large open areas
- Building features that may confuse or trap firefighters during a fire, such as:
  — Dead-end corridors or hallways
  — Open vats, pits, or shafts

Chapter 2 • Prefire Planning

— Openings into underground utility shafts or tunnels

— Multilevel floor arrangements

— Mazelike room divisions or partitions

— Alterations that disguise the original construction

Floor Collapse at Brick House

Firefighters in Arkansas responded late in the afternoon to a call of a fire that started in a house's basement. Noticing flames coming from the one-story brick house, personnel entered the structure to try to attack the fire. However, they quickly noticed that the floor system had a "spongy" feel to it.

Firefighters were unable to carry hoselines into the house because of the soft floor, and moments later, they fled the structure when a portion of the floor collapsed. All of the firefighters got out of the house before the floor collapsed, and the Incident Commander was forced to adjust his plan. The IC told his personnel to knock down most of the fire from the outside before re-entering the house to fully extinguish the fire.

The firefighters followed the new plan, and the fire was put out minutes later without anyone suffering an injury. Their observation that the floor felt "spongy" underneath their boots prevented any of them from staying in the house too long and falling through the floor when it collapsed.

Building Contents

Everything within a structure that is not a part of the structure can be considered its contents. The amount of contents is used to determine the fuel load in the structure. The intensity of a content fire will depend on the quantity and characteristics of the contents. Large quantities of combustibles will generate high temperatures within the structure, which will ignite or weaken the structure. Highly combustible contents, such as pallets, or flammable contents, such as paint thinner in a warehouse or store room, and the use of plastics in furniture and decorations will also generate high fire temperatures. Materials with high surface to mass ratios, such as saw dust in a lumber mill, will also ignite and burn more rapidly than solid pieces of lumber **(Figure 2.15)**.

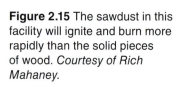

Figure 2.15 The sawdust in this facility will ignite and burn more rapidly than the solid pieces of wood. *Courtesy of Rich Mahaney.*

When company officers observe and record the fuel load of buildings during preincident surveys, they are primarily addressing the fire-control considerations of preincident planning. Subsequently, they devise plans for dealing with fires that may feed on this load. The materials used in the construction of most modern commercial and mercantile buildings contribute relatively little fuel to a fire. Fires in structures will more commonly start in the contents than the structure. Once a contents fire flashes over, it will eventually spread out of the compartment of origin and into the structure. In many cases, the only evidence that a fire exists within concealed structural spaces is discoloration of wall surfaces, noise, radiant heat, or light smoke in the air.

Structure fires can, however, be more severe due to the amount of time it takes to locate and extinguish the fire within the concealed space. Undetected fires in concealed spaces can result in the complete loss of a structure.

Ventilation Systems

Most structures have some form of climate control or HVAC system. These systems range from small window-mounted units to huge commercial units. While the potential hazards associated with large commercial units are widely recognized, even the small window-mounted units can be hazardous to firefighters under certain conditions **(Figure 2.16)**

Figure 2.16 Under certain conditions, the small window-mounted air conditioning units in this building can be hazardous to firefighters. *Courtesy of Matthew Daly/MMattyPhoto, Bronxville, NY.*

Some HVAC systems can be used during emergency operations to remove contaminated atmospheres from a structure if the system itself and natural ventilation is not helping. Company officers must have a thorough knowledge of buildings that incorporate these types of HVAC systems. Company officers should consult with building engineers to ensure the system's emergency operations are not overridden.

The survey should also identify any built-in ventilation devices that can be used to control a fire or remove hazardous atmospheres from the structure. The company officer should be aware of any underfloor air distribution systems that may be present.

> **Draft Curtains —** Noncombustible barriers or dividers hung from the ceiling in large open areas that are designed to minimize the mushrooming effect of heat and smoke and impede the flow of heat. *Also known as* Curtain Boards *and* Draft Stops.

Built-In Ventilation Devices

Some structures are equipped with built-in ventilation devices that are designed to limit the spread of fire, release heated fire gases, or control smoke and contaminated atmospheres. NFPA 204, *Standard for Smoke and Heat Venting*, provides guidelines for the design and installation of smoke and heat venting equipment and recommends using automatic heat-activated roof vents and **draft curtains** (curtain boards). Roof and wall vents and curtain boards are most common in large buildings having wide, unbroken expanses of floor space. These built-in ventilation devices may, in some cases, adversely affect the travel of heat and smoke to portions of compartments or attics not immediately involved in a fire.

The presence of these devices should be noted in the preincident survey. Company officers need to become familiar with the specific types in use in their areas. The various types of vents and curtain boards include:

Figure 2.17 Some vents may operate manually or automatically.

- **Automatic roof and wall vents** — Release heat and smoke to the outside through vents that work automatically and are placed at the highest point of a roof or wall to limit the spread of fire within a building. Smoke detectors may activate some automatic roof vents; however, most operate through the use of fusible links connected to spring-loaded or counterweighted cover assemblies. Operating sprinklers may slow or prevent the activation of automatic roof vents. If vents do not open automatically, firefighters will have to open them manually with manual-release mechanisms **(Figure 2.17)**.

- **Atrium vents** — Release heat and smoke from atriums (large, vertical openings in the center of structures) to the outside. Building codes in most areas require that atriums be equipped with automatic vents.

- **Monitors** — Release heat and smoke to the outside from square or rectangular structures that penetrate a building's roof. They may have metal, glass, wired glass, or louvered sides. Monitors with solid walls should have at least two opposite sides hinged at the bottom and held closed at the top with a fusible link that allows gravity to open them in case of a fire. Those with glass sides rely upon the glass breaking to provide ventilation in case of a fire. If a fire does not break the glass, firefighters will have to remove it.

Figure 2.18 Skylights may be appropriately placed to allow for easier ventilation of a roof. *Courtesy of McKinney (TX) Fire Department.*

- **Skylights** — Skylights with **thermoplastic** panels or ordinary window glass act as automatic vents when a fire's heat melts the plastic or breaks the glass **(Figure 2.18)**. Skylights without thermoplastic panels or automatic venting will have to be removed or glass panes will have to be broken. In skylights equipped with wired glass, the panes have to be removed from their frames or cut with saws.

- **Curtain boards** — Fire-resistive half-walls (also known as draft curtains) extend down from the underside of a roof to limit the horizontal spread of heat and smoke, which confines them to a relatively small area directly over their sources. Curtain boards also concentrate heat and smoke directly under automatic roof vents to accelerate the vents' activation. They may also accelerate the activation of automatic sprinklers in the area.

Underfloor Air Distribution Systems

Underfloor air distribution (UFAD) systems, a recent advance in HVAC system design, introduce thermostatically controlled air into the space through openings in the floor. The absence of overhead ducts or return air plenums allows an increase in ceiling heights. Return air passes through sidewall vents located adjacent to the HVAC system mechanical room.

Concerns have been raised that UFADS pose a life-safety risk for two reasons. First, if smoke develops under a floor, it will be distributed into the space at floor level. Second, water may enter the underfloor area and result in a short circuit that could cause a fire in the UFAD's electrical system. As a deterrent, model building codes require smoke detectors in each space as well as in the mechanical rooms. Smoke detectors are not currently required in the UFAD distribution system. In addition, design and installation must meet all existing code requirements to reduce electrical hazards in the system.

> **Thermoplastic** — Plastic that softens with an increase of temperature and hardens with a decrease of temperature but does not undergo any chemical change. Synthetic material made from the polymerization of organic compounds that become soft when heated and hard when cooled.

Fire Protection Systems

Any built-in fire protection equipment or system should be checked during the preincident survey. Fire companies should not test this equipment, but rather merely note its presence and condition and evaluate its usefulness during a fire on the premises. If the team observes some condition that would reduce the effectiveness of such equipment, it should be reported to the owners/occupants with suggestions for corrective actions. During the survey, personnel should identify the need for specialized hose adapters or possible obstructions that may require greater lengths of fire hose.

During a preincident survey, the survey team should pay particular attention to the absence or locations and conditions of the following systems:

- **Fixed fire-extinguishing systems** — Automatic sprinklers, carbon dioxide, dry chemical, halon-substitute flooding systems, which may reduce the need for interior attack hoselines, but may increase the need for system support.

- **Standpipe systems** — All classes of wet- and dry-pipe systems that permit the use of hoselines on upper floors and in remote areas of large-area structures; may allow firefighters to carry hose packs into a building rather than lay long attack hoselines from outside **(Figure 2.19)**.

- **Fire detection and alarm systems** — All types of automatic detection systems for smoke, carbon monoxide, low oxygen content, and other situations that result in a toxic atmosphere and all types of systems used to alert occupants to the need to evacuate a structure or area.

- **Smoke, heat, or alarm activated doors** — Alarm initiation or fusible link operation closes the doors to prevent

Figure 2.19 Standpipes provide water access to higher levels of a building.

the spread of smoke and heat. The activation of these assemblies may potentially restrict the ability for people to move throughout a structure, and door activation may damage, pinch, or block attack lines. Interior operating crews should consider alternate routes whenever possible.

For facilities protected with sprinkler or standpipe systems, the required water supply should have been determined during the design and installation of the systems. Changes in water demand, such as the construction of additional buildings using the same supply line, can reduce the actual water supply available.

Determining the availability and reliability of water supplies is critical to the development of any preincident plan. The preincident survey of any given occupancy should gather the following information:

- Locations of all water supplies
  — Auxiliary water supplies
  — Private water supply systems, such as impounded bodies of water or wells
- Locations of water-system interconnections
  — Hydrants, including hydrant main intake facing roadway **(Figure 2.20)**
  — Fire protection system flow meters and alarms

Figure 2.20 A preincident survey should include the location of hydrants.

> **Pitot Gauge** — Instrument that is inserted into a flowing fluid (such as a stream of water) to measure the velocity pressure of the stream; commonly used to measure flow; functions by converting the velocity energy to pressure energy that can then be measured by a pressure gauge. The gauge reads in units of pounds per square inch (psi) or kilopascals (kPa). *Also known as* Pitot Tube.

  — Water-demand systems such as high-water demand processes connected to the supply system
- Required fire flow based on construction type and fuel load information or on calculations that owners/occupants provide
- Water supply system pressure (reading the pressure at hydrants with a **pitot gauge** while flowing water from them)

- Available fire flow (flowing a hydrant and determining water flow rates based on the pressure readings at the hydrant)
- Reliability of water supplies (reading the water pressure at a series of hydrants while simultaneously flowing water from them)
- Water supply utilization methods (how water is used and distributed within the facility)

The public works department can provide information regarding the sizes and locations of water mains as well as pressures that can be anticipated serving the occupancy based on the municipality's water atlas, a map book or online database of all waterlines in the jurisdiction.

Available Resources

Regardless of your response area, you must be familiar with your local public water supply distribution system, the location of operating hydrants, and their ability to supply a sufficient quantity and pressure of water. This knowledge extends to static water supplies, such as swimming pools, ponds, and lakes **(Figure 2.21)**. While the first arriving unit may not have the time to establish

Figure 2.21 Personnel should be aware of static water supplies, such as lakes, that could be helpful at certain incidents.

a drafting operation, you must realize that other units will require time and personnel to do so. Where private water supply systems exist, you must be aware of their capacity and pressure as well as the location and operation of control valves. Establishing a primary and secondary water supply will typically guarantee sufficient water to extinguish most fires.

CAUTION
Always consider the potential need for a secondary water supply.

Water Supply

Your initial visual size-up, your preincident plan, and your experience should tell you if your apparatus water supply is sufficient to control the fire. At the same time, you must consider the probability that you will need a continuous water supply provided from a municipal water system, a private water supply, or static water. The preincident plan should indicate the location of each of these sources. Hydrant spacing, which will determine the length of supply hoseline that you will need, is generally constant in most urban areas and some rural areas. However, the condition of hydrants may make them unserviceable or inadequate to supply the amount of water you will need. You must be prepared to assign units to perform relay pumping or request water tenders to shuttle water to the site.

Your ability to control a fire will depend mainly on proper tactics and the amount of water that you have available. The preincident plan or knowledge of the area will provide a basic idea of the available water sources, and the following will need to be determined:

- What is the size of the fire? Can it be managed with the water tank?
- How much fire growth can occur before getting water on the fire?
- Would the tank supply be better suited to protect exposures and disregard the original structure if there is no life hazard?

If you will need to depend on a water tender or water shuttle, you must have an idea of how long it will take to set up the water supply. Many districts have structures that will present water supply challenges. Reasons for this situa-

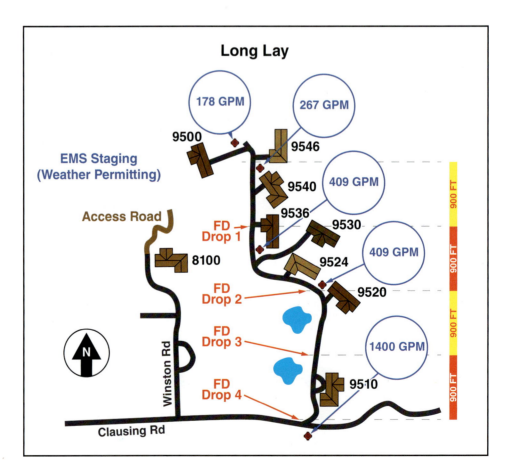

Figure 2.22 A diagram illustrates how personnel use a long lay of hoseline to get water resources to a structure.

tion could include difficult access, terrain, or distance from an adequate water source. As water supply is critical to an incident, these situations should be recognized and pre-planned.

For structures that would require an extended length supply line, this situation can be mapped out ahead of time. Generally, if the length of the lay line would require more than one engine's supply line, this would be considered an extended length supply line.

Staging pumpers at strategic intervals within the supply line lay can boost pressure and help overcome friction loss that distance can cause. This practice is commonly referred to as *relay pumping*.

With permission from the property owner, signs can be placed to mark premeasured distances showing designated hose drop off points to show engine companies where to lay their supply line. There would be a sign for each engine needed to make up the distance between the structure and the water source **(Figure 2.22)**.

Required Fire Flow Calculations

Part of the preincident survey involves calculating the required fire flow for the structure. The term *required fire flow* is used to describe the estimated uninterrupted quantity of water expressed in gallons per minute (gpm) (liters per minute [L/min]) that is needed to extinguish a well-established fire. You should be able to calculate the quantity of water that will be needed to extinguish a compartment or structure fire. Knowing the amount of water needed will help determine if you have the resources available to apply the water, including staffing and pump capacity **(Figure 2.23)**.

Figure 2.23 The Passaic (NJ) Fire Department deployed hoselines that stretched down the street to battle a large structure fire. *Courtesy of Ron Jeffers, Union City, NJ.*

You should calculate the fire flow for each type of structure within your response area during the preincident survey of target hazards or other structures. Some structures, such as single-family dwellings of similar floor space (area), can be based on one general fire flow requirement because they are similar in construction and fuel load.

Numerous formulas exist for calculating required fire flow, and some are more intricate than others. Those formulas based on detailed mathematical formulas, such as the Insurance Services Office (ISO), International Code Council (ICC), and Factory Mutual (FM) formulas, are intended for insurance and building officials to use. In addition, NFPA 1142, *Standard on Water Supplies for Suburban and Rural Fire Fighting*, contains a fire flow formula that includes occupancy and construction types as well as considerations for exposures.

However, simpler formulas are available that will provide an adequate estimate, and fire fighters should know the formulas used to calculate and determine the needed fire flow for any given structure. Taking the time to calculate the required fire flow during the preincident survey and including that information in the preincident plan will save critical time during an emergency incident.

In the 1980s, the National Fire Academy (NFA) developed a formula for estimating fire flow based on the percent of involvement of the structure or compartment. This formula was intended for both preincident planning and for on-scene calculations at the incident:

Needed Fire Flow (GPM) = (Length x Width)/3 x % involvement

This formula is intended for use with an offensive interior fire attack in a compartment or structure that is no more than 50 percent involved. The formula is less accurate, and therefore less effective, at involvement greater than 50 percent or when flows are determined to be greater than 1,000 gpm (3 785 L/min). Preincident estimates using this formula are generally calculated for 25-percent, 50-percent, 75-percent, and 100-percent involvement.

Calculating Fire Flow

Although there are numerous formulas to calculate fire flows, the initial IC usually will not have time to use them during a structure fire. For situations where you do not have a fire flow estimate or where the fire has exceeded the estimate, apply the simple rule of thumb. You can also apply the statement, "Big Fire, Big Water." That is, the larger the fire, the more water, stated in GPMs (L/min), that will be required to extinguish it. In any case, it is better to be prepared with sufficient supply hoselines and attack lines in place than to not have enough capacity available.

Resource Needs

The most basic resources at your disposal at the incident are the personnel, tools, equipment, water supply, and apparatus assigned to you. You should be aware of their capabilities and limitations. While the apparatus and its contents may remain constant, your personnel may not.

The capabilities of your personnel will vary. You must take into consideration how variables, such as staffing levels, temporary transfers and replacement personnel, training, and experience, will affect the safe and efficient operations at the incident **(Figure 2.24)**. Establishing an Incident Action Plan will help to determine the best use of personnel resources.

Figure 2.24 Firefighters respond to three-alarm fire in Jersey City (NJ). *Courtesy of Ron Jeffers, Union City, NJ.*

Your ability to control an emergency also depends on the condition and capabilities of the apparatus, tools, equipment, and water supply that you bring to the incident scene. For example, before making an offensive interior attack with the water available on your apparatus, you must consider how much time will pass before you have a constant water supply from a water distribution system, a relay from a static source, or a water tender. You must also consider whether you have sufficient personnel to perform the required hoseline attack.

Automatic Aid

Automatic aid is a formal, written agreement between jurisdictions that share a common boundary. The authority to request automatic aid is understood, and the dispatch or communication center alerts the correct unit or jurisdiction. Automatic aid occurs whenever certain predetermined conditions occur.

Fire officers who respond to automatic aid incidents operate in the same manner as when they respond to any emergency incident. If the incident involves activities that place them under the operational control of another jurisdiction, officers should perform the following:

- Report to the IC
- Determine the proper operational radio frequency **(Figure 2.25)**
- Determine the location of the Command Post
- Adhere to the personnel accountability system that the primary agency uses
- Adhere to the established procedures
- Maintain situational awareness

When an officer is in charge of an incident that will involve automatic aid units from other jurisdictions, the officer is responsible for the following:

Figure 2.25 Firefighters providing automatic aid at an incident must be informed of the proper operational radio frequency. *Courtesy of Ron Jeffers, Union City, NJ.*

- Requesting the dispatcher to instruct all responding units to acknowledge that they are on the assigned radio frequency
- Assigning units based on arrival time and capabilities
- Establishing and communicating the location of staging areas

All fire officers should be familiar with the resources that will automatically respond to various target hazards or facilities and the circumstances under which automatic aid will be activated. It is a good idea to include these resources in the preplan for a particular building or target hazard.

Mutual Aid

Mutual aid is a reciprocal agreement between two or more fire and emergency services organizations. The agreement may be between local, regional, statewide, or interstate agencies that may have contiguous boundaries. The agreement identifies the resources that will be provided, the types of incidents where they will be used, and how the actions of the resources will be monitored and controlled. Responses under a mutual aid agreement are usually based on an on-request basis. These arrangements do not guarantee a response from outside organizations. For example, if a disaster such as a tornado or hurricane affected the jurisdiction that received the request, the request can be denied.

Additional Resources

Additional resources are similar to mutual aid, except that payment rather than reciprocal aid is made by one agency to the other. Outside aid is typically addressed through a signed contract under which one agency agrees to provide aid to another in return for an established payment. This payment is typically an annual fee, but it may be on a per-response basis. Otherwise, the outside aid agreement differs little from the mutual aid agreement.

Understanding a Building's Construction

Personnel must understand building construction and the building codes that regulate construction in the jurisdiction. Building codes define the type of construction materials and techniques used for particular purposes. These specific purposes determine the type of occupancy classification that will be assigned to the completed structure. Company officers develop preincident plans based on the completed structure and its use.

NOTE: For more information on structural specifications and their effect on a building's purpose, refer to IFSTA's **Building Construction for the Fire Service** manual.

Company officers should survey buildings while they are under construction. These surveys provide opportunities to view and discuss various construction techniques and building components that will be hidden once those structures are complete **(Figure 2.26)**. The company officer should always obtain permission from the project manager or job superintendent before entering a construction site. The company officer should also speak with the appropriate governmental officials, such as the fire marshal or building officials, to learn about any special issues or considerations applicable to the site. Head, eye, and hearing protection may be required, and all safety regulations must be followed during the survey.

Figure 2.26 Shear walls constructed early in the process of erecting a building support structures against lateral loads. *Courtesy of Rich Mahaney.*

Because each type of building construction behaves differently under fire conditions, company officers must be able to identify the various types during the surveys. Knowing how stable different materials and assemblies are under fire conditions allows appropriate plans and procedures to be developed that will allow firefighters to operate with greater safety and efficiency. The sections that follow provide a brief overview of the primary structural building assemblies, lightweight construction, and other building components.

Lightweight/Engineered Construction

In lightweight/engineered construction, **purlins** support plywood panels (called *panelized roofing*) between laminated wooden beams or gusseted wooden trusses that span from outside wall to outside wall. Open web (diagonal member) trusses or wooden I beams have replaced conventional subfloor construction and contain gusset plates that are known to prematurely fail during fire conditions **(Figure 2.27)**. Similar conditions can develop from excessive or accumulated rain or snowfall. Roof or floor systems that open web trusses support are prone to sudden and unexpected collapse, if the unsupported bottom chord is subjected to downward force. An example of this is when firefighters inadvertently pull on them while pulling ceiling panels with pike poles. Extreme care should be taken when working above or below this type of construction.

Purlin — Horizontal member between trusses that support the roof.

United States Construction

Both the *International Building Code* and the NFPA recognize five types of construction, designated as Type I through Type V. Each type is further divided into

Figure 2.27 Gusset plates are steel connectors used on trusses. *Courtesy of Colorado Springs (CO) Fire Department.*

Load-Bearing Wall — Wall that supports itself, the weight of the roof, and/or other internal structural framing components, such as the floor beams and trusses above it; used for structural support. *Also known as* Bearing Wall.

Nonload-Bearing Wall — Wall, usually interior, that supports only its own weight. These walls can be breached or removed without compromising the structural integrity of the building. *Also known as* Nonbearing Wall.

Noncombustible — Incapable of supporting combustion under normal circumstances.

subcategories, depending on the code and construction type. Although the building codes have subtle differences regarding construction types, there are significant common areas.

The construction materials and their performance under fire conditions determine the construction type of the structure. Every structure is composed of basic building elements:

- Structural frame
- **Load-bearing walls**, both interior and exterior
- **Nonload-bearing walls** and partitions, both interior and exterior
- Floor construction
- Roof construction

Type I

In Type I construction, all structural members are composed of only **noncombustible** materials that possess a high fire-resistance rating. Type I construction can be expected to remain structurally stable during a fire for the duration of the structural members' fire resistance rating. Reinforced and precast concrete, masonry, and **protected steel**-frame construction meet the criteria for Type I construction **(Figure 2.28)**.

Figure 2.28 Steel is often used to increase the structural strength of concrete.

Figure 2.29 Typical Type I construction building. *Courtesy of Ron Moore.*

Protected Steel — Steel structural members that are covered with either spray-on fire proofing (an insulating barrier) or fully encased in an Underwriters Laboratories Inc. (UL) tested and approved system.

Specific combustible materials in small quantities are occasionally allowed for use in Type I construction. These materials include some types of roof coverings, wood trim, finished flooring, and wall coverings. Examples of these exceptions are described in detail in each of the model building codes.

Structures that are Type I construction are often referred to incorrectly as being *fireproof* **(Figure 2.29)**. This perceived characteristic is frequently cited as justification for reducing automatic sprinklers or other fire suppression provisions. Although the use of Type I construction provides structural stability during a fire and limits fire spread because of fire barriers, it may not offer greater life safety or loss reduction. Combustible materials in the structure, such as furniture, wall and window coverings, and merchandise, may compromise the structure's fire resistance and structural integrity.

Type I construction will generally have limited interior vertical fire extension. Most common fires in this type of construction are room and contents. Some of the factors that will affect strategy and tactics with fires in Type I buildings constructed using protected steel frames or reinforced concrete include:

- High occupancy load and occupancy type
- Location of fire and number of floors involved
- Access/egress
- Building safety features
- Ventilation difficulties
- Resource needs

Type II

Buildings classified as Type II construction are composed of materials that will not contribute to fire development or spread. This construction type consists of noncombustible, or protected noncombustible, materials that do not meet the stricter requirements of materials used in the Type I building classification. The most common form of Type II construction are structures with metal framing members, metal cladding, and concrete block or tilt-slab walls with metal deck roofs that unprotected open-web steel joists support **(Figure 2.30)**. Type II construction is normally used when fire risk is expected to be low or when fire suppression and detection systems are designed to meet the hazard load.

The personnel must keep in mind that the term noncombustible does not always reflect the true nature of the structure. Noncombustible buildings often incorporate combustible materials into their construction. This practice is most notable with some combustible roof systems, flooring, and display areas. Additionally, combustible features can be included on the exterior of Type II structures, such as balconies or wall coverings added for aesthetic purposes.

Some of the factors that will affect strategy and tactics with fires in Type II structures with metal framing concrete block walls include:

- Confirming the actual Type II construction
- Mixed occupancy types within the structure
- Likely areas of collapse due to construction features
- Potential large area and volume for fire growth
- Contents and processes
- Hoseline deployment and search tactics

Figure 2.30 Unprotected steel is most commonly used in unprotected, noncombustible construction. *Courtesy of McKinney (TX) Fire Department.*

Figure 2.31 Type III construction may survive many decades. *Courtesy of Dave Coombs.*

Type III

Type III construction is common in older churches, schools, apartment buildings, and mercantile structures. This construction type requires that exterior walls be constructed of noncombustible materials, typically masonry **(Figure 2.31)**. Interior elements can be constructed of any material that the adopted code permits. Brick, concrete, and reinforced concrete are typical materials used in exterior walls and interior loadbearing walls. Floors, roofs, and interior non-loadbearing framing and partitions are constructed of small-dimension wood or metal stud systems.

Unprotected steel and aluminum nonload-bearing wall framing members are also found in Type III construction. It is common to find older buildings of Type III construction with wood or steel trusses, while new buildings tend to have more wood trusses and floor joist systems.

Tie Rods

Some Type III buildings have tie rods and anchor plates that span from one side of a building to the opposite side. These rods are connected to either ornamental stars or plates on the exterior of the building. When conducting a walk around or pre-planning a structure, personnel should make note of this construction feature. When exposed to heat in a fire, the tie rod will expand and allow the exterior walls to move outward and potentially collapse

Fire Stop — Solid materials, such as wood blocks, used to prevent or limit the vertical and horizontal spread of fire and the products of combustion; installed in hollow walls or floors, above false ceilings, in penetrations for plumbing or electrical installations, in penetrations of a fire-rated assembly, or in cocklofts and crawl spaces.

Some of the factors that will affect strategy and tactics with fires in Type III structures, such as older churches and apartment buildings, include:

- Voids exist inside the wooden channels that roof and truss systems create that will allow fire spread unless proper **fire-stopping** is applied.
- Renovations in older Type III structures may have resulted in greater fire risk due to the creation of large voids above ceilings and below floors.
- New construction materials may have been substituted for original materials during renovations. This substitution may result in reducing the load-carrying capacity of the supporting structural member.
- The original use of the structure may have changed to one that requires a greater load-carrying capacity than that of the original design.

Type IV

Type IV construction is often referred to as heavy-timber construction. This construction type uses large-dimension timber (greater than 4 inches [100 mm]) for all structural elements. The dimensions of all structural elements, including columns, beams, joists, girders, and roof sheathing (planks), must adhere to minimum dimension sizing.

In Type IV construction, exterior walls are constructed of noncombustible materials. Interior building elements are solid or laminated wood with no concealed spaces. Fire-retardant-treated wood framing is permitted within interior wall assemblies. Floors and roofs are constructed of wood and generally have no void or concealed spaces that could provide a means for fire to travel **(Figure 2.32)**. Any other materials used in construction and

Figure 2.32 In Type IV construction, concealed spaces are not permitted between structural members. *Courtesy of McKinney (TX) Fire Department.*

not composed of wood must have a fire-resistance rating of at least one hour.

When involved in a fire, the heavy-timber structural elements form an insulating effect derived from the timbers' own char that reduces heat penetration to the inside of the beam. Therefore, these types of structures are more resistant to collapse.

Modern Type IV construction materials may include smaller-dimension lumber that is glued together to form a strong **glue-laminated beam**. They are used for many types of buildings, most often in churches, auditoriums, and other large, vaulted facilities.

Some of the factors that will affect strategy and tactics with fires in Type IV (heavy-timber) structures, such as old factories and mills converted for residential use include:

- Occupancy type and use (including any changes in use)
- Specific requirements for apparatus placement due to potential collapse of structure
- Lack of sufficient fire protection systems
- Conditions that require high fire flow/supply water
- Potential limited access

> **Glue-Laminated Beam** — (1) Wooden structural member composed of many relatively short pieces of lumber glued and laminated together under pressure to form a long, extremely strong beam. (2) Term used to describe wood members produced by joining small, flat strips of wood together with glue. *Also known as* Glued-Laminated Beam *or* Glulam Beam.

Type V

Type V construction is commonly known as *wood frame* or *frame* construction **(Figure 2.33)**. The exterior bearing walls may be composed of wood and other combustible materials. A brick or stone veneer may be constructed over the wood framing. The veneer offers the exterior appearance of masonry-type construction, but provides little additional fire protection to the structure. A single-family dwelling or residence is perhaps the most common example of this type of construction.

Most often, Type V wood-frame construction consists of framing materials that include wood 2 × 4-inch (50 mm by 100 mm) studs or wood sill plates. Sheathing, which is applied directly to the stud face, may consist of the following:

Figure 2.33 The definitive characteristic of Type V construction is the wood framing components.

- Wooden boards
- Plywood
- **Oriented strand board (OSB)**
- Rigid insulation boards
- Any combination of the above materials

The exterior surface is then covered with a covering material, such as shingles, shakes, wood clapboards, sheet metal, plastic siding, and stucco. Nails, screws, or glue attaches the exterior siding. Stucco is spread over a screen lattice that is attached to the framing studs.

Over the years, Type V construction has evolved to include the use of more lightweight engineered lumber systems to support both roof and floor loads. The

> **Oriented Strand Board (OSB)** — Wooden structural panel formed by gluing and compressing wood strands together under pressure. This material has replaced plywood and planking in the majority of construction applications. Roof decks, walls, and subfloors are all commonly made of OSB.

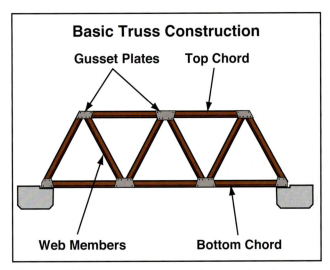

Figure 2.34 Trusses are constructed as a series of triangles arranged in a plane.

lightweight wood truss can be assembled in many configurations to support large loads over relatively long spans. Trusses typically use a triangular configuration to provide stability and efficiently transfer the applied loads **(Figure 2.34)**.

These trusses are an efficient use of lumber materials and provide many open areas to install mechanical, plumbing, and electrical systems throughout a structure. The disadvantage with truss construction is that structural redundancy is reduced. All truss members are critical to maintain the strength and stability of the truss, and the loss of any member can be catastrophic.

Like the lightweight wood truss, wood I-joists are efficient at spanning large distances with little material. This building element is an assembly of two wood flanges that a wooden web connects. The web originally was 3/8-inch or 7/16-inch plywood; in recent years OSB of a similar thickness has become more common. The flanges can be made of either solid-sawn dimension lumber joined using finger joints or may be an engineered wood product such as laminated veneer lumber.

Energy efficient construction works to keep the heat and combustion products from a fire inside the structure. The insulating system also works to keep air from infiltrating into the building. Therefore, a fire in the structure can quickly become ventilation-limited. Most energy efficient homes are typically associated with Type V construction.

Floor Protection
Since 2012, the International Residential Code has required that ½-inch gypsum board or equivalent protection is to be provided on the underside of floor framing members that are less than 2 inches x 10 inches nominal or larger framing members. This code change was driven because of the hazard firefighters risked from sudden and unexpected floor failures during basement fires.

Risks of Floor Collapse
In the afternoon of a late summer day in 2006, a firefighter was killed and two others injured when a floor collapsed into a basement fire. Further complicating the incident was a report of the possibility of someone trapped inside. The floor collapse was a result of the failure of a parallel chord truss floor support system. The fire in the basement had direct access to a system of parallel chord trusses that supported the very floor the firefighters had used to access the fire. The construction of the structure itself contributed to the tragedy as the basement fire had unchecked access to the floor support system.

Basement fires are challenging and dangerous situations. They become significantly more dangerous when firefighters are placed on the "roof" of the basement – the floor above the fire. This issue is further exacerbated when that ceiling/floor assembly is constructed with an unprotected engineered structure where fire has had the ability to degrade its strength. Support systems – whether in a roof or flooring system – are quick to degrade in direct flame / high heat exposure situations; even for relatively short periods of time.

Preincident surveys or direct size-up observations should identify structures with lightweight trusses. Firefighters should be prevented from entering such structures if those structures have had direct flame contact for any significant amount of time, as established by the AHJ.

Some of the factors that will affect strategy and tactics with fires in Type V (wood frame) structures, such as single-family dwellings and commercial buildings, include:

- High life hazard
- Structure contributes to the fire with early collapse potential
- Rapid fire spread due to the construction and age of the building
- Non-permitted remodeling and alterations to the building
- Open floor plans and void spaces do not limit the ability of fire and smoke to travel through the structure

Improper Use of Materials

Firefighters must be aware that owners/occupants might install combustible materials without the knowledge of the building or fire department. Personnel should be aware that buildings may not fit exactly into the definitions above and that different construction types can be combined for mixed construction. Personnel must understand that the unregulated use of certain materials in all types of construction may be prohibited because the materials contribute to an unacceptable increase in risk.

Unclassified Construction Types

Unclassified construction types include factory-built homes (also referred to as *manufactured, prefabricated,* and *industrialized housing*). Factory-built homes do not conform to the model building codes **(Figure 2.35)**. However, the U.S. Department of Housing and Urban Development (HUD) regulates their manufacturing. Detailed requirements are included in the *Code of Federal Regulations — Title 24: Housing and Urban Development; Chapter XX, Part 3280*. Like the model building codes, this federal standard describes the fire-resistance requirements of materials used in the construction of these buildings.

Figure 2.35 Typical manufactured or mobile home found throughout the United States and in some parts of Canada.

There are five categories of factory-built homes:

- **Manufactured:** Of all the types of factory-built homes, manufactured homes are the most common type, almost completely prefabricated prior to delivery and the least expensive. The HUD code preempts all local building codes and is more stringent than model building codes. Because the HUD code is based on performance standards, it tends to encourage construction innovations. Manufactured homes usually have a permanent steel undercarriage and are delivered to the site on wheels. They normally range from one-section, single-wide homes to three-section, triple-wide homes.

- **Modular:** Modular or sectional homes must comply with the same local building codes as site-built homes. Modular homes comprise only about 6 percent of all factory-built housing. Modular sections can be stacked vertically and connected horizontally in a variety of ways. The modular sections are constructed and then transported to the site. From there, they are attached to each other and a permanent foundation, which may include a full basement.

- **Panelized:** Panelized homes are assembled on-site from preconstructed panels made of foam insulation sandwiched between sheets of plywood. The individual panels are typically 8 feet (2.44 m) wide up to 40 feet (12.2 m) long. The bottom edges of the wall panels are recessed to fit over the foundation sill. Each panel includes wiring chases. Because the panels are self-supporting, framing members are unnecessary.

- **Precut:** Precut homes come in a variety of styles, including pole houses, post-and-beam construction, log homes, A-frames, and geodesic domes. The precut home consists of individual parts that are custom cut and must be assembled on-site.

- **Hybrid modular:** One of the most recent developments in factory-built homes, the hybrid modular structure includes elements of both the modular design and the panelized design. Modular core units, such as bathrooms or mechanical rooms, are constructed in the factory, moved to the site, and assembled. Preconstructed panels are then added to the modules to complete the structure.

Canadian Construction

The *National Building Code of Canada (NBC)* defines the following three types of building construction:

1. **Combustible** — Construction that does not meet the requirements for noncombustible construction.

2. **Noncombustible** — Construction in which the degree of fire safety is attained using noncombustible materials for structural members and other building assemblies.

3. **Heavy timber** — Combustible construction in which a degree of fire safety is attained, placing limitations on the sizes of wood structural methods and the thickness and composition of wood floors and roofs; also it avoids concealed spaces under floors and roofs.

To enable Canadian code users to understand these definitions, the NBC identifies specific requirements and limitations on materials used for each type of construction within the code. These requirements are listed in table formats that are easy to read and understand based on the occupancy classification and construction type.

In recent years, both the harsh winter climate and attempts at increasing energy efficiency have combined in a new construction standard primarily used in Quebec, Canada. Residential dwellings may be built under the design requirements of the *Novoclimat standard* **(Figure 2.36)**. This standard is designed to make new homes more energy efficient and better insulated. The result is an almost airtight structure constructed with smaller dimensional lumber. Air is heated and circulated through a closed system of HVAC units, thermo-pumps, and other devices. The result for firefighters is a structure more likely to fail rapidly under fire conditions as thermal heat and fire gases are contained within the compartments and structure. The design also makes vertical ventilation more difficult.

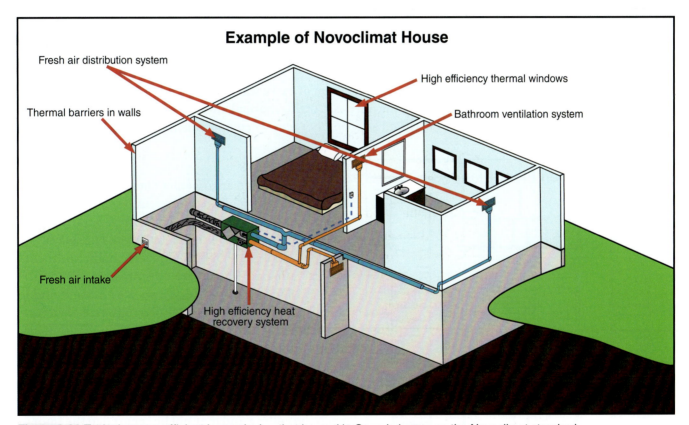

Figure 2.36 Typical energy efficient home design that is used in Canada known as the Novoclimat standard.

Interior Building Arrangement

Most buildings are simply boxes that are designed to include different types of safety features and may be built to accommodate specific functions. With the exception of common residential structures, the actual interior arrangement of most buildings can be difficult to determine from the outside. Even visual clues, such as the location and arrangement of doors, windows, chimneys, wall height, and the number of stories, cannot provide you with the exact indication of the internal floor plan. However, it does provide a general idea of the room arrangement **(Figure 2.37, p. 112)**.

Figure 2.37 The location of doors, windows, and a chimney could provide visual clues of a room's arrangement.

The interior arrangement of a structure influences fire behavior within the structure. For the purposes of this manual, interior arrangement of a structure can be classified as *open floor plan or compartmentalized*.

Open Floor Plans

An open floor plan arrangement lacks interior floor-to-ceiling walls to break up the area into smaller compartments. Many of the exterior walls are generally visible from any point inside the open floor plan. The spaces that the open floor plan creates may have high, multistory ceilings, such as "Big Box" retail stores, or ceilings that are 8 to 10 feet (2.44 to 3.05 m), such as those found in office spaces on a single floor of a high rise structure. The open floor plan can be found in warehouses, offices, auditoriums, and houses of worship **(Figure 2.38)**. Open floor plan offices may be subdivided into individual work spaces that use 4 to 6 feet (1.2 to 1.83m) high partitions. The partitions, which create cubicles, generally do not reach the ceiling and may not be secured to the floor or a permanent wall assembly.

In single-family residences built using the open floor plan design, there will be no walls between the kitchen, family, dining, and living rooms. High ceilings are a common design feature of new construction residences of this type. Walls will usually be provided for bathrooms, bedrooms, utility rooms, and storage spaces. While the interior walls of residences are not always fire resistant, fire codes require that the wall between the garage and the dwelling is fire resistant.

Fires that begin in an open floor plan structure will likely be initially fuel controlled because they will have adequate oxygen available. These open floor

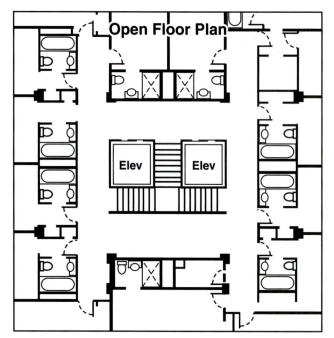

Figure 2.38 An open floor plan arrangement can be found in warehouses, offices, auditoriums, "Big Box" retail stores, and houses of worship.

plans can complicate search and rescue as well as occupant survivability. Ceiling height will influence the increase in temperature at the ceiling level. Air currents can influence fire spread that exterior openings or HVAC systems create. Conduction from one fuel item to the next will permit fire to spread due to the lack of fire barrier walls. If the fire is allowed to spread undetected or uninterrupted, the interior temperature will increase to the point where it can weaken the roof structure and cause it to collapse.

While the fire may initially be fuel controlled where high ceilings exist, its ability to remain this way through decay is dependent on the amount and type of fuel available. In most houses and offices, the amount of modern fuel available to burn is capable of consuming the oxygen and becoming ventilation limited. This is because of the nature of the fuel and its need for oxygen to burn.

NOTE: For more information about fire dynamics, refer to the IFSTA manual, **Essentials of Fire Fighting**.

Compartmentalization

A compartmentalized structure contains interior walls that create small boxes or spaces within the confines of the exterior walls and roof or ceiling. Compartmentalized layouts are typical for residential-type structures built before the 1980s where the structure is divided into multiple rooms or compartments. This arrangement is also common in commercial structures, office buildings, schools, and institutions **(Figure 2.39)**.

Figure 2.39 Compartmentalized offices have corridors connecting work areas and entrance lobbies along with other function areas.

A fire in a compartmentalized structure will generally be ventilation-controlled based on the amount of oxygen that is available within the compartment. The fire resistive quality of the compartment's wall, ceilings, and doors will limit fire spread. If the compartment contains sufficient fuel and oxygen, however, the fire can reach a magnitude that will breach even fire resistive barriers and extend into other compartments.

Basements, Cellar, and Crawl Spaces

While many buildings are built on slab foundations, some new buildings and many older ones are built with a space between the first floor and the soil. This space may be large enough for human occupancy, such as a basement or cellar, or simply a space for ventilation, such as a crawlspace.

Depending upon the geographic region, the terms *basement* and *cellar* are interchangeable. Both terms are used to describe a room or space beneath a structure wholly or partially below ground level that may be used for a living space. The space may be unfinished, with the foundation forming the exterior walls and the first floor joists visible, or covered with an interior finish such as Sheetrock®. The floor may be dirt, concrete, or stone. Basements may contain the structure's HVAC system, including fuel storage such as coal or fuel oil. Cellars are generally smaller unheated spaces used to store food or wine at cooler ambient temperatures. Cellars may also be designed to provide protection from tornadoes or other storms. In either case, access may be gained via interior stairs or an exterior entrance.

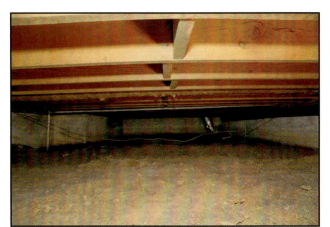

Figure 2.40 Crawl spaces consist of a perimeter wall foundation with a sub-floor that is framed with wood beams and floor joists.

Crawl spaces are found in single-family dwellings and small commercial structures. Crawl spaces consist of a perimeter wall foundation with a sub-floor that is framed with wood beams and floor joists **(Figure 2.40)**. In recent years, concrete slab construction has replaced this type of construction. Crawl spaces were originally used to provide ventilation for the structure, allowing cool air to circulate beneath the ground floor and separate the floor from the ground below. Floor furnaces fuelled with natural gas may be found in crawlspaces. This design uses the crawlspace to funnel oxygen for the unit's combustion chamber.

Basement and cellar fires are particularly challenging to control. Conditions that make these fires difficult to control include:

- Most basements have only one entry point that will funnel heat, fire, and smoke into the path of fire crews **(Figure 2.41)**.
- The ceiling above the basement is generally exposed and the ceiling height is approximately 8 feet (2.4 m) high, allowing the fire to rapidly spread upward.
- Basements under balloon construction buildings permit the fire to spread up the insides of exterior walls as far as the roof.
- Fires in crawl spaces are difficult to contain due to the lack of access to the area.
- Ventilation options are limited in basement/cellar and crawlspace fires, contributing to extinguishment difficulties.

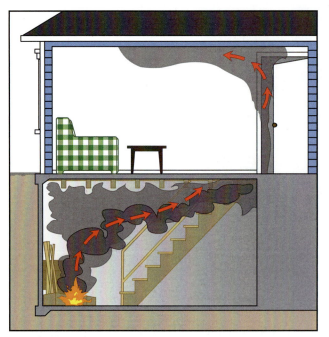

Figure 2.41 Smoke from a basement fire will flow up a stairwell in much the same way as it flows up a chimney.

Deadly Basement Fire in Washington D.C. Townhouse

On May 30, 1999, two firefighters died and two others were injured as the result of a fire in the basement of a multistory brick townhouse in Washington, D.C. The casualties were located on the first floor of the structure. The firefighters had entered through the front door, believing that the fire was located on that level. An engine and a ladder responding to the back of the structure reported seeing the fire when they opened the grade level door. This door actually led into the basement and not the first floor as the IC believed. The IC did not give permission to attack the fire because he did not want opposing fire streams. Opening the basement door resulted in a rapid increase in the magnitude of the fire, forcing it up the interior stairs and onto the crews operating on the first floor. The first floor crews made a rapid withdrawal, and two firefighters became lost while others exited with burns. When it was determined that two men were missing, rescue personnel reentered the building and located one firefighter using his activated Personal Alert Safety System (PASS) device. The other firefighter was located 4 minutes later; his PASS device had not been turned on and did not activate.

NIOSH firefighter fatality report (FACE#99F21) contains a number of recommendations that could have prevented these deaths. Among these are:

- Ensure that the department's SOP/Gs are followed and refresher training is provided
- Provide the Incident Commander with a Command Aide
- Ensure that firefighters from the ventilation crew and the attack crew coordinate their efforts
- Ensure that when a piece of equipment is taken out of service, appropriate backup equipment is identified and readily available
- Ensure that personnel equipped with a radio position the radio to receive and respond to radio transmissions
- Consider using a radio communication system that is equipped with an emergency signal button, is reliable, and does not produce interference
- Ensure that all companies responding are aware of any follow-up reports from dispatch
- Ensure that a Rapid Intervention Team (RIT) is established and in position immediately upon arrival
- Ensure that any hoseline taken into the structure remains inside until all crews have exited
- Consider providing all firefighters with a PASS integrated into their Self-Contained Breathing Apparatus (SCBA)
- Develop and implement a preventive maintenance program to ensure that all SCBAs are adequately maintained

Attics and Cocklofts

The space between the top floor of a structure and the roof is referred to as the **attic** or cockloft. While both terms refer to similar spaces, an attic is usually found in residential structures and is large enough for a person to walk upright.

> **Attic** — Concealed and often unfinished space between the ceiling of the top floor and the roof of a building. *Also known as* Cockloft *or* Interstitial Space.

Attics may be finished with flooring and interior finish on the underside of the roof, or they may be unfinished with the roof rafters and ceiling joists exposed **(Figure 2.42)**. They are generally used for storage but may also contain an HVAC unit to supply the upper floor of the structure. A cockloft is a space 2 to 3 feet (0.6 to 0.9 m) in height that is found over commercial building spaces and is not designed for human habitation. It is often found beneath flat roofs and may be accessed through a hatch from below. In strip malls, row houses, and garden apartments, the cockloft may extend throughout the structure, connecting individual occupancies.

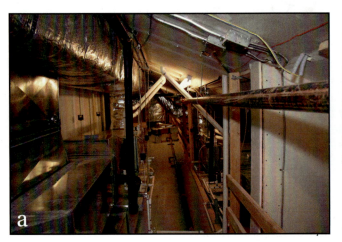 

Figure 2.42 a and b The space between the uppermost ceiling and the roof may be referred to as (a) an attic or (b) as a cockloft.

Attic and cockloft fires are difficult to access and control. These fires create unsafe conditions for firefighters working on the roof attempting ventilation. Most of today's strip mall, row houses, and garden apartment buildings have truss roof assemblies that create a dangerous condition when exposed to fire. In strip malls, row houses, and garden apartments, fires may extend across several occupancies and spread rapidly without the original point of origin being evident. Locating cockloft fires may require both opening the roof and pulling ceilings down to determine the extent of fire spread. In occupancies that have high ceilings, this activity may require the use of a combination of tools, including long pike poles, ground ladders, and solid bore streams. The addition of drop ceilings installed under existing ceilings and rain roofs over existing roofs can make cockloft fires difficult to locate and extinguish. Controlling the attic in structures that share a common attic is vital to controlling fire spread and fire extinguishment.

Concealed Spaces

The term **concealed space** *(void space)* is used to describe any area between wall surfaces, over ceilings, and under floors that are not visible from the normal occupied area. Concealed spaces are generally designed to provide passive insulation and sound barriers. They may also provide paths for return air from the compartments to the HVAC system called **plenums**, kitchen exhaust ducts, or as spaces for electrical and plumbing systems to connect multiple floors of a structure. Trash and laundry chutes may also exist in concealed spaces. A concealed space is not visible unless an access panel is provided to it.

Concealed Space — Structural void that is not readily visible from a living/working space within a building, such as areas between walls or partitions, ceilings and roofs, and floors and basement ceilings through which fire may spread undetected; also includes soffits and other enclosed vertical or horizontal shafts through which fire may spread.

Plenum — Open space or air duct above a drop ceiling that is part of the air distribution system.

Concealed spaces create pathways for fire spread from one floor to the next or from compartment to compartment **(Figure 2.43)**. Although rare, an electrical short circuit in the wiring or a leak in a natural gas line within the space may create fires in a concealed space. Controlling a fire spreading through a concealed space requires gaining access to the space to apply water directly to the fire.

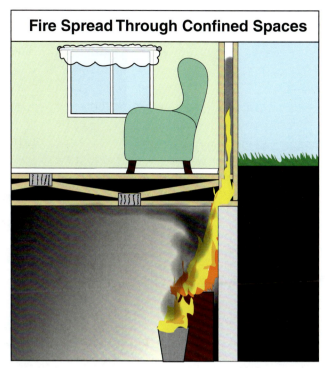

CAUTION
Keep a charged hoseline available when exposing a concealed space with suspected fire conditions.

Figure 2.43 Large or small voids can contribute to the spread of fire between floors.

Structural Collapse Potential

The structural failure of a building or any portion of it resulting from a fire, snow, wind, water, or damage from other forces is referred to as **structural collapse**. Structural collapse can also be the result of an explosion, earthquake, flood, or other natural occurrence. However, natural or explosion caused collapses usually occur without warning prior to an emergency response while fire-caused collapses often occur during emergency operations. The ability to understand how fire can cause building elements to deteriorate resulting in a collapse is extremely important for fire officers and firefighters. Collapse potential should be considered during preincident surveys and throughout the size-up process until the situation is mitigated and control of the property is returned to the owner or other authorities.

Factors that should be considered when trying to determine the potential for structural collapse include:

- Construction type
- Age of structure
- Renovations, additions, and alterations
- Contents
- Length of time the fire has been burning
- Stage of the fire
- Amount of water used to extinguish the fire
- Weather **(Figure 2.44, p. 118)**

A collapse or safety zone must be established adjacent to any exposed exterior walls of the structure. Apparatus and personnel operating master stream appliances must not be positioned in the collapse zone. Traditionally, taking the height of the structure and multiplying it by a factor of 1 1/2 has been used to estimate the **collapse zone** (Figure 2.45, p. 118). For example, a 3-story structure that is 30 feet (9 m) tall would require a collapse zone no less than 45 feet (13.5 m) from the base of the structure. Structures over a certain height do not lend themselves to this practice because of the impractical use of accessibility space for defensive fire fighting operations.

Structural Collapse — Structural failure of a building or any portion of it resulting from a fire, snow, wind, water, or damage from other forces.

Collapse Zone — Area beneath a wall in which the wall is likely to land if it loses structural integrity.

Figure 2.44 Snow and ice load can add enough weight to collapse a large building. *Courtesy of West Allis (WI) Fire Department.*

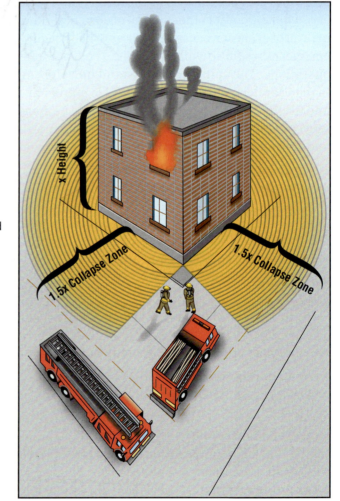

Figure 2.45 A simple equation traditionally used to estimate the collapse zone is the height of the structure multiplied by a factor of 1 ½.

118 Chapter 2 • Prefire Planning

The following guidelines should be considered when determining a collapse zone:

- **Type I** construction high-rise buildings are not as likely to collapse, making the primary concern the hazard of flying glass from windows or curtain walls **(Figure 2.46)**. In Type I construction, it is the contents of the building burning, not the structure itself. Collapse zones must be determined considering the direction and velocity of wind currents that can carry the glass shards. Structural collapse, if it does occur, may be localized and not structure wide.

Figure 2.46 The primary concern with high-rise buildings is flying glass from windows and walls, not collapse.

- **Type II** construction consists of unprotected steel or noncombustible supports, such as I-beams. When exposed to temperatures above 1,000°F (538°C), unprotected steel will expand and twist, pushing out walls, and when cooled will slightly constrict. These movements will cause floors and walls to collapse. Any type of construction that includes brick and block walls supporting unprotected steel bar joists and I-beams are involved in a large number of these collapses.

- **Type III** construction multistory buildings should have a collapse zone of 1½ times the height of the structure. For example, a building 7 stories (70 feet [21m]) will require a collapse zone of approximately 105 feet (31.5 m) during defensive operations. In Type III ordinary construction, exterior load bearing walls are made of concrete, brick, or masonry, while wood, masonry, or unprotected steel carry interior loads. Masonry construction walls can collapse in one piece or crumble in many parts. When the debris strikes the ground it can travel a long distance and even cause the collapse of other structures or objects.

- **Type IV** heavy timber or mill construction is the least likely to collapse. The weight-bearing capacity of the large dimension wood members will resist collapse unless a large volume of fire has affected them for an extended period of time. A collapse zone should be established if the fire is intense or repeated fires over time have weakened the structure.

- **Type V** construction collapses are influenced by the style of construction. That is, a multistory platform structure will generally burn through and collapse inward while a balloon structure can have full walls fall outward in a single piece. Exterior masonry and veneer walls that are not load-bearing are placed over load-bearing wood walls. Brick veneer attached to the frame can fall straight down *(curtain collapse)* into a pile or fall outward as a unit as the ties and supports fail. Although it is rare for a Type V building to collapse outward, there is a great danger to firefighters due to interior collapses. Lightweight structural material may fail within minutes when exposed to direct flame or high heat.

In North America, examples of structural collapse involving high rise buildings or Type I construction buildings is limited. Strict building codes have ensured that structural members exposed to fire and high temperatures will remain sound until the fire is extinguished. Structural collapses due to earthquakes generally involve smaller buildings, such as the buildings in the Marina Section of San Francisco in 1989, with heights ranging from one to four stories. Some of these buildings simply fell over or against another building.

Collapse zones should be established when:

- There is an indication that prolonged exposure to fire or heat has weakened the structure
- A defensive strategy has been adopted
- Interior operations cannot be justified

Figure 2.47 Regardless of materials, loss of interior supports can lead to collapse of walls.

The size of the collapse zone must include factors such as the type of building construction, other exposures, and the safest location for apparatus and personnel. Church steeples, water tanks, chimneys, walkways, fire escapes, and false facades that extend above the top of the structure must be viewed as a potential collapse hazard even if the structure is not. Most collapses usually involve brick or masonry block and may be structural components or veneer **(Figure 2.47)**. Structural collapses are not limited to the actual emergency and can occur well after the fire is extinguished. Fire personnel must ensure the structural stability of the site before entering it.

Because the collapse zone extends the full length of all of the affected walls, the safest location for defensive operations is at the corner of the building. Master streams and apparatus can be located in the area that a 90-degree arc forms from the wall intersection as long as they are far enough away that flying debris will not strike them.

The stage of the fire can easily indicate the quantity of heat that the structure has been exposed to and the potential for structural collapse **(Figure 2.48)**. Fires in the incipient stage will not have generated sufficient heat or flame to cause unprotected steel or wood frame construction to collapse. However, collapse potential increases in the growth stage as heat increases in the upper levels of the space and flame spreads to and consumes the combustible structural members. In the decay stage, and during post-suppression safety, collapse becomes likely due to the weakened state of structural members and the buildup of water.

Figure 2.48 A burning grain silo was on the verge of collapse as firefighters responded to a large fire inside of it. *Courtesy of McKinney (TX) Fire Department.*

Another factor that can contribute to structural collapse is the contents within the structure or on the roof. The contents may contribute to collapse in three ways:

- Adding to the fuel load in the building and generating higher temperatures and rapid combustion that will weaken the structure
- Adding weight to the weakened structural members, causing them to more rapidly collapse
- Ability of the contents to retain water increases their weight and the stress on the structural members.

Contents include things such as stored materials, furniture, and machinery. Like knowledge of the construction type, knowledge of the contents is gained through preincident surveys and inspections.

While contents within a structure are visible during preincident surveys, storage in concealed spaces and attics of residences is not. Attic storage is often heavier than the ceiling joists have been designed to carry. When fire weakens the joists, storage increases the potential for ceiling joists to fail, putting firefighters at risk. Such storage is common in residential dwellings and has been the cause of firefighter fatalities in industrial fires.

Finally, the quantity of water that is used to suppress the fire can have a direct effect on an unstable structure. Every U.S. gallon (SI liter) of water that is used to suppress the fire adds 8.33 pounds (4 kilograms) (Imperial gallon 10 pounds) of weight to floors that may already be weakened. The added weight may cause floors to pancake down or push walls out, resulting in a complete failure of the structure. As an estimate, 250 gpm (1 000 L/min) adds 1 ton (900 kg) of water per minute to the structure.

In addition to the factors previously listed, indicators of potential or imminent collapse include:

- Roof sagging, pulling away from parapet walls, or feeling spongy (soft) under foot
- Floors sagging or feeling spongy (soft) under foot
- Chunks of ceiling tiles or plaster falling from above
- Movement in the roof, walls, or floors
- Noises from structural movement
- Little or no water runoff from the interior of the structure
- Cracks appearing in exterior walls with smoke or water appearing through the cracks
- Evidence of existing structural instability such as the presence of tie rods and stars that hold walls together **(Figure 2.49)**
- Loose bricks, blocks, or stones falling from buildings
- Deteriorated mortar between the masonry
- Walls that appear to lean
- Structural members that appear distorted
- Fires beneath floors that support heavy machinery or other extreme weight loads
- Prolonged fire exposure to the structural members (especially trusses)
- Structural members pulling away from walls **(Figure 2.50)**
- Excessive weight of building contents

Figure 2.49 Symbolic of reinforced walls in older structures are cast iron stars on exterior walls.

Figure 2.50 Stone masonry veneers may pull away from their anchors as a result of fire damage. *Courtesy of Ed Prendercast.*

Collapse Times for Floor Systems

Until recently, there has not been any documented evidence of how long structural members will remain intact when exposed to fire. Underwriters Laboratories (UL), in conjunction with the National Institute of Standards and Technology (NIST), conducted a series of 17 full-scale fire experiments to examine the impact of floor system types on firefighter safety.

Four types of floor systems were examined during these experiments to include dimensional lumber, engineered wood I-joists, steel C-joists, and parallel chord trusses. The collapse times of the various floor systems ranged from 3 minutes, 28 seconds to 12 minutes, 45 seconds, depending on the type of flooring system, ventilation present, simulated floor loading, and fuel load. Dimensional lumber floor systems collapsed at an average of 11:57 from ignition. Engineered floor systems collapsed at an average of 7:00 from ignition.

In the 17 iterations, several variables were tested to determine their influence on the outcome, such as:

- The fuel load was arranged to replicate a typical basement fuel load.

- The ventilation was varied to examine different conditions. Max Vent indicates a scenario in which all vents are open. No Vent indicates a scenario in which all vents are closed. Sequenced Vent indicates a more realistic scenario in which some vents are open and others are closed.

- Floor loading was also arranged to replicate that of a typical home in addition to a lighter load in which the perimeter was loaded, simulating furniture, with two 300-pound firefighters in the center of the floor.

Experiment results showed that the floor system itself was a significant contributor to the fire growth and flashover conditions present within the structure. Additionally, varying the fuel load resulted in collapse times that were within 100 seconds of each other. Limiting ventilation within the structure only slowed collapse times at a maximum of 2:40. Results also indicated that the loading on the floor system did not play a significant role in determining the collapse time. The collapse of the floor system was due to the degradation of the components as the fire consumed and weakened them.

Separate structural collapse experiments were also conducted in acquired structures. A legacy bungalow home and a two-story modern colonial home were examined. Basement fires were ignited in both structures. With no ventilation present, the fires became ventilation-limited and entered the decay stage. With ventilation present, the fires continued to grow, create flashover conditions in the basement with a flow path of heated fire products up the stairs to the first floor, and caused the collapse of the flooring assemblies in both structures. In the legacy home, the flooring assembly collapsed 18 minutes from ignition. Meanwhile, in the modern home, the flooring assembly collapsed 15 minutes from ignition.

With collapse times of no more than 20 minutes from ignition in all experiments, personnel need to be aware of the potential for sudden and unpredicted structural collapse when faced with a basement fire. Since the fire department is unable to know the time from ignition, there is no safe operational time above a basement fire. Considerations should be made to avoid operating above a basement fire. Crews should also avoid operating within the flow path present in the structure. This situation places personnel in a high risk area for potential rapid fire development.

Source: UL/NIST

Collapse of structures using lightweight construction can occur earlier in the incident and may not provide the warning indicators listed above. A thorough preincident survey and size-up of the incident scene will provide you with some indication of the presence of lightweight construction.

> **WARNING!**
> Structural collapse can occur with little warning. If indicators start to appear, collapse is imminent and personnel must be withdrawn from the structure and the collapse zone.

Safety Alert

Firefighters may be injured or killed when fire-damaged roof and floor truss systems collapse, sometimes without warning. Firefighters should take the following steps to minimize the risk of injury or death during structural firefighting operations involving roof and floor truss systems:

- Know how to identify roof and floor truss construction.
- Report immediately the presence of truss construction and fire involvement to the IC.
- Use a thermal imager as part of the size-up process to help locate fires in concealed spaces.
- Use extreme caution and follow SOP/Gs when operating on or under truss systems.
- Open ceilings and other concealed spaces immediately whenever a fire is suspected of being in a truss system. *Other guidelines to follow:*
  — Use extreme caution because opening concealed spaces can result in backdraft conditions.
  — Always have a charged hoseline available.
  — Position between the nearest exit and the concealed space to be opened.
  — Be aware of the location of other firefighters in the area.
- Understand that fire ratings may not be truly representative of real-time fire conditions and that fire severity may affect the performance of truss systems.
- Take the following steps to protect firefighters before emergency incidents:
  — Conduct preincident planning and inspections to identify structures that contain truss construction.
  — Ensure that firefighters are trained to identify roof and floor truss systems and use extreme caution when operating on or under truss systems.
  — Develop and implement SOP/Gs to safely combat fires in buildings with truss construction.
- Use the following procedures to protect firefighters at the emergency incident:

- Ensure that the IC conducts an initial size-up and risk assessment of the incident scene before beginning interior fire fighting operations.
- Evacuate firefighters performing operations under or above trusses as soon as it is determined that the trusses are exposed to fire, and move to a defensive mode.
- Use defensive overhauling procedures after extinguishing a fire in a building containing truss construction.
- Use outside master streams to soak smoldering trusses and prevent rekindles.
- Report any damaged sagging floors or roofs to Command.

Source: *National Institute for Occupational Safety and Health (NIOSH): "Preventing Injuries and Deaths of Fire Fighters Due to Truss System Failures," NIOSH Publication No. 2005-123.*

Roof Types and Hazards

A number of factors must be considered for roof operations. These factors can be tied directly to the construction of the roof, loads placed on roofs, and other hazards.

Roof types include the following:

- Flat
- Pitched **(Figure 2.51)**
- Arched

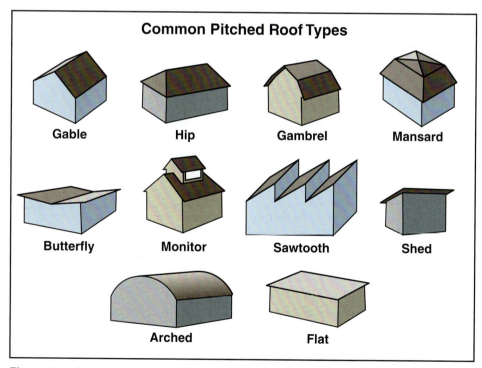

Figure 2.51 Common pitched roof types have unique features that make them ideal for some applications.

Flat

Flat roofs are commonly found on commercial, industrial, and apartment buildings and may have a slight slope to facilitate water drainage. Chimneys, vent pipes, shafts, scuttles, and skylights often penetrate flat roofs. Parapets may surround and/or divide these roofs. Roofs may also support water tanks, air-conditioning equipment, antennas, and other objects that add to a building's dead load. The structural part of a flat roof consists of wooden, concrete, or metal joists covered with sheathing. The sheathing is often covered with a layer of insulating material, plus a finish layer of some weather-resistant material **(Figure 2.52)**.

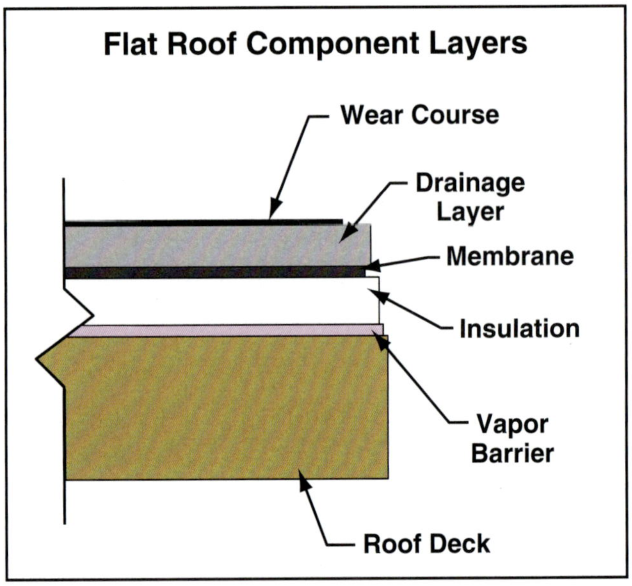

Figure 2.52 Flat roofs often include several layers of materials, each intended to mitigate specific conditions.

Tactical considerations with this type of roof include but are not limited to:

- Conventional vs. lightweight
- Attic void space
- Roof decking type (asphalt, rubber, metal, stone/gravel)
- Parapets/facades
- Skylights/scuttles/vents
- Dead loads
- Ventilation method and the tool selection based on this type of roof

Pitched

Pitched roofs have a peak along one edge or in the center and a deck that slopes downward from the peak. Pitched roofs consist of timber rafters or metal trusses that run from the ridge to a wall plate on top of the outer wall **(Figure 2.53)**. Sheathing boards or panels are usually applied directly onto the rafters. Pitched roofs usually have a covering of roofing paper (felt) applied before final weather coverings are laid. The final roof covering may be made of wood, metal, composition, asbestos, slate, rubber, concrete, or tile.

Tactical considerations with this type of roof include:

- Steepness of pitch
- Attic/cocklofts
- Type of decking (metal, slate, concrete/clay tile, asphalt, wood shingles)
- Lightweight/engineered or conventional
- Ventilation method and the tool selection based on this type of roof

Arched

One form of arched roof construction uses **bowstring** (bow-shaped) **trusses** as the main supporting members. The lower chord (bottom longitudinal member) of the truss may be covered with a ceiling to form an enclosed cockloft or roof space. *Trussless arched roofs* (sometimes called *lamella roofs*) are composed of relatively short timbers of uniform length. These timbers are beveled and bored at the ends, where they are bolted together at an angle to form an interlocking network of structural timbers. This network forms an arch of mutually braced and stiffened timbers in which the roof exerts a horizontal reaction in addition to the vertical reaction on supporting structural components. **(Figure 2.54)**

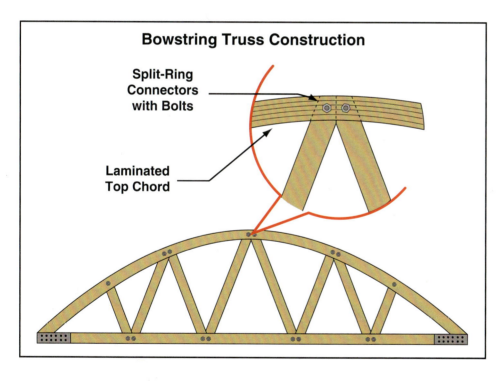

Figure 2.53 The roofing structural supports of a pitched roof place an outward load against the wall supports. *Courtesy of Wil Dane.*

Bowstring Truss — Lightweight truss design noted by the bow shape, or curve, of the top chord.

Figure 2.54 Bowstring trusses use split-ring connectors instead of gang-nailed gussets at most of the intersections within the truss.

Tactical considerations with this type of roof include but are not limited to:

- Heavy timber vs. lightweight/engineered or metal
- Parapets
- Attic space/storage
- Type of decking (rolled composition or fiberglass)
- Ventilation method and the tool selection based on this type of roof

Figure 2.55 A six-story parking garage is immediately under this park. *Courtesy of Ed Prendergast.*

Roof Hazards

The primary concern for firefighter safety is a roof's susceptibility to sudden and unexpected collapse because fire has weakened its supporting structure. The combustibility of the surface of a roof is a basic concern to the fire safety of an entire community. Flaming embers frequently cause major fires. A rooftop garden also constitutes a dead load on the roof structural system, which must be capable of supporting the load. A **green roof** can take several forms, ranging from the use of potted plants and flower boxes to a layer of earth with growing plants covering a large area of a roof **(Figure 2.55)**.

Photovoltaic (Solar) Panels

It has become more common to find photovoltaic (solar) panels on roofs. These panels are designed to convert solar energy into electricity for use within a structure or structures **(Figure 2.56)**. These panels and the other components within photovoltaic systems (wiring harnesses, inverters, batteries, and disconnect switches) can become involved in structural emergencies and create additional hazards to emergency responders. When dealing with photovoltaic systems, company officers and firefighters should:

- Never walk on these panels as they will not support the weight of personnel walking on them and can be a tripping hazard.
- Always wear full protective clothing and SCBA because these systems give off toxic vapors when on fire.
- Consider these systems to be charged or "hot" at ALL times.
- Shut off and Lock Out/Tag Out system disconnects.
- Use insulated tools around photovoltaic system components.
- Consider that photovoltaic system components may remain energized after regular power service has been shut off.

Figure 2.56 Solar panels may not be readily visible in low-visibility conditions. *Courtesy of McKinney (TX) Fire Department.*

Firefighters who touch or come into contact with solar (photovoltaic) panels are at risk of receiving a serious electrical shock. The electrical current could be strong enough to cause a firefighter to fall off the roof, fall into a solar panel, or be strong enough to cause death. The panels generate electricity from exposure from sunshine, streetlights, the moon, and the lights used during nighttime emergency response. Covering panels with salvage covers can reduce the energy to very-low levels, but they will still have a charge that can seriously injure personnel.

Occupancy Types

Occupancy types are the classifications of the use a structure is designed to contain. Structures may be divided into either single use occupancies or multiple use occupancies. These two types are described in the following sections.

Single Use

A structure that is designed for a single use must meet the building code requirements for that purpose. For instance, an office building must meet the requirements found in the Business Occupancy Classification, while an elementary school must meet the requirements of an Educational Occupancy. Requirements include exit access, emergency lighting, fire protection systems, construction type, and fire separation barriers.

In many cases, a structure such as an industrial facility will contain multiple types of uses, including storage, processing, and office. If the structure has one owner and occupant, then the structure is generally classified by its primary function. For example, a restaurant located in a single structure would be classified as an Assembly Occupancy, even though it also contains areas for cooking, washing, and multiple types of storage **(Figure 2.57)**. Fire separations and fire protection systems may be required to protect one area from another.

Green Roof — Roof of a building that is partially or completely covered with vegetation and a growing medium, planted over waterproof roofing elements. Term can also indicate the presence of green design technology including photovoltaic systems and reflective surfaces.

Figure 2.57 A freestanding restaurant is an example of a single-occupancy structure found in most communities.

Figure 2.58 A strip mall may contain different occupancy classifications, and each space is classified according to its use. *Courtesy of Ron Moore.*

Multiple Use

Structures that contain multiple-use occupancies must meet the requirements for each individual occupancy classification. In a strip mall, each space is classified according to its use and a fire-rated assembly or a wall separates it from other units as the building code requires **(Figure 2.58)**. Within the strip shopping mall, there may be retail outlets (Mercantile Occupancy), offices (Business Occupancy), and small restaurants (Assembly Occupancy). While the concept that each space will be separated from the others and meet its own requirements may have been applied when the structure was built, this may not be the case after units are sold and occupancy types change. For instance, a strip mall that was initially intended to contain only Mercantile Occupancies may have been built with limited fire separations due to the similarity of hazards in the structure. Over time, other types of occupancies may have moved in, such as a restaurant. The fire separation wall may not have the required rating, and the wall may not extend through the roof to form a complete barrier. In addition, unauthorized and non-code compliant penetrations may have been made in fire walls that will permit the spread of fire and smoke.

Chapter Review

1. Why are prefire plans and surveys important?
2. What components are commonly found in a preincident survey?
3. What resources may be available to a fire department?
4. What types of building construction might firefighters encounter?
5. Why should firefighters understand building construction?

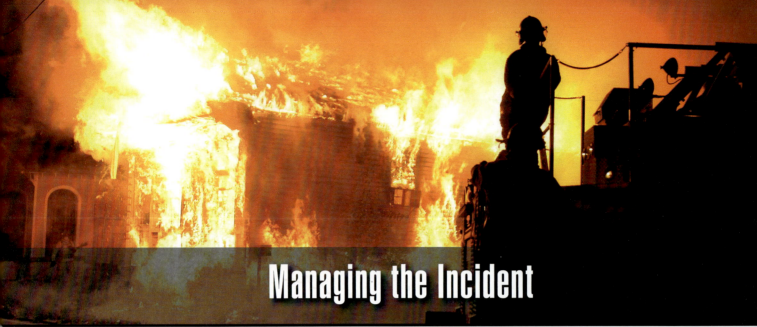

Managing the Incident

Photo courtesy of Chris Mickel, New Orleans (LA) F.D. Photo Unit.

Chapter Contents

Decision-Making **135**	Common Terminology 144
Identify and Prioritize the Problems 139	Common Communications 146
Determine a Solution 139	Unified Command Structure 148
Implement the Solution 140	Incident Action Plan 149
Monitor the Results 141	Manageable Span of Control 150
Adjust Plan as Necessary 141	Comprehensive Resource Management....................151
Result of Indecision..141	Personnel Accountability 151
National Incident Management System-Incident Command System (NIMS-ICS) **142**	Resource Tracking..152
	Chapter Review **155**

chapter 3

Key Terms

After Action Reviews 137
First Alarm Assignment 139
Freelance 141
Immediately Dangerous to Life and
  Health 152
Incident Action Plan 136

Jargon 147
National Incident Management System -
  Incident Command System 142
Span of Control 150
Personnel Accountability Report 151

Managing the Incident

FESHE Learning Outcomes

After reading this chapter, students will be able to:

4. Describe the steps taken during size-up.

Chapter 3
Managing the Incident

This chapter provides the basic information you will need to manage a structural fire incident as a fire officer:

- Decision-making
- National Incident Management System-Incident Command System (NIMS-ICS)

Decision-Making

As a fire officer, you are responsible for managing any emergency incident assigned to you. Your success as an IC will require having a strong command presence with good decision-making **(Figure 3.1)**. Applying the knowledge gained from preincident surveys, you will implement an Incident Command System (ICS) and select the best incident priorities, strategy, and tactics for the incident. The following are challenges for the Incident Commander (IC):

Figure 3.1 Considering the uncertainty that personnel may face when they arrive on-scene, an Incident Commander must show strong leadership, command presence, and solid decision-making. *Courtesy of Ron Jeffers, Union City, NJ.*

- Incident size and complexity
- Managing the safety of personnel, victims, and bystanders **(Figure 3.2, p. 136)**
- Inadequate resources
- Communication concerns
- Infrastructure, road systems, and mapping

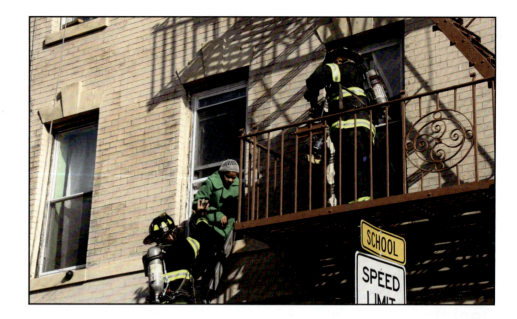

Figure 3.2 The IC must take into the account the safety of all individuals at an incident, including victims and the firefighters needed to rescue them. *Courtesy of Ron Jeffers, Union City, NJ.*

Incident Action Plan (IAP) — Written or unwritten plan for the disposition of an incident; contains the overall strategic goals, tactical objectives, and support requirements for a given operational period during an incident. All incidents require an action plan. On relatively small incidents, the IAP is usually not in writing; on larger, more complex incidents, a written IAP is created for each operational period and disseminated to all units assigned to the incident. When written, the plan may have a number of forms as attachments. *Also known as* Building Emergency Action Plan.

Incident scene management applies to all types of emergency responses and all levels of resource commitment, from single-resource situations to multijurisdictional and multiagency disasters requiring many resources. Learning and applying incident scene management at single-resource situations will help the IC perfect the skills that can later be applied to more complex situations.

A sound decision-making process is required while developing and implementing a written or nonwritten **Incident Action Plan (IAP)**, which includes the incident priorities, strategies, and tactics, and assigning resources. During an emergency incident, the IC is under pressure to quickly make the right decisions; having a pre-existing tactical worksheet greatly aids in that process **(Figure 3.3)**. Applying the tactical worksheet on routine incidents will help you become more confident and able to apply it in an emergency.

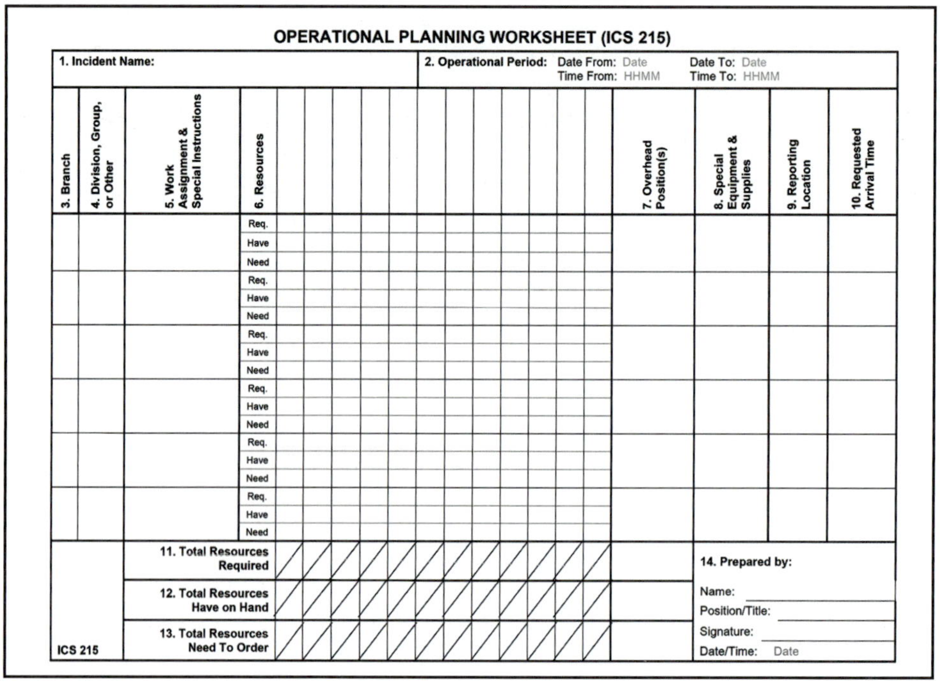

OPERATIONAL PLANNING WORKSHEET (ICS 215)

1. Incident Name:

2. Operational Period: Date From: Date Time From: HHMM Date To: Date Time To: HHMM

3. Branch	4. Division, Group, or Other	5. Work Assignment & Special Instructions	6. Resources	7. Overhead Position(s)	8. Special Equipment & Supplies	9. Reporting Location	10. Requested Arrival Time
			Req.				
			Have				
			Need				
			Req.				
			Have				
			Need				
			Req.				
			Have				
			Need				
			Req.				
			Have				
			Need				
			Req.				
			Have				
			Need				
			Req.				
			Have				
			Need				

11. Total Resources Required

12. Total Resources Have on Hand

13. Total Resources Need To Order

ICS 215

14. Prepared by:
Name:
Position/Title:
Signature:
Date/Time: Date

Figure 3.3 An example of a tactical worksheet. *Courtesy of FEMA.*

Recognition Primed Decision Making (RPDM)

Firefighters, law enforcement, and military personnel utilize a form of the Recognition Primed Decision Making (RPDM) process to make decisions on time critical and emergency situations because they seldom have time and all of the information to make decisions. This is even more critical in low-frequency, high-risk situations where the officer may have limited experience with the parameters involved.

In essence, the mind works like a computer with "files" categorized by "file folders." When faced with a decision, the human mind wants to look for familiarity. The brain starts searching previous experiences (or "files") for familiar and matching circumstances to suggest a course of action. Once a match is found, or one that the brain feels is as closely matched as possible – the experience is retrieved and becomes a conscious (or in some cases) an unconscious choice for the officer.

So where do those "files" come from? They come from a variety of sources that include past on-scene experiences, training drills, classes, books, magazine articles or even informal discussions with someone who just experienced an unusual circumstance. This last one is why **After Action Reviews (AAR)** are so important. The first part is that decisions are dissected to determine if they were correct or not for the circumstances. Good decisions are reinforced and poor decisions are evaluated to determine a more correct path. A final advantage with AARs is if they are published and discussed, the wealth of experience contained within is disseminated even further.

A problem exists when a mental "file" is corrupted either because it is outdated or was in error to begin with. For instance, as technology or science advances, better ways to approach the problem, sometimes with improved equipment, makes the previous decision invalid in today's world. Another way is if the person witnessed or experienced what was really a poor decision but just had a good outcome through luck. If the decision is not recognized as flawed, the person may recall it, try to implement it, and be faced with consequences that did not happen before.

The key to RPDM is to understand the process and continue to fill the "computer" with "files" that are appropriate and correct. This process also reinforces the need for continued and varied experience and continued education to maintain appropriate, up to date and recent "files" to enhance the officer's ability to have experiences relevant to the situations they may be faced with and the knowledge to discard "corrupted files."

After Action Reviews — Learning tools used to evaluate a project or incident to identify and encourage organizational and operational strengths and to identify and correct weaknesses.

The steps of a decision-making process are:

- Identify and prioritize the problems.
- Define the best solution.
- Implement the solution.
- Monitor the results.
- Adjust the plan as necessary.

As the situation changes, it may be necessary to again go through the cycle and identify any new problems that have resulted from implementation of the solution **(Figure 3.4)**. Decision-making is a dynamic process that continues until the incident is terminated. The examples in the following sections address only one problem. You must recognize that multiple problems may exist, requiring multiple solutions.

Figure 3.4 Incident commanders should go through a step-by-step process to help them assess a situation and then devise plans based on these observations. If new problems arise, ICs should revisit this decision-making process. *Courtesy of Chris Mickal.*

Deadly Fire Inside a Three-Bedroom House

Firefighters responded to an early morning call of a fire in a small, three-bedroom house that was built in the 1950s. The fire had originated in the center of the three bedrooms. While on scene, personnel learned that two occupants had gone back into the house after initially exiting it.

Two firefighters entered the house to perform an interior search for the two occupants. It was a wood-framed structure with a stucco exterior, and security bars covered nearly all of the windows. However, arriving crews failed to notice or report that the structure had been altered like other houses in the neighborhood to include a rain roof that may have hidden an original flat roof.

After firefighters knocked down the fire inside the house, the hoseline was abandoned and the two firefighters performing the interior search went into the structure without it. However, the altered roof contributed to the crew performing improper ventilation and heat quickly mixed with unburned fuels inside the small house. Rapid gas ignition trapped the firefighters performing the search in a bedroom, where their PPE could not protect them from the extreme fire conditions. Both firefighters died.

An investigation determined that the crew made several key mistakes, including the Command structure failing to properly manage the incident. There was confusion early on as to who was in Command within the first few minutes of the fire. The Incident Commander (IC) also failed to have a Rapid Intervention Crew/Team (RIC/T) ready to assist the firefighters performing the search inside the house.

Identify and Prioritize the Problems

To identify the problem, you must first:

- Gather facts based on the dispatch report
- Check the preincident survey
- Assess your personal observations **(Figure 3.5)**
- Check information relayed
- Assess your experiences

Once gathered, the facts must be analyzed. The analysis will identify and define the problem. For example, your company has been dispatched to a report of smoke in a single-story, single-family dwelling at 3 a.m. on a Monday. Your company is fully staffed with four people, and a complete **first alarm assignment** has been dispatched. According to the dispatch report, the occupants have evacuated the structure and reported a fire in the master bedroom. Upon arrival, you identify the problem as a working fire. At this point, you must consider the incident priorities of life safety, incident stabilization, and property conservation.

Figure 3.5 Since new problems can quickly arise at a fire, the IC must factor his or her observations into the problem-solving process. *Courtesy of Ron Jeffers.*

Determine a Solution

When the problem has been identified, you need to determine a solution. If resources and time permit, alternate solutions should be determined and the best solution chosen. In most instances, alternative solutions may be limitless. You may have to establish limits on the alternate solutions based on:

- Knowledge of local SOP/Gs
- Available resources **(Figure 3.6)**
- Time available to implement the solution

> **First Alarm Assignment** — Initial fire department response to a report of an emergency; the assignment is determined by the local authority based on available resources, the type of occupancy, and the hazard to life and property.

Figure 3.6 In order to find a solution to a problem at a scene, the IC must take into account the availability of resources, such as the equipment and personnel needed to provide a water supply. *Courtesy of Ron Jeffers.*

The best solution results in the best outcome. The following questions should be considered when making your decision:

- Has the safety of everyone, including firefighters, been considered?
- Are there enough resources to correct the problem?
- What will happen if I do not do anything?

In the example given above about the early morning house fire, your first action is to establish command and give a brief initial radio report to the dispatch center and the other units that are part of the assignment. Based on the information that you have from a 360-degree building assessment or size-up, you are reasonably certain that occupant lives are not in danger and rescue is not needed. Your primary incident priority becomes incident stabilization, which will also accomplish property conservation through an offensive strategy. Your best solution is to apply water to extinguish the fire in the bedroom while coordinating ventilation of the structure.

Implement the Solution

Your resources will implement the solution by performing tasks, such as:

- Coordinating horizontal or vertical ventilation
- Setting ladders **(Figure 3.7)**
- Advancing hoselines

Figure 3.7 Firefighters are assigned to set ladders outside a house fire. *Courtesy of Bob Esposito.*

In fire fighting terms, the decision becomes an action when the tactics are accomplished by performing the necessary tasks. In the above example of the single-story house fire, an interior attack crew controls the interior environment using hose streams to cool fire gases and extinguishing the fire while additional personnel perform ventilation. Through coordination of ventilation and cooling the interior environment, firefighters limit fire growth, which reduces both the temperature of the heated fuels and the amount of oxygen introduced during ventilation.

Monitor the Results

You determine whether the decision has effectively solved the original problem through feedback from radio reports, personal observation, and face-to-face communication from other officers or firefighters. Feedback and observation will also tell you whether the initial decision has created other problems. In the example, you are monitoring the effectiveness of the hose stream in controlling the fire and of ventilation through the removal of smoke from the compartment or structure.

Adjust Plan as Necessary

During emergency operations, factors constantly change. An IC must always evaluate the changing information and make adjustments to the IAP if they are necessary. While monitoring the results of fire fighting efforts, you may determine that the current actions are not effectively mitigating the emergency **(Figure 3.8)**.

Conditions may change to the point where it is determined that the fire attack must be changed from an offensive strategy to a defensive strategy to better protect your personnel. You must constantly evaluate the ongoing operations and make adjustments to best mitigate the current emergency. It is common for ICs to make necessary adjustments to an IAP throughout an emergency.

Figure 3.8 Three firefighters descend a ladder when the fire they were fighting intensified, putting them in danger while standing on the roof. *Courtesy of Ron Jeffers, Union City, NJ.*

Figure 3.9 Incident commanders must make decisions quickly to keep incidents from escalating. *Courtesy of Bob Esposito.*

Result of Indecision

While some types of problems can be resolved by making no decision, indecision is not an option at an emergency incident. The lack of action can and likely will result in greater damage, a potential for loss of life, and/or rapid fire spread **(Figure 3.9)**. Indecision on the part of an IC can result in **freelancing**. Freelancing occurs when individuals or crews decide on their own to take action that they believe will help to control the situation. Such actions effectively remove them from the IC's command and control. Their actions may even increase the hazard to other personnel and themselves as well as decrease the effectiveness of the actions ordered by the IC.

Freelance — To operate independently of the incident commander's command and control.

The Cedar Fire

The Cedar Fire was reported at approximately 5:37 p.m. on October 25, 2003. The fire, burning under a Santa Ana wind condition, consumed 280,278 acres and destroyed 2,232 residential structures, 22 commercial buildings, and 629 outbuildings. In addition, the fire was responsible for the death of 13 civilians and one firefighter as well as more than 100 injuries. The fire was managed under a Unified Command structure with the United States Forest Service, the California Department of Forestry and Fire Protection, and local government.

Four days later, on October 29, 2003, four personnel from Engine Company 6162 (Engine 6162) of the Novato Fire Protection District were overrun by fire while defending a residential structure located on Orchard Lane in the community of Wynola in rural San Diego County.

Shortly before the burnover, the Liaison Officer from the nearby Paradise Fire, without communication or coordination with any personnel assigned to the Cedar Fire, began lighting ring fires around the homes protected by the engine companies on the Strike Team. The freelancing actions of the Paradise Fire Liaison Officer resulted in driveways and roads becoming impassable due to the firing operations, and ignited vegetation around the house that was being protected by Engine 6162 without the knowledge of the company members.

National Incident Management System-Incident Command System (NIMS-ICS)

National Incident Management System - Incident Command System (NIMS-ICS) — The U.S. mandated incident management system that defines the roles, responsibilities, and standard operating procedures used to manage emergency operations; creates a unified incident response structure for federal, state, and local governments.

Most levels of government recognize the importance of a standardized Incident Management System. The United States government mandates that all fire and emergency service organizations that accept federal funds adopt the **National Incident Management System (NIMS)**. NIMS provides a model Incident Command System (ICS) structure that can be implemented for emergency incidents of all types and sizes.

The ICS is the basis for safe and efficient incident scene management. As a result of Presidential Directive 5, NIMS is required to be used. ICS is a component of NIMS. The first-arriving emergency services personnel should:

- Establish the NIMS-ICS
- Make decisions
- Take actions that will influence the rest of the operation

The initial decisions must be based on the organization's incident scene management procedures. Essential to all emergency incident scene management is the management of emergency response resources:

- Apparatus
- Personnel
- Equipment
- Materials

NIMS-ICS establishes an organizational structure for all types of emergency incidents, regardless of size **(Figure 3.10)**. Every organization member, especially ICs, must be familiar with the system and trained in its application. All agencies with mutual or automatic aid agreements must know and use the same system. This system may require extensive cross-training at all organizational levels among units of the participating agencies. These levels may include independent EMS providers, law enforcement agencies, and public works.

Common characteristics of the NIMS-ICS are as follows:

- Common terminology for functional structure
- Modular organization
- Common communications
- Unified command structure
- Incident Action Plan (IAP)
- Manageable span of control
- Predesignated incident facilities
- Comprehensive resource management
- Personnel accountability

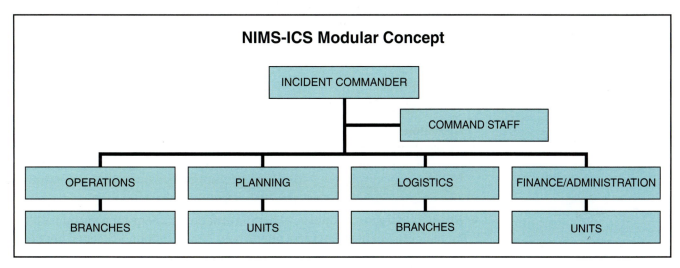

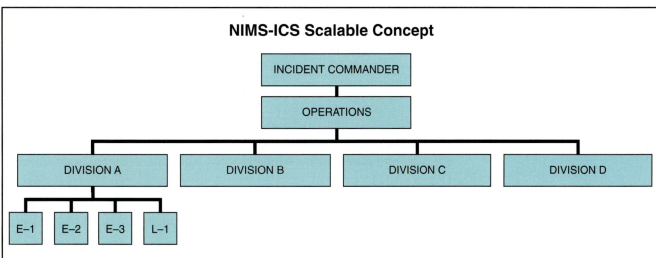

Figure 3.10 a. The modular scalable concept. B. The scalable concept.

Common Terminology

Common terminology for functional elements, position titles, facilities, and resources is essential for any command system, especially one that units from multiple agencies will use. All fire and emergency service responders must understand the terms in the following lists. A quick review of the most frequent tools of ICS is listed below.

Organizational Levels

- **Command** — Act of directing, ordering, and/or controlling resources by virtue of explicit legal, agency, or delegated authority; also denotes the organizational level that is in command (Incident Commander) of the incident. Lines of authority must be understood by all personnel involved. Lawful commands by those in authority must be immediately followed.

- **Command Staff** — Incident management personnel who report directly to the incident commander; includes the public information officer, safety officer, and liaison officer.

- **General Staff** — Incident management personnel who represent the major functional Sections.

- **Section** — Organizational level having responsibility for a major functional area of incident management includes:
  - Operations
  - Planning
  - Logistics
  - Finance/Administration
  - Information and Intelligence

  NOTE: Information and Intelligence may be designated as a Section, a Branch within Operations, or part of the Command Staff.

- **Branch** — Organizational level having functional/geographic responsibility for major segments of incident operations; organizationally located between Section and Division or Group. Branches are identified by Roman numeral or functional area (such as Command or Operations).

- **Division** — Organizational level having responsibility for operations within a defined geographic area; organizationally between Branch and single resources, task force, or strike team (which are described in the following section). Resources assigned to a Division report to that Division supervisor.

- **Group** — Organizational level, equal to Division, having responsibility for a specified functional assignment at an incident (such as ventilation, salvage, water supply) without regard to a specific geographical area. When the assigned function has been completed, it is available for reassignment **(Figure 3.11)**.

- **Unit** — Organizational level within the sections that fulfill specific support functions such as the resources, documentation, demobilization, and situation units within the Planning Section.

Figure 3.11 A group of responders will be assigned a specific task, such as ventilate a roof. *Courtesy of Bob Esposito.*

Uncertainty at a House Fire

After responding to a car accident that involved minor injuries, an engine company was dispatched to a house fire in which two children were believed to be trapped inside the structure. As the firefighters approached the house, they requested additional crews after noticing smoke with colors ranging from white to dark brown coming from the house. Upon arrival, they saw a woman trapped on the porch roof with a young child.

A police officer who was already at the scene grabbed a ladder and climbed to the roof to rescue the woman and child from the porch. At the same time, a firefighter entered the burning house to search after a neighbor confirmed that two children were still inside the structure.

A fire chief assumed the role of Incident Commander and instructed a firefighter to pull a 1 1/2-inch handline through the front door and into the house. Two other firefighters were told to don their personal protective equipment and enter the house to assist the first firefighter with the search of the two children.

A few moments later, one of the firefighters emerged from the house with one of the children. Because of the limited crew at the scene, the IC turned his attention away from the fire to perform CPR on the child. As the firefighter prepared to re-enter the house to continue with the search-and-rescue, she noticed water was flowing from the hoseline, which had been burned through.

A thermal blast of heat ripped through the house before the three firefighters conducting the search could exit, killing them. The child was trapped inside the structure with injuries, but survived. An investigation of the incident determined that a RIC/T had not been established and in position before the search began. The report also called for firefighters to be trained in conducting defensive searches and that Command always maintain close accountability for all personnel at a scene.

Resources

Resource designations provide an indication of the type and capability of the resources that may be available to the IC for assignment. It is important to track the statuses of the following resources so that they may be assigned without delay when and where they are needed.

- **Crew** — Specified number of personnel assembled for an assignment such as search, ventilation, or hoseline deployment and operations. The number of personnel assigned to a crew should be within span-of-control guidelines. A crew operates under the direct supervision of a crew leader.

- **Single resources** — Individual pieces of apparatus (engine, ladder/truck, water tender, bulldozer, air tanker, and helicopter) and the personnel required to make them functional **(Figure 3.12)**.

Figure 3.12 An engine is an example of a single resource. *Courtesy of Ron Jeffers, Union City, NJ.*

- **Task force** — Any combination of resources (engines, ladders/trucks, bulldozers) assembled for a specific mission or operational assignment. All units in the force must have common communications capabilities and a designated leader. Once a task force's tactical objective during an incident has been met, the force is disbanded; individual resources are reassigned or released.

- **Strike team** — Set number of resources of the same type (engines, ladders/trucks, bulldozers) that have an established minimum number of personnel. All units in the team must have common communications capabilities and a leader in a separate vehicle. Unlike task forces, strike teams remain together and function as a team throughout an incident.

Common Communications

A common means of communication is essential to:

- Maintaining control
- Coordination
- Safety at any incident

Experience at emergency incidents has shown that background noise and personal safety equipment can significantly affect the ability to hear and understand radio transmissions **(Figure 3.13)**.

INCIDENT RADIO COMMUNICATIONS PLAN (ICS 205)

1. Incident Name:	2. Date/Time Prepared: Date: Date Time: HHMM	3. Operational Period: Date From: Date Date To: Date Time From: HHMM Time To: HHMM

4. Basic Radio Channel Use:

Zone Grp.	Ch #	Function	Channel Name/Trunked Radio System Talkgroup	Assignment	RX Freq N or W	RX Tone/NAC	TX Freq N or W	TX Tone/NAC	Mode (A, D, or M)	Remarks

5. Special Instructions:

6. Prepared by (Communications Unit Leader): Name: Signature:

ICS 205 | IAP Page | Date/Time: Date

Figure 3.13 An example of an incident radio communications plan. *Courtesy of FEMA.*

ICS requires that a common communication system provides the ability to be understood and to contact all units or agencies that are assigned to the emergency incident. To ensure effective communications, all units must use clear text (specified phrases in plain English) rather than the numeric 10-codes, other agency-specific radio codes, or **jargon**.

Jargon — The specialized or technical language of a trade, profession, or similar group.

Communications between units will be through the jurisdiction's emergency radio communication system or via face-to-face communication. As units are assigned to the incident, the central communications center may announce the radio frequency in use for Command or automatically place all radios on that frequency. Units assigned to the incident must ensure that they have complete communication with the Command Post. Face-to-face communication is always the preferred method of exchanging information, but is not always practical on the fireground **(Figure 3.14)**. Some communications should be announced over the radio so that all personnel on the fireground are aware of the information.

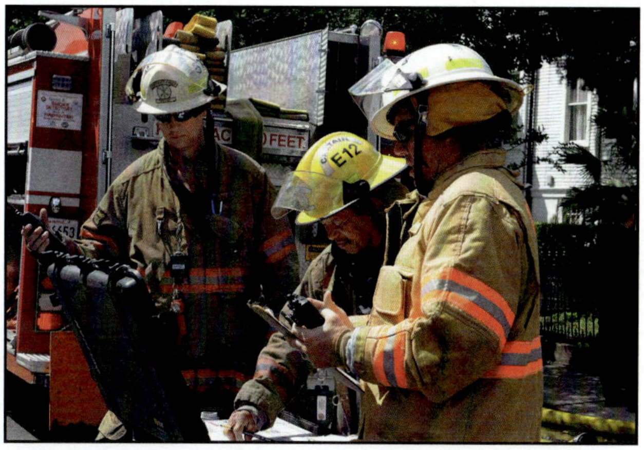

Figure 3.14 Units should maintain communication with the Command Post, preferably face-to-face communication if possible. *Courtesy of Chris Mickel, New Orleans (LA) F.D. Photo Unit.*

The ICS also requires the establishment of a common incident communications plan that identifies radio frequencies or channels to be used exclusively for specified organizational functions. To avoid the chaos that would result from all units attempting to receive and transmit on the same channel, the Incident Communications Plan, consistent with departmental SOP/Gs, assigns specific channels to specific functions or units.

While most modern mobile and portable radios are capable of scanning, receiving, and transmitting over dozens of channels, not every organization is equipped with the latest communications equipment. If mutual aid units are not equipped with radios that can receive and transmit on the channels assigned to them in the plan, they must be issued portable radios that will function on those channels.

Everyone at the emergency scene must follow two basic communications rules. First, units or individuals must identify themselves in every transmission as outlined in the local radio protocols. The protocol may be to first identify yourself and then the receiver as in *"Engine One to Dispatch"* or to identify the person or unit you are calling first as in *"Dispatch from Engine One."* Second, the receiver must acknowledge every message by repeating the essence of the message to the sender.

Example:

Engine 4: *Communications from Engine 4. We are on scene and have a trash fire. Engine 4 can handle. Return all other units.*

Communications: *Engine 4, I copy you have a trash fire, and you will handle. Other units can be canceled.*

Requiring the receiver to acknowledge every message ensures that the message was received and understood. This feedback can also tell the sender if the message was not correctly understood and further clarification is necessary.

Unified Command Structure

A unified command structure is necessary when an incident involves or threatens to involve more than one jurisdiction or agency. These multijurisdictional incidents are not limited to fires. For example, hazardous materials releases may spread from one jurisdiction to another **(Figure 3.15)**. Large-scale natural disasters, such as Hurricane Katrina, can easily affect multiple jurisdictions. A unified command allows multi-juridistictional operations to be conducted using the following:

- Common approaches
- Communications
- Procedures

Figure 3.15 The release of hazardous materials can evolve into a multijurisdictional incident that requires a unified command structure. *Phil Lindner/Canada Transportation Accident Specialist.*

Incident Action Plan

NFPA 1021 requires the Fire Officer I to be able to develop an Incident Action Plan (IAP). According to NFPA 1561, *Standard for Fire Department Incident Management System*, an IAP establishes the overall strategic decisions and assigned tactical objectives for an incident.

The IC should develop the IAP with a focus on actions that can be implemented during the initial phase of the incident. Examples may include the following:

- Fire suppression
- Rescue
- Water supply
- Ventilation

The number and capability of resources initially at the scene will determine the exact implementation. The IC must prioritize response actions based on available resources **(Figure 3.16)**. The initial IC should not hesitate to call for additional resources if the judgment is made that the resources on-scene or dispatched are inadequate to deal appropriately with the incident.

Figure 3.16 The IC must base the response to an incident on the resources available and their capabilities. The IC might need to call for more resources than those initially available at the scene. *Courtesy of Chris Mickel, New Orleans (LA) F.D. Photo Unit.*

Organizational policies will determine the transfer of command process. For example, when the IC transfers command to a higher ranking officer, the receiving IC will review, evaluate, and revise the IAP (if necessary) to determine the effectiveness of the actions already taken.

A formal written IAP is required for long duration events. In addition, the Incident Command System will likely expand to include functions of the Planning Section to assist with the development of the IAP. Based on their training and experience, fire officers may be asked to serve in this role or reassigned to other operational aspects of the incident.

Manageable Span of Control

Span of control is the number of direct subordinates that one supervisor can effectively manage. An effective span of control ranges from three to seven subordinates per supervisor, depending upon the variables, with five considered the optimum number **(Figure 3.17)**. Supervisors can more easily keep track of their subordinates and monitor their safety if an effective span of control is maintained.

> **Span of Control** — Maximum number of subordinates that that one individual can effectively supervise; ranges from three to seven individuals or functions, with five generally established as optimum.

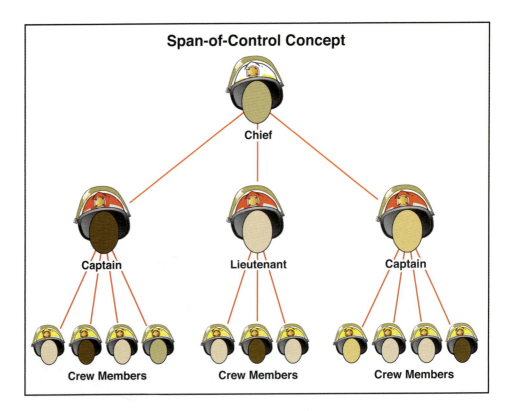

Figure 3.17 Span of control is a tool used to promote organization during the activities required at an incident.

Based on the following variables, the supervisor may want to increase the number of subordinates in the span of control **(Figure 3.18)**:

Figure 3.18 A supervisor must consider several variables when evaluating his or her subordinates to determine the span of control at a scene. *Courtesy of Chris Mickel, New Orleans (LA) F.D. Photo Unit.*

- Subordinates perform similar functions as other members of the team
- Subordinates work within sight of the supervisor for monitoring/evaluation
- Subordinates are able to communicate with other members of the team
- Subordinates are skilled in performing the assigned task
- Supervisor judges the incident to be relatively simple

Comprehensive Resource Management

The purpose of comprehensive resource management is to provide the IC with access to and control over all available resources. Resource management involves maintaining an accurate and up-to-date awareness of resources needed for utilization, including:

- Personnel
- Teams
- Equipment
- Supplies
- Facilities

> **Personnel Accountability Report (PAR)** — Roll call of all units (crews, teams, groups, companies, sectors) assigned to an incident. The supervisor of each unit reports the status of the personnel within the unit at that time, usually by radio. A PAR may be required by standard operating procedures at specific intervals during an incident, or may be requested at any time by the incident commander or the incident safety officer.

Personnel Accountability

Accountability is vital during an incident, particularly when the hazards change. At a structure fire, that change might be the extension of the fire through a concealed space or the rapid increase in the volume of fire creating a flashover or backdraft situation or a change in strategy. SCBA can malfunction or run out of air and firefighters can get lost in mazes of rooms and corridors. Without having an accountability system, it is impossible to determine who and how many firefighters may be trapped inside or injured **(Figure 3.19)**. Too many firefighters have died because they were not discovered missing until it was too late. An accountability system should include a Command Officer requiring constant feedback, plus updated **personnel accountability reports (PAR)** to track and maintain firefighter safety.

The IC must know which firefighters are at the incident, where each one is located, and his or her assignment. To help with this task, the IC is responsible for managing the personnel accountability system employed by the organization. The IC may delegate the authority by assigning an accountability or Incident Safety Officer to monitor the system. The system should indicate the individuals assigned to each apparatus, the names of people responding individually (such as staff personnel and volunteers) the time of arrival, the assigned duty or unit, and the time of release from the scene. Common types of personnel accountability systems used at the emergency incident include:

Figure 3.19 While several personnel accountability systems are available, a department must utilize at least one of them to ensure that all the firefighters responding to a call are accounted for and not accidentally overlooked. *Courtesy of Ron Jeffers, Union City, NJ.*

- Tag/passport systems **(Figure 3.20)**
- SCBA tag systems
- Bar code readers

Each organization should employ a standardized accountability system at every incident. This system should also be compatible with any mutual/auto aid departments. The system should identify and track all personnel working at the scene. All personnel must be familiar with the system and participate in it. The system must also account for personnel who respond to the scene in personally owned vehicles.

Personnel operating in the **Immediately Dangerous to Life and Health (IDLH)** environment must be assigned in crews of two or more and be close enough to each other to provide immediate assistance in case of need. Crew members must maintain constant communication with each other through visual, audible, or physical means and have a portable radio to keep in contact with the IC. Company officers are responsible for keeping track of assigned resources.

The initial interior attack shall not be initiated in an IDLH environment until there are sufficient personnel on scene to meet the two-in, two-out rule per OSHA 1910.134 The only exception to this rule listed in the OSHA mandate applies to firefighters performing a known rescue. See NFPA 1500 for further details.

Figure 3.20 Accountability systems provide a mechanism for ensuring the safe return of responders from a fire scene. *Courtesy of Ron Jeffers, Union City, NJ.*

Immediately Dangerous to Life and Health (IDLH) — Description of any atmosphere that poses an immediate hazard to life or produces immediate irreversible, debilitating effects on health; represents concentrations above which respiratory protection should be required. Expressed in parts per million (ppm) or milligrams per cubic meter (mg/m^3); companion measurement to the permissible exposure limit (PEL).

Resource Tracking

In most career departments, units responding to an incident arrive fully staffed and ready to be assigned an operational objective **(Figure 3.21)**. In volunteer and combination departments, personnel may arrive individually or in small groups. Departments should have in place SOP/SOGs to guide personnel in establishing a tracking and accountability system at incidents.

Figure 3.21 An apparatus races toward a large structure fire to join other units already on-scene. *Courtesy of Ron Jeffers, Union City, NJ.*

The SOP/SOGs must incorporate the following elements, which must be included in the IAP:

- Procedure for checking in at the scene
- System for tracking the location and assignments of each unit and all personnel on scene
- Determination of apparatus staging **(Figure 3.22)**

Figure 3.22 A department's SOP/SOG must determine a tracking system for staging apparatus at an incident. *Courtesy of Chris Mickel, New Orleans (LA) F.D. Photo Unit.*

Apparatus Staging Protocols

Many fire departments use two different staging protocols. Level I staging is applied to the initial response of more than one fire department unit. Level II staging is enacted when a large number of units are responding to an incident. The IC or Operations Section Officer initiates this level when requesting additional resources.

Level I staging is often used on any response of two or more units. During Level I staging operations, later-arriving apparatus stop (stage) approximately one block away from the scene in their direction of travel and await further instructions If the first-arriving unit has no immediate orders for later-arriving apparatus, the officer may call for Level I staging to be implemented while the incident is investigated. Engine companies in this scenario typically stage near a hydrant or water source. While staged, driver/operators should not allow their engines to get blocked in by other resources.

Level II staging is implemented when numerous units are responding to the same incident, particularly those situations that require mutual aid or result in the transmittal of multiple alarms. The Operations Section Officer may designate one or more apparatus staging areas from which the IC can draw additional resources. Units responding to the incident are advised of the staging area location when dispatched, and they respond directly to that location. A parking lot or other large open area, such as a field, may be designated for use as a staging area so long as it can be secured and is free of civilian traffic.

Though not necessarily part of the IAP, procedures for releasing resources no longer needed should be included in department SOP/SOGs. Personnel must be trained to check out when leaving the incident.

The IC must be able to locate, contact, deploy, and reassign the units that are assigned to an emergency incident. These tasks are accomplished through the ICS procedures that assign units to locations within the operating area. As units arrive on the scene, the IC assigns them to locations or functions as needed. The units may be held in a staging area until needed or until they are released from the incident **(Figure 3.23)**. If staging has not been implemented, company officers must check in with the IC at the Command Post for an assignment upon their arrival.

Figure 3.23 Personnel wait on the scene of a four-alarm fire in New Orleans. *Courtesy of Chris Mickel, New Orleans (LA) F.D. Photo Unit.*

CAUTION
Personnel must not *freelance* or assign themselves to a task.

The IC can use a number of visual aids to help manage and track resources assigned to the incident. The visual aid can be as simple as a preprinted form, called *Tactical Work Sheets*, to sketch the incident scene and the location of units as they arrive. For tracking personnel, the IC could use simply white boards with a grease pencil or Velcro® or more elaborate tracking board with magnetic symbols identifying each of the units **(Figure 3.24)**. Regardless of the tracking device or system, it should be simple to read and contain as much information as necessary about the activities of all the units on scene. The visual aid may contain the following information:

- Assigned radio frequencies
- Assigned, available on-scene, and requested units

Figure 3.24 Fire personnel in Passaic, New Jersey, use a tracking board with magnetic symbols identifying individual units at a structure fire. *Courtesy of Ron Jeffers, Union City, NJ.*

- Activated ICS functions
- Site plan
- Staging areas
- Logistics location
- Control zones
- Unit incident assignment and location

Chapter Review

1. What steps are included in the decision-making process?
2. What are the elements of NIMS-ICS?

Size-Up: Evaluation and Assessment

Photo courtesy of Chris Mickel, New Orleans (LA) F.D. Photo Unit.

Chapter Contents

Incident Size-Up Considerations 159	**Firefighter Survivability Approaches 181**
While Responding 159	Occupant Survival Profile 181
Facts, Perceptions, and Projections and Probabilities 162	Crew Resource Management 183
On Arrival 167	Rules of Engagement 184
Reading Smoke 172	**Critical Fireground Size-Up Factors 185**
During the Incident 179	Building Characteristics 186
Decision-Making 180	Life Hazard via Occupancy Condition 191
Plan of Operation 180	Arrival Condition Indicators 200
	Chapter Review 202

Chapter 4

Key Terms

Crew Management Resource 183
End-of-Service-Time Indicator 192
Ignition Source 173
International Association of Fire Chiefs (IAFC) 184
Laminar Flow 178
Life Safety, Incident Stabilization, and Property Conservation (LIP) 180
Means of Egress 188
Metal-Clad Doors 189
Mobile Data Computer 159
Occupant Survival Profile 181

Plan of Operation 180
Rapid Intervention Crew or Team 167
Rehabilitation 201
Risk-Benefit Analysis 199
Situational Awareness 164
Shelter in Place 196
Size-Up 159
Turbulent Flow 178
Velocity 162
Ventilation-Controlled 177

Size-Up: Evaluation and Assessment

FESHE Learning Outcomes

After reading this chapter, students will be able to:

1. Discuss fire behavior as it relates to strategies and tactics.
3. Identify the basics of building construction and how they interrelate to prefire planning and strategy and tactics.
4. Describe the steps taken during size-up.
5. Examine the significance of fire ground communications.

Chapter 4
Size-Up: Evaluation and Assessment

To perform an effective and thorough size-up, personnel should consider factors that are relevant throughout the incident **(Figure 4.1)**:

- Incident size-up considerations
- Firefighter survivability
- Critical fireground size-up factors

Figure 4.1 Personnel, especially those in a Command role, must consider factors such as firefighter survivability while performing the size-up of a scene. *Courtesy of Ron Jeffers, Union City, NJ.*

Incident Size-Up Considerations

Size-up is a continuous process that actually begins before an incident is reported and continues throughout the incident. This section discusses the application of size-up theory to three specific time intervals: while responding, on arrival, and during the incident.

While Responding

The size-up process begins when the alarm is sounded or your pager is activated. Depending on the dispatch system that your department uses, you may receive information via a range of channels, including a verbal account over a radio broadcast or a written account on a **mobile data computer** in the apparatus **(Figure 4.2, p. 160)**. This information contains a description of the current conditions as reported by witnesses at the scene. While paying close attention to the message, you must also remember that the information provided may be inaccurate and incomplete.

> **Size-Up** — Ongoing evaluation of influential factors at the scene of an incident.
>
> **Mobile Data Computer (MDC)** — Portable computer that, in addition to functioning as a Mobile Data Terminal, has programs that enhance the ability of responders to function at incident scenes.

Figure 4.2 Information about an incident may come over a radio.

You must be prepared for any type of situation. For example, reported vehicle fires have been found to be well-involved vehicles parked in garages under apartment buildings. Prior to the arrival of fire personnel, some of the things you can verify immediately include:

Figure 4.3 Firefighters arriving at this scene can easily determine that the structure fire is occurring at nighttime. *Courtesy of Bob Esposito.*

- Time of day **(Figure 4.3)**
- Weather conditions
- Capabilities of your department's response

Local SOP/SOGs will generally establish the minimum information that is given during a dispatch broadcast, including:

- **Time** — Exact time of dispatch, not the time the fire started
- **Situation** — Type of emergency reported to the dispatch center
- **Location** — Address of the emergency situation
- **Resources dispatched** — Units responding

- **Pre-plan resources** — Information sources available such as running card, Mobile Data Computer/Terminal (MDC/T), mapbook

Pay particular attention to nonstandard alarm assignments. Notice if units are being called that are not typically on your first assignment or if some units are missing that may indicate that resources are unavailable.

Time of Day

Consider the time of day when the alarm is received and how it will affect the incident by answering the following questions:

- Is a commercial property occupied during normal business hours?
- Is it during evenings or weekends when residential properties are more likely to be occupied?
- Is it the middle of the night when residential occupants may be sleeping?
- Is it during school hours on a weekday?
- How are the month, day of the week, and time of day likely to affect traffic congestion along the response route?

In addition to the time of day, other time factors can help you calculate details at the scene. For example, although you do not initially know the time the fire started and how long it has been burning, being able to calculate your response time and the visual clues of the fire upon arrival will help determine the stage of the fire.

Weather Conditions

Continue to evaluate weather-related variables as you respond to the reported incident by asking the following questions:

- How will the weather affect response times? **(Figure 4.4)**
- How will the weather affect apparatus safety during response and set-up?
- How will the weather affect firefighter safety and fire conditions?

Figure 4.4 Firefighters must consider how weather could slow their response time, such as trying to advance hoselines in heavy snow. *Courtesy of Ron Jeffers, Union City, NJ.*

Chapter 4 • Size-Up: Evaluation and Assessment   **161**

Capabilities of Your Department's Response

Based on your knowledge of the response area, the units assigned, and the radio communications, you should be able to determine which unit will arrive first and establish the ICS. Consider the answers to the following questions to evaluate the resources your department will be able to contribute to the response:

- What are the capabilities of my personnel?
- How long before additional resources arrive at the scene?
- Have the correct resources been dispatched to meet the current hazard?
- Are my resources trained to meet the actual hazard?
- Where is my water supply and is it sufficient to control the situation?

Additionally, consider any information that you have available, such as:

- Review the preincident plan for the building (if available) to more thoroughly prepare for what responders may encounter. Some fire departments require stations to have preincident plans of high hazard occupancies easily accessible to the company officer on the apparatus.
- Be prepared to observe the color, thickness, volume, **velocity**, density, pressure, and movement of any smoke as soon as it is visible **(Figure 4.5)**.

> **Velocity** — Rate of motion in a given direction; measured in units of length per unit time, such as feet per second (meters per second) and miles per hour (kilometers per hour).

Figure 4.5 The Incident Commander will be able to collect information from the color, thickness, volume, velocity, density, pressure, and movement of the smoke. *Courtesy of Bob Esposito.*

This information, combined with knowledge of fire behavior and the building where the fire is burning, can help you better assess resource needs. You should also ask for and evaluate any additional information that the dispatch center provides over the radio during the response.

Facts, Perceptions, and Projections and Probabilities

When considering the foundation for your decision-making process, think about four components: facts, perceptions, and projections and probabilities. When you learn how to assign priority to different types of information, you will be able to make sound decisions when you arrive at the incident scene.

Facts

The facts of the situation are things that you know to be true. The dispatch center should provide the facts as they are known based on the report of the emergency. However, not all the information given when dispatched may be incomplete. All of this information can and should be factored into your thought process regarding the emergency. Facts are also based on the following:

- Preincident survey of the site
- Knowledge of building construction and occupancy
- The current time of day and the day of week and on-scene observations, including a 360-degree survey
- Knowledge of basic fire development

You should also consider that a preincident plan may not exist and that you may not have made a survey of the site. In these cases, your facts may be reduced to a minimum and based on the information provided in the alarm dispatch and what you see upon arrival. Refer to Chapter 2, Prefire Planning, for more information on preincident plans.

Construction and occupancy. Knowing the type of construction can help you project potential fire growth, potential directions of fire travel, and ways to use the structure to protect occupants. Knowledge of the type of occupancy or use of the structure can tell you the potential number and location of occupants. It can also alert you to the ability of the occupants to self-evacuate or whether they will require assistance. Construction and occupancy types can also give the first-arriving officer an idea of the dangers and obstacles that the unit may face in trying to attack the fire and perform rescue. Steel bars in windows and doors create problems for occupant egress and firefighter access **(Figure 4.6)**. Occupancy information may indicate possible hazards while searching. Construction type and occupancy dictates strategy and tactics. For example, a concrete roof delays or eliminates the possibility of vertical ventilation.

Figure 4.6 Firefighters responding to a structure fire encounter metal bars over a window. *Courtesy of Chris Mickel, New Orleans (LA) F.D. Photo Unit.*

Time of day/day of week. Although some types of occupancies are inhabited twenty-four hours a day, seven days a week, most structures are not. The presence of security personnel or a cleaning crew is possible and must be considered. The preincident survey may include this type of information.

Fire behavior. The ability to estimate how fires will grow, spread, and generate toxic atmospheres within a given structure or occupancy will help determine how much danger the occupants are in and the level of risk to which your crew will be exposed. In some incidents, fatalities caused by smoke or fire gas inhalation occurred in locations far away from the actual seat of the fire. The only real assumption you can make is that the fire will spread regardless of the construction if water is not applied correctly, sufficiently, and swiftly. Fire will also grow significantly slower by limiting the oxygen available.

Resources. You must have appropriate resources to attempt any action. Resources include sufficiently trained personnel with appropriate equipment and water supply. The authority having jurisdiction (AHJ) may include the following resources to be dispatched or called: utility companies, law enforcement, and more.

Weather. You must be aware of how weather conditions can influence fire behavior and growth, such as wind-driven fires. The weather at the scene can affect your choice in tactics and your location and plan of attack.

Perceptions

Your perceptions are based on observations influenced by your knowledge, biases, beliefs, and past experiences. Maintain **situational awareness** and continually reassess the situation to prevent from developing tunnel vision. The presence of lights on or vehicles in the parking lot of a commercial building at 2 a.m., children's toys near the front door of a residence, or evidence of forced entry to a vacant building may be indications that the building is occupied **(Figure 4.7)**. These indications can be used to justify implementing a rescue tactic. Your emotions may also influence your perceptions. In a small community, you may know the people who live or work in the structure. Neighbors or witnesses who are upset, regardless whether they have actual knowledge of who is in the structure, may influence your decisions.

> **Situational Awareness** — Perception of the surrounding environment and the ability to anticipate future events.

Figure 4.7 The graffiti inside this vacant building could be an indication that people sometimes occupy it. *Courtesy of Chris Mickel, New Orleans (LA) F.D. Photo Unit.*

Sometimes, you may be misled to accept as a fact what is really only a perception. To counteract misperceptions, you must rely on additional sources for your gathering of facts. The observations of your crew members are some of the best resources. You should encourage your crew to apply situational awareness during emergencies. Communicate with and listen to the observations provided by them throughout the size-up. The synergy created by the shared knowledge and experience of the entire unit should provide a balanced approach to decision-making.

Projections and Probabilities

Probabilities are things that are likely to occur based on a given situation. To assist in making decisions, the following questions must be answered regarding the probabilities of a fire-emergency situation:

- Where is the fire located and in which direction is the fire likely to spread?

- Are exposures at risk? **(Figure 4.8)**

- Have explosions occurred or are they likely? Is a secondary explosion likely? Is structural collapse likely?

- Do occupants need to be evacuated?

- Are additional resources likely to be needed? If so, what types and how many?

Figure 4.8 A probability is something that is likely to occur based on a given situation, such as the risk of a fire spreading to exposures. *Courtesy of Ron Jeffers, Union City, NJ.*

Combustible Furnishings

Research by the National Institute of Science and Technology (NIST) and Underwriters' Laboratory (UL) has shown that with modern construction and furnishings many fires are oxygen-lean and fuel-rich. This result can lead to *rapid fire development* if ventilation and fire attack are not closely coordinated. This phenomenon can occur without warning. Fire officers and Incident Commanders should consider this hazard prior to sending crews inside a structure for search or fire attack operations. The IFSTA Manual, **Building Construction Related to the Fire Service**, includes research data on modern furnishings and interior finishes as relevant to their effect on a structure fire.

Your ability to predict what might happen next at a scene is based on your knowledge and experience of the following:

- *Fire behavior* — Rapid fire development recognized, in part, from reading the smoke, fire growth, and fire spread may threaten emergency responders and occupants in other portions of the structure and adjacent structures. Have resources available to address changes in fire growth. Resources that are not committed to a task should be staged for quick deployment if conditions change.

- *Building construction* — Knowledge of building construction can help you predict how long a fire can be contained within a compartment and how rapidly it can spread through concealed spaces. This knowledge also helps to estimate when the structure, or a portion of it, has the potential to collapse.

- ***Fire fighting activities*** — Improper application of fireground procedures can increase fire growth. Improperly performed tasks, such as nozzle pattern selection, inadequate fire flow, and uncoordinated ventilation, will cause the fire to spread, endanger lives, and may cause a flashover or backdraft. Fire attack must be closely coordinated with ventilation.

In addition to the construction types, materials, and structural weaknesses of a building, knowing the layout of the building will also improve the success and safety of rescuers at an incident. For example, knowing the location of access and egress doors and stairways will assist in the following:

- Predicting areas to expect victims
- Searching for and removing victims
- Protecting yourself and your crew during the search operation
- Advancing hoselines/performing ventilation **(Figure 4.9)**

Figure 4.9 Firefighters perform ventilation on a house fire. *Courtesy of Bob Esposito.*

The IC must be aware of work cycles and air management in initially determining resource needs. Firefighter fatalities at structural fire incidents tend to be the result of personnel becoming disoriented and running out of air or being involved in a structural collapse. The following contribute to firefighter safety:

- Creating and maintaining multiple means of egress
- Using hoselines or tag lines
- Performing proper ventilation
- Maintaining awareness of proper air management

Methods of preparing for a potential rescue include:

- Knowing where crews are at all times through an effective accountability program

- Having an established **rapid intervention crew/team (RIC/RIT)** on site
- Keeping in constant communication with all crews by requesting continuous feedback reports
- Being aware of how the fire is developing

The Occupational Safety and Health Administration (OSHA), some state regulations, and NFPA standards mandate the presence of an RIC/RIT outside of the IDLH (immediately dangerous to life and health) environment for the potential rescue of firefighters **(Figure 4.10)**. While the regulation only requires two trained personnel as RIC/RIT, it is highly recommended that the crew be more than two members and include an officer. In a true RIC/RIT deployment, it will take several crews to facilitate a rescue. The IC must use the same care in planning a rescue effort as planning for fire control. Sufficient and appropriate resources must be assigned to safely carry out the rescue assignment. While the NFPA standards and federal/state/provincial regulations allow the RIC/RIT personnel to have concurrent outside assignments, those assignments may not be such that leaving them to assist the firefighters inside would increase the jeopardy of any firefighters on the scene.

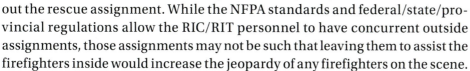

Figure 4.10 Basic equipment that a RIC/RIT carries might fit into a duffel bag.

> **Rapid Intervention Crew or Team (RIC/RIT)** — Two or more firefighters designated to perform firefighter rescue; they are stationed outside the hazard and must be standing by throughout the incident.

On Arrival

The most intense part of the size-up process occurs when you are the first resource to arrive at the emergency incident scene. You may arrive on a scene of chaos or one with no visible clues that an emergency exists. In addition to the emergency situation and those individuals directly involved in it, numerous spectators may be gathered at the scene, making it difficult to distinguish them from occupants or victims. These bystanders may be upset, and possibly irrational, screaming for responders to do something **(Figure 4.11)**. Some

Figure 4.11 First responders could arrive at a chaotic scene that includes occupants or victims standing nearby, some upset and demanding that personnel do more to rectify the situation. *Courtesy of Ron Jeffers, Union City, NJ.*

individuals may attempt to extinguish the fire, assist victims, or perform a rescue. These actions may place them in direct danger.

In the midst of this scene, you must assess and communicate information to other responding units and the dispatch center. Your scene assessment begins before the apparatus comes to a stop and you step from the apparatus. The incident may be categorized by the severity, extent, and dynamics of the incident. One of two general situations are encountered upon arrival:

- There may be nothing evident that will prompt an investigation.
- There may be smoke or fire showing.

These two situations are the core of the arrival report or conditions report (conditions on arrival or size-up report in some jurisdictions) that you relay over the radio. Many fire and emergency services organizations mandate this report in their operations SOP/SOG. This initial report is an essential communication that will inform other responders and establish a foundation upon which to build the incident.

According to the AHJ, the arrival report may contain:

- Unit number and update of address if necessary
- Brief description of the scene, including the extent of fire conditions
- Status of water supply
- Establish Command and Command mode as well as define the strategy

A situational update report during a 360-degree survey may include:

- Special considerations and hazards
- Entrance and egress points
- Intended initial actions and assignments for incoming units
- Requests for any additional resources

The report provides other responding units with an idea of what they will encounter. You are describing the fire scene for crews who have not arrived yet. It should clearly portray what is to be expected by giving a description of the scene, including:

- Building type and size
- Extent and location of fire
- Life safety issues
- Occupancy type
- Water supply needs or availability
- Type and location of any special hazards:
  — Barriers that could impede access
  — Location and condition of victims in need of assistance
  — Downed power lines
  — Other observations

The arrival report also includes:

- Command statement
- Location and type of the Command Post (fixed or mobile)

Figure 4.12 The arrival report should provide a description of the initial action at an incident. *Courtesy of Bob Esposito.*

- Description of initial action **(Figure 4.12)**
- Command option (investigation, fast-attack/offensive, or defensive, rescue mode)

If the first person who takes the IC role is a company officer, he or she must determine whether his or her direct involvement with the crew will make a significant difference, or should the IC separate and establish command outside. The stationary Command post is proven based on NIOSH reporting to be a safer Command option, but other AHJ give other types of Command operations, such as mobile. If the operation will be managed over a specific radio frequency (some jurisdictions assign a tactical channel on dispatch), then this information is also provided to responding units and they are directed to move to that frequency.

Depending upon the organization's SOP/SOG, if responders from the first-arriving engine do not deploy a supply hoseline from the nearest accessible hydrant (or otherwise ensure a water supply), responders from the next arriving engine may be assigned this operation. As with any assignment, the initial IC must be specific about assigning water supply as in ordering, "Engine Two, connect to a hydrant and supply Engine One." Depending upon policy, this procedure may be automatic or the IC may assign it. In the latter case, providing the water supply would be part of the IAP and one more decision that the first-arriving officer must assign.

Personnel should be assigned tasks upon arrival to prevent units from self-assigning tasks, and to ensure personnel are given assignments that will result in saving lives, limiting fire growth, and preserving property. The assigning of tasks typically occurs just after the arrival report is transmitted. The idea is that the first-arriving crews are told what to do as the IC begins obtaining information from the 360-degree building assessment.

Following the arrival report, perform a visual size-up by looking at the scene from all sides. If you cannot complete a full 360-degree survey or walk around the structure, you must use other means to learn conditions on all sides of the incident.

While checking all sides of the structure, the information you gather includes:

- Victim survivability based on conditions
- Any indicators of fire location
- Lowest floor of fire involvement
- Smoke conditions
- Forcible entry requirements
- Special hazards such as propane, chemicals, electrical lines, or elevation changes **(Figure 4.13)**
- Building construction features such as basement access and balloon construction

Figure 4.13 Electrical lines could pose a safety hazard to firefighters at an incident and their location should be noted during a 360-degree survey. *Courtesy of Chris Mickel, New Orleans (LA) F.D. Photo Unit.*

When the structure is small, such as a single-family dwelling, you may be able to walk all the way around it to view it from four sides. In the case of large structures, strip malls, row houses, structures with limited access, or big box retail stores, performing a 360-degree survey may not be possible. In these instances, it will be necessary to have other members of your crew or other arriving units provide information from the other sides, including a view from above. As soon as possible, have an individual provide an ongoing assessment of all four sides of the structure. In commercial buildings, the roof is often included in this request and aerial apparatus or ground ladders are used to view the roof area. Other incoming units may be assigned to various sides of the structure and supply a report back to the IC.

You must focus on the situation and answer the question, *Can the resources at the scene and en route handle this situation?* If the answer is *no* or even *maybe*, then additional resources must be requested immediately, even while en route. Addressing this question is why the initial size-up sets the tone for the balance of the incident.

If there are not enough resources initially available and the arrival of additional ones may be delayed, personnel are likely to start the incident at a disadvantage. This type of situation will take significant work and organization to overcome. As a result, the initial size-up must be done with attention to detail and focus on the desired outcome and the activities and resources needed to accomplish that outcome **(Figure 4.14)**. Considerations must be given to the level of acceptable personnel risk (risk/benefit analysis) and which operational strategy to initiate. Properly interpreting condition indicators can help you make the correct decisions.

Figure 4.14 A 360-degree survey should be made of the structure when possible.

Once the 360-degree survey is performed, determine the appropriate course of action, develop an IAP, and assign duties to your crew and the remainder of your resources. Keep in mind that additional units may not arrive for some time, and it may be necessary for you to prioritize the activities that your plan requires. You must also consider that while you are setting up operations, the fire is increasing in size and extension. A risk assessment is necessary to ensure that you are making the right choice and doing it safely.

> **CAUTION**
> While your crew may assist with information collection and offer suggestions, the IC is responsible for making all decisions.

Before you commit any resources to an interior fire attack, you must have the resources to adhere to the two-in, two-out rule. For the first two firefighters to enter the IDLH, you must have two other firefighters on the outside prepared to respond to a firefighter rescue. The two personnel on the exterior must be fully trained and prepared to enter the structure in full personal protective equipment (PPE), including SCBA. These individuals may be the pump operator, yourself, or another firefighter. Performing a rescue is the only exception to the two-in, two-out rule. However, for non-OSHA states, the two-in, two-out rule is highly recommended, but not required. Refer to Chapter 5, Strategy, for more on the two-in, two-out rule.

Two-In, Two-Out
The presence of four firefighters at a structure fire does not automatically ensure that the two-in, two-out rule has been met. If it is necessary for one or both of the outside members to leave their primary duty, they may not be considered available for rescue. They may not leave their post if it will further endanger the incident scene. A member performing a task critical to the safety of other members on the scene should not be considered available for the initial RIT if by leaving that assigned task to make a rescue, other on-scene members are placed at greater risk. Your decisions must be driven by your local SOP/SOG and by best industry practices if national, state/provincial, or local laws/ordinances do not apply.

Reading Smoke

Firefighters should learn how to "read" smoke at a fire. The quantity and movement of smoke will indicate factors, including:

- Potential fuels involved and potential hazards to responders
- Location and stage of the fire, and its direction of travel

Smoke color can differ for various types of fuels, such as legacy versus modern. Tar, soot, and carbon are the most common heated particles found in smoke, giving it the black color. Water moisture and heated gases give the smoke its white color.

Large particles cool quickly. Because of their size, the particles drop out or are filtered out of the smoke over time, changing the black smoke to gray and then white **(Figure 4.15)**. Responders must remember that visible smoke conditions may be misleading. Location of smoke may be an indicator of the location of the fire or a false indicator caused by movement of smoke through a structure. Light-colored smoke or slow-moving smoke can indicate if smoke has traveled distances from the origin. A small amount of smoke from a large structure could easily be an indicator of a well-seated fire inside.

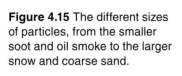

Figure 4.15 The different sizes of particles, from the smaller soot and oil smoke to the larger snow and coarse sand.

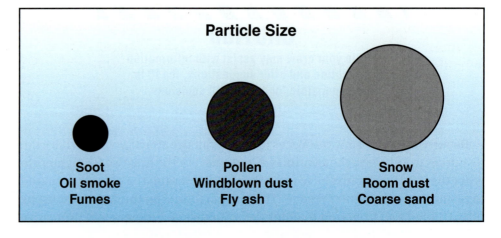

> **WARNING!**
> Visible smoke may be unreliable as an indicator of the conditions inside a structure.

Smoke color may be used to troubleshoot the types of materials included in a fire. The following sections discuss characteristics unique to the different colors that may be seen in smoke.

White Smoke

Light white smoke indicates that pyrolysis (chemical change by heat) is occurring in areas adjacent to the main body of fire. The light white color indicates moisture and gases are being released from the product **(Figure 4.16)**. White smoke has lost particles because the particles have cooled from travel, water, and the addition of cold air. White wispy smoke can indicate pyrolysis in a compartment fire and should be considered unburned fuel. A deep-seated fire, where the smoke has been filtered or cooled from travel, will also be white. On cold days when the temperature is below freezing, smoke immediately turns white and turbulent on leaving the structure. When smoke is forced from a structure through cracks, it filters large particles and will be white.

Figure 4.16 White smoke may have been filtered or may be an indication that water is being applied to the fire.

When a product is heated at a consistent temperature, it can pyrolize and release types of flammable gases. When the gases accumulate in the right mixture with air, they can explode on contact with an **ignition source**. White smoke explosion is rare.

Thick, white smoke was black but has lost the solid particulates that gave it its dark color. This change can happen when smoke has chilled to cold temperatures or has travelled a long distance inside a structure while losing heat. Its consistency is fluffy as compared to a steam cloud, which is more opaque and mostly consists of condensation or water mist.

Ignition Source — Mechanism or initial energy source employed to initiate combustion, such as a spark that provides a means for the initiation of self-sustained combustion.

Brown Smoke

Brown smoke is common in mid-stage heating as moisture mixes with gases and carbon as pyrolysis increases. It is also common in mid-to late-stage heating to see the caramel color brown smoke. Brown smoke is an indication of burning wood, whether unfinished found largely in attics, or finished as found in structural wood members **(Figure 4.17)**.

Gray Smoke

Gray smoke indicates a combination of mixing. It can be mid-stage heating with white, brown, or black, or it can be when different smoke areas combine. It can indicate smoke production changes from mid-stage heating to high heat.

Figure 4.17 The brown smoke escaping from the upper windows of this brick structure is an indication of burning wood. *Courtesy of Ron Jeffers, Union City, NJ.*

Black Smoke

Black smoke contains high quantities of carbon particles and is also an indicator of the amount of ventilation available at the seat of fire. The thicker the smoke, the less clean burning and the less oxygen available, as smoldering fires produce massive amounts of black smoke. In the past, hydrocarbon fires were considered the source of black smoke. Now, fires involving synthetics, plastic, resins, polymers and products made from hydrocarbon derivatives will also give off large quantities of black smoke. This black smoke contains unburned fuels and is a good indicator of carbon monoxide and many other flammable and toxic gases.

WARNING!
Smoke is a combustible by-product of a fire and will burn rapidly when exposed to enough heat.

Thin, black smoke is the direct result of heat from a flame. This smoke is black, but possible to see through. When thin, black smoke is fast-moving or otherwise active, it indicates that the fire is nearby. Thin, black smoke with smooth lines exiting high in an opening and going straight up indicates flame-driven smoke with good ventilation.

In addition to indicating the materials it is burning, thick, black smoke suggests the fire is in the late stages of pyrolysis, which produces large amounts of carbon as unburned product **(Figure 4.18)**. This characteristic indicates a high percentage of carbon monoxide gas in smoke, creating a highly flammable and toxic atmosphere.

Figure 4.18 Thick, black smoke comes from a large structure fire in New Orleans. *Courtesy of Chris Mickel, New Orleans (LA) F.D. Photo Unit.*

Black fire refers to dense, black smoke (fuel) that is ready to ignite, possibly at the vent point. An occupant would not be able to survive this area if the smoke layer extends from the floor to the ceiling. This thick black smoke can act like flames, cause pyrolysis, and even char. Black fire can be seen traveling quickly out of openings in hallways and other channels. Once the ventilation plane is unable to feed oxygen to the large growing fire, it will become ventilation controlled and fill the area, becoming a backdraft or smoke explosion hazard. Under these conditions, black fire can reach autoignition temperatures, and cause the compartment to flashover. In this type of environment, the best way to reduce the potential for a flashover is to apply water to cool the ceiling area.

WARNING!
Black fire can autoignite.

Unusual Colors in Smoke

Unusual color smoke should give personnel an indication that different extinguishing agents may be needed. Flammable metals and chemicals will give off uncharacteristic colors of smoke as they burn.

Neutral Plane

The first responders at a scene of a structural fire should note the height of the neutral plane. This level will indicate several conditions, including the stage of the fire. Also known as thermal balance, the neutral plane lowers as a fire develops and density of fire gases (smoke) increases **(Figure 4.19)**. Interior attack teams must observe the level of the neutral plane carefully and prepare to withdraw when there is a rapid change in conditions causing the neutral plane to lower.

A high neutral plane may indicate that a fire is in the early stages of development. A higher neutral plane can indicate you are further away from the seat of the fire. High ceilings can mask the stages of fire growth. A high neutral plane can also indicate a fire above your level.

When the neutral plan is roughly centered between the ceiling and floor, the fire is beginning to be ventilation controlled, and flashover conditions are developing. A neutral plane close to the floor may indicate that the compartment (room) is reaching flashover conditions, or that the fire is at a grade below the room being observed.

Smoke Volume

The quantity of smoke visible at an incident may provide some clues to the stage of the fire, but those clues should not be acted on without further information. For example, a small residence may fill with smoke from a small kitchen fire. In contrast, a large retail building with an open floor plan may only show small indicators of white smoke from a large fire. However, a large building

Figure 4.19 A laboratory test illustrating the neutral plane of a fire. *Courtesy of National Institute of Standards and Technology.*

Chapter 4 • Size-Up: Evaluation and Assessment

pushing white smoke from openings is a sign of a large fire. As smoke travels farther from the seat of the fire, it will lose its black color and become volume pushed, meaning that its own mass is the primary energy causing it to move. Volume pushed smoke will usually flow, neither smooth nor turbulent. It floats out of openings, rising slowly.

Smoke Density

The darker and more turbulent the smoke is, the closer you are to a rapid fire event. Indicative of a large fire or an under ventilated fire is lots of suspended particles of tar, carbon, ash, and soot. Thick black smoke has lots of heat; thick, white smoke has traveled and lost particles but still may have plenty of heat and fuel to burn **(Figure 4.20)**. The denser the smoke, the lower the visibility, and the more likely that heat buildup indicates a pending flashover. Fast-moving dense black smoke (black fire) can create fast-moving fires with rapid fire spread.

Figure 4.20 Thick, white smoke can still have heat and fuel to burn despite losing particles as it travels. *Courtesy of Ron Jeffers, Union City, NJ.*

WARNING!
The denser the smoke, the more likely flashover conditions are developing.

Types of Smoke Activity

The movement of smoke can be described in terms of speed and direction. When those two factors are considered at the same time, they are discussed in terms of velocity. Changes in the velocity of smoke over time can indicate changes in condition, and the stage of the fire generating the smoke.

The flow path is the volume between an inlet and an outlet that allows the movement of heat and smoke from the higher pressure within the fire area towards the lower pressure areas accessible via doors and window openings. Based on varying building configurations, there may be several flow paths within a structure. Additionally, flow paths and flows can change dependent on fire department ventilation and fire ground tactics employed by the crews on scene. Understanding the flow path can help in tactical decision making on the fire ground for both the placement of suppression crews and the timing and location of ventilation.

Velocity

Slow movement of fresh air in the direction of a fire can indicate that the fire is still in the early stages and still fuel-controlled. In contrast, that velocity of air flow can indicate that the fire is severely ventilation-limited. These differences can be identified by examining the level of the neutral plane within the flow path. Air movement is rapid and turbulent as a fire becomes **ventilation-controlled**. Additionally, a rapid intake of fresh air can be an indicator of a fire reaching flashover conditions. A sudden and total rush of fresh air into a compartment can indicate that a backdraft condition is imminent. The direction of airflow can indicate the location of the seat or base of the fire.

> **Ventilation-Controlled —** Fire with limited ventilation in which the heat release rate or growth is limited by the amount of oxygen available to the fire. (NFPA 921).

> **Need for Coordinated Ventilation**
> Opening access to ventilation must be coordinated carefully. Depending on the stage of fire development, a rapid intake of fresh air can lead to rapid fire growth. Current research indicates that the rapid intake of air can be indicative of an imminent flashover.

Pulsations

Ventilation-limited conditions can result in smoke pulsing out of openings within a structure. Opening the structure improperly or at the wrong location can add fresh air to the compartment, resulting in rapid fire growth leading to flashover conditions. The fire can also be drawn to an undesired location. With these conditions, ventilation operations should not be initiated until suppression crews are prepared to react according to the changing conditions.

Whistling noise created by the movement of air into the structure can indicate fire conditions approaching or at the ventilation-limited phase. Take caution opening a structure with this indicator.

Smoke Movement

Three common types of movement in smoke are: floating or hanging, volume pushed, and heat pushed. The heat in the smoke will determine the smoke's speed.

Floating or hanging (lost its heat) smoke is the same temperature as the air around it. This characteristic is often found in air conditioned buildings, fires that are sprinkler controlled, or where smoke particles are filtered and cooled by passing through cracks in walls. Floating or hanging smoke will move according to air currents **(Figure 4.21)**. Lazy smoke indicates a small, early stage

Figure 4.21 Floating or hanging smoke looms over a street near a structure fire. *Courtesy of Ron Jeffers, Union City, NJ.*

fire, mostly containing moisture from the first stage of pyrolysis. It could also mean a deep-seated fire in a large building, where the smoke is cooling off as it travels significant distances.

Once the area fills up with smoke, the speed it pushes out of the fire room will indicate some factors of the fire. For example, the faster it exits, the more the conditions indicate a large fire and the speed of possible growth. Slower smoke exits in slow rolls that are not turbulent or active.

When smoke fills the compartment, it becomes pressurized. It can reduce flame action if under ventilated. If confined, it can be seen forced out of cracks around windows, doors, and eaves. The color and volume can indicate room location and contents or structure fire. The area from which the smoke comes will indicate passage for smoke, which means passage for fire. Black smoke means the fire has an open area of travel, while white indicates a filtered area of travel.

Turbulent Flow — Movement of a liquid or gas at a high rate of speed and no definite pattern to the movement of the particles.

Laminar Flow — Movement of a liquid or gas at a low rate of speed and in a predictable direction.

> **CAUTION**
> Black smoke indicates that a fire has an open area of travel; white smoke indicates a filtered area of travel.

Heat pushed smoke is characterized by its speed and direction, which can be described as a **turbulent flow** (bubbling, boiling, chaotic) or **laminar flow** (smooth, straight). Fast, turbulent, or active smoke means there is an established working fire. Turbulent black smoke contains lots of particles and, indicative of ventilation-controlled smoke, has heat. It pushes from openings coming out rolling up and to sides as more smoke pushes out into it. Laminar smoke is heat pushed and not obstructed or ventilation-limited. Laminar smoke flow exits from openings near the active flaming fire.

Heat

Heat is a form of energy transferred from one body to another as a result of a temperature difference. Indicators:

— Blackened or crazed (patterns of short cracks) glass: Indicates a fire in the room or nearby as hot smoke condenses on a cooler window. While crazing indicates high interior temperatures, this is an indicator mostly seen on single pane windows. Take caution when opening a structure with these indicators.

— Blistered paint: Indicates both temperature extreme and location of the neutral plane. It may also indicate fire behind the wall **(Figure 4.22)**.

Figure 4.22 Blistered paint in a kitchen destroyed by a house fire. *Courtesy of Ron Moore/McKinney (TX) Fire Department.*

— Sudden heat buildup: Indicates flashover conditions are present. When operating inside a structure, personnel must be aware of a rapid increase in temperature and act immediately to apply water or exit the structure.

Flame Color

Color is usually an indicator of the oxygen supply and the extent of fuel-oxygen pre-mixing, which determines the rate of combustion.

During the Incident

After the IAP is implemented, emergency responders will be busy performing their assignments and making progress toward resolving the incident. This phase (between arrival and termination) can be relatively short or it can last for a considerable length of time. When the initial assignment is small and/or you as the IC make good decisions, the incident may be resolved in a few minutes. If the situation is complex, resolution may require additional resources and time.

During this phase, the situation will change by either improving or worsening. As a result, you must constantly reassess conditions and the effects of current operations. You must continue to size up the situation; validate or adjust the incident objectives, strategies, and tactics; and make changes to the IAP, as needed. By getting continuous feedback from crews, you can follow the progress of your IAP and continue to size up the scene.

An acronym for this part of the size-up is CARA, which stands for **(Figure 4.23)**:

- *Conditions* — An assessment of the interior conditions and the visible indicators on each side and roof of the structure
- *Actions* — Specific activities being performed by each crew inside and outside the structure that are engaged in fire fighting operations
- *Resources* — Includes additional personnel, apparatus, equipment, ventilation, a second hoseline, or an exchange of personnel
- *Air* — Refers to the breathing air supply available for interior fire attack as well as salvage and overhaul operations

Variations on CARA

Many regions, states/provinces, and local departments use variations on the CARA acronym. What you use should be based on your own SOP/SOG. Two other variations include:

CAN	CAAN
Conditions	Conditions
Actions	Actions
Needs	Air
	Needs

Figure 4.23 While sizing up a scene, an Incident Commander can rely on the four aspects of the acronym CARA, which stands for Conditions, Actions, Resources, and Air. *Courtesy of Chris Mickel, New Orleans (LA) F.D. Photo Unit.*

Figure 4.24 Incident Commanders must make determinations at a scene, including whether first responders have the appropriate resources needed to extinguish a fire. If not, the IC must request more resources. *Courtesy of Chris Mickel, New Orleans (LA) F.D. Photo Unit.*

Decision-Making

Using the information gathered prior to and upon arrival and from the 360-degree survey, you have established an action plan. You have identified the problem and selected the incident priorities based on **life safety, incident stabilization, and property conservation (LIP)**. Your IAP contains a clear and precise outline of the actions to be taken and the objectives to be achieved.

As the incident progresses and the situation changes, supplemental decisions will have to be made. For example, the IC needs to decide whether the initial deployment of resources is still producing the desired results or if the deployments need to be changed **(Figure 4.24)**. As the fire scene progresses and the situation changes, new objectives are formed or lower priority ones can move up the list. On large incidents, consideration must be given to relief personnel, additional supplies, and rehab.

> ### A Structure Collapse During a Search Results in LODD
>
> On May 20, 2013, a firefighter with more than 28 years of experience was killed while conducting a primary search inside a burning three-story apartment building. The firefighter and his partner were making their way through the first-floor hallway, knocking on doors in search of any occupants trapped by the large structure fire. As they reached a set of doors, the second-floor walkway and perhaps the third-floor walkway collapsed, killing the firefighter and trapping the partner. The partner was rescued after he called a MAYDAY and activated his personal alert safety systems (PASS) devise.
>
> A NIOSH investigation into the incident identified several issues with the Incident Command that contributed to the line-of-duty death (LODD):
>
> - The initial Incident Commander never conducted a 360-degree size-up of the scene.
>
> - A stationary Command post was never established and personnel were not working in coordination with each other at the scene. As a result, there was a lack of accountability and Command did not evaluate the risk-versus-gain of conducting another primary search.
>
> - The fireground was never notified that another primary search was being conducted, so the seven master streams being used as part of defensive strategy continued flowing water on the building.

> **Life Safety, Incident Stabilization, and Property Conservation (LIP)** — Three priorities at an incident, in order of importance.

> **Plan of Operation** — Clearly identified strategic goal and the tactical objectives necessary to achieve that goal; includes assignments, authority, responsibility, and safety considerations.

Plan of Operation

The **plan of operation**/IAP does not need to be in writing on relatively small, routine incidents involving only an initial assignment. However, there should be a written IAP for long duration or complex incidents. A written IAP is useful when command changes or multiple jurisdictions are involved. How the plan is implemented is the operational phase. After defining the strategy, the plan is started by assigning tactical objectives to company officers in Divisions or Groups. Select the highest priority assigned first and work down your list as more crews arrive or on-scene crews complete tasks.

Firefighter Survivability Approaches

To reduce the number of firefighters injured or killed during structural fire fighting operations, a series of approaches to improving firefighter survivability has been developed and applied to these operations **(Figure 4.25)**. These approaches are explained in the following sections.

Figure 4.25 Life safety is always the top priority at an incident, and efforts are being made to try to limit the number of firefighters who suffer injuries or are killed in the line of duty. *Courtesy of Bob Esposito.*

While these approaches were primarily developed with structural fire fighting operations in mind, they may also be applied in other situations, such as hazardous materials incidents or wildland/urban interface fires where emergency responders may be endangered.

Occupant Survival Profile

Modern fire conditions are increasingly more hazardous compared to legacy construction. This shift is primarily due to the prevalence of plastics and the increasing use of lightweight trusses, because fires burn hotter, with more toxins, and structural supports have less mass. Upon arriving at a structure fire, responders should examine the situation and make a rational, informed decision based on the known events or circumstances. This data assists personnel in determining whether building occupants can survive the fire and smoke conditions that are present and whether to commit personnel to interior operations, including search and rescue.

An **occupant survival profile** is a type of size-up that should be employed to evaluate the potential of an occupant being alive within a structural fire environment. In conducting occupant survivability profiling, the IC asks the following questions:

1. Are occupants suspected of or known to be trapped?

2. Is it reasonable to assume that the occupants are still alive?

The environment within a modern structural fire can quickly exceed 500° F (260° C) with the potential for flashover (approximately 1,110° F [599° C]) to

Occupant Survival Profile — Type of incident size-up that evaluates whether an incident should be treated as a rescue or recovery.

occur within minutes. Firefighters wearing full personal protective equipment (PPE) and self-contained breathing apparatus (SCBA) will have some protection in this environment, but even that protection is limited. In contrast, civilian occupants may not have any protection against fire conditions.

The IC should take the entire structure into consideration when determining survivability whether with full protective equipment or without. If the answers to these above questions are "no," then the responders should take a different approach. They should stop what they are doing, analyze the bigger picture, and gather additional information on the situation. The firefighters should focus on fighting the fire first, whether directly or indirectly, and search for victims later when it is safer to do so. This approach may contradict the long-held practices of entering a burning building as quickly as possible to attempt rescue, but the use of occupant survivability profiling can be paramount to saving the lives of firefighters **(Figure 4.26)**.

Figure 4.26 Firefighters have to resist the urge to charge into a burning structure. They instead might need to fight the fire from the outside before searching for victims. *Courtesy of Chris Mickel, New Orleans (LA) F.D. Photo Unit.*

CAUTION
The Incident Commander's action plan and modes of attack should be followed to ensure incident safety.

Cleveland Fire at Vacant Building

When Cleveland firefighters received an early morning call about a fire in a vacant building just down the block, they had an idea about what to expect. They were familiar enough with the abandoned two-story building to know that it probably was not empty.

182 Chapter 4 • Size-Up: Evaluation and Assessment

> A locksmith had operated out of the building at one time, but like many other structures in the neighborhood, it had been vacant for several years. However, firefighters had recently looked around their response area for buildings that could pose a problem because homeless people were known to sleep in them. Personnel had noted that the former locksmith's shop was problematic because it was a frequent hangout for squatters, who had stripped all of the copper pipes and wire out of the building.
>
> When firefighters arrived at the structure fire, they saw heavy smoke and flames coming from the building's second floor. Knowing that homeless people could be trapped inside the building, firefighters entered the structure to search for survivors. However, after not finding anyone inside, the firefighters were exiting the building when one of them injured his leg when it went through the floor.
>
> The Incident Commander shifted to a defensive strategy and determined that it was best to contain the fire as fast as possible. The fire department's familiarity with the abandoned building influenced how they attacked the fire.

Crew Resource Management

Crew resource management (CRM) is a system that optimizes the utilization of all available resources, personnel, procedures, and equipment in order to promote safety and improve operational efficiency. Originated by the air transportation industry, crew resource management has been adopted by other career fields to include the fire and emergency services. (CRM) training should become a part of each fire and emergency services organization's training schedule for all personnel.

> **Crew Resource Management (CRM)** — Training procedures intended to improve communications, leadership, and decision making to reduce human error.

CRM is designed to create a culture or climate of freedom in which personnel are encouraged to contribute to the safety and goals of mitigating the incident. Using CRM assists personnel in identifying the first indicator of errors occurring; the discrepancy between what is happening and what *should be* happening. Personnel can then communicate the discrepancy to the supervisor or incident commander in a manner that is forceful, yet respectful, in order to overcome the problem and save firefighter lives.

CRM training achieves its goals of increasing scene safety and responder effectiveness by emphasizing the following topics:

- **Communications** — Using CRM, personnel focus on communications, speak directly and respectfully, and communicate responsibly.

- **Situational awareness** — Personnel need to recognize that emergency situations are dynamic and require each individual's full attention **(Figure 4.27)**.

- **Decision making** — The availability of little or no information can result in a poor risk assessment while too much information can overload the decision maker and interfere with making a decision.

- **Teamwork** — The efficient and safe interactions among responders and others can significantly improve an incident response.

Figure 4.27 Considering how unpredictable some incidents can be, personnel must maintain situational awareness and pay even more attention during emergency situations. *Courtesy of Bob Esposito.*

- **Barriers** — Factors that interfere with communication, situational awareness, decision making, and teamwork. CRM training helps to identify these barriers and prevent them from impeding an operation.

Rules of Engagement

In 2010, the Safety, Health, and Survival (SHS) Section of the **International Association of Fire Chiefs (IAFC)** released its initial draft of *Rules of Engagement for Structural Fire Fighting – Increasing Firefighter Survival*. This document includes two sets of rules for structural fire incidents, one intended for use by Incident Commanders and another for firefighters. Through the application of these rules of engagement, the IAFC/SHS Section stresses that firefighter line-of-duty injuries and deaths can be reduced.

All fire and emergency services personnel should be trained on the *Rules of Engagement for Firefighter Survival*. All personnel who may serve as an Incident Commander at an emergency scene should be trained on the *Incident Commander Rules of Engagement for Firefighter Safety*.

> **International Association of Fire Chiefs (IAFC)** — Professional organization that provides leadership to career and volunteer chiefs, chief fire officers, and managers of emergency service organizations throughout the international community through vision, information, education, representation, and services to enhance their professionalism and capabilities.

International Association of Fire Chiefs Rules of Engagement

Rules of Engagement for Firefighter Survival

1. Size-up your tactical area of operation.
2. Determine the occupant survival profile.
3. **DO NOT** risk your life for lives or property that cannot be saved.
4. Extend **LIMITED** risk to protect **SAVABLE** property.
5. Extend **VIGILANT** and **MEASURED** risk to protect and rescue **SAVABLE** lives.
6. Go in together, stay together, come out together.
7. Maintain continuous awareness of your air supply, situation, location, and fire conditions.
8. Constantly monitor fireground communications for critical radio reports.
9. You are required to report unsafe practices or conditions that can harm you. Stop, evaluate, and decide.
10. You are required to abandon your position and retreat before deteriorating conditions can harm you.
11. Declare a MAYDAY as soon as you **THINK** you are in danger.

Incident Commander Rules of Engagement for Firefighter Safety

1. Rapidly conduct, or obtain, a 360-degree size-up of the incident.
2. Determine the occupant survival profile.
3. Conduct an initial risk assessment and implement a **SAFE ACTION PLAN**.
4. If you do not have the resources to safely support and protect firefighters – seriously consider a defensive strategy.
5. DO NOT risk firefighter lives for lives or property that cannot be saved – seriously consider a defensive strategy.
6. Extend **LIMITED** risk to protect **SAVABLE** property.

7. Extend **Vigilant** and **Measured** risk to protect and rescue **SAVABLE** lives.

8. Act upon reported unsafe practices and conditions that can harm firefighters. Stop, evaluate, and decide.

9. Maintain frequent two-way communications and keep interior crews informed of changing conditions.

10. Obtain frequent progress reports and revise the action plan.

11. Ensure accurate accountability of all firefighter locations and status.

12. If, after completing the primary search, little or no progress towards fire control has been achieved – seriously consider a defensive strategy.

13. Always have a rapid intervention team in place at all working fires.

14. Always have firefighter rehab services in place at all working fires.

Source: © 2012, The International Association of Fire Chiefs.

Figure 4.28 An IC must understand factors such as building characteristics and the resources available in order to make sound decisions on the fireground. *Courtesy of Ron Jeffers, Union City, NJ.*

Critical Fireground Size-Up Factors

What you look for in your size-up is determined by the factors required to make your decisions **(Figure 4.28)**. Common considerations are described in the following sections. Many departments include a list of factors in their SOP/SOG for fireground operations. These factors include:

- Building characteristics
- Life hazard
- Resources
- Miscellaneous conditions
- Arrival condition indicators

There is no specific order that you should follow as you consider these factors. Some factors will be based on your preincident plan, personal knowledge

Chapter 4 • Size-Up: Evaluation and Assessment

of the site, and risk-based analysis. Each factor, however, will have an effect on the decisions you make and the actions you take.

Building Characteristics

Building characteristics should be contained in your preincident plan, and supplemented by your knowledge of building construction and through area familiarization. If a preincident plan is unavailable, your initial information will come from your first view of the exterior of the structure. As you evaluate building characteristics, you should learn:

- **Access points** — Types and location of doors, windows, exterior stairs, basement entrances, garage doors, skylights, and other openings **(Figure 4.29)**
- **Barriers to access or egress** — Security bars, fences, gates, and other physical barriers to access
- **Barriers to aerial operations** — Overhead wires, trees, or porches will prevent or obstruct aerial apparatus operations
- **Topography** — Slopes that may indicate grade-level access onto different floors from alternate sides of the structure
- **Roof design** — Type of roof, peaks, dormers, overhangs, potential concealed spaces, passive ventilation systems, skylights, chimneys to indicate location of furnace, and vent pipe to indicate kitchen and bathroom locations
- **Construction type** — Type of exterior wall construction that can help determine the overall construction type
- **Number of stories** — Floors above and below grade
- **Exposures** — Exterior proximity to other structures, existence of fire separation walls
- **Area** — Approximate floor space
- **Height** — Ceiling height of ground level floor
- **Occupancy** — Determined by building appearance or signage

Figure 4.29 Firefighters could have trouble reaching this burning apartment because its window is not easily accessible. *Courtesy of Ron Jeffers, Union City, NJ.*

- **Fire protection systems** — Signs indicating location of FDC, fire control room, or private water supply system

Knowledge of local building styles can also assist in determining the interior layout of the structure. Refer to Chapter 2, Prefire Planning, for information on building construction. Examples of similarities within a jurisdiction may include:

- Neighborhoods with two-story homes all designed and built alike will generally have similar interior floor plans **(Figure 4.30)**.
- Commercial occupancies in a strip mall will generally have a show room or display area in the front, storage at the rear, and small areas used for toilets, basement access, and offices near the back.

Figure 4.30 The two-story homes in this neighborhood are designed and built alike, which could assist an IC in determining the interior layout of a structure if a preincident plan is unavailable. *Courtesy of Ron Jeffers, Union City, NJ.*

- Strip malls usually have glass store fronts and a steel door without exterior door handles in the rear.

You can learn a lot from the building access shown on the preincident survey or visible to you during your approach and 360-degree survey. The first-in company officer should consider pulling past the address to get a three-sided view of the structure to assist with both the size-up and the initial arrival report.

CAUTION
RIC/RIT companies should be advised of the Incident Action Plan and any access issues.

Street Access

Your first assessment is of the street access as you approach the incident scene. By being aware of what is going on in the neighborhoods in your response area, you will know more about the general road conditions in the area. Streets may be closed due to construction or occasional block parties that will delay arrival time. Parking patterns may change to accommodate alternating parking, snow removal ordinances, or day/night parking requirements. Parking patterns may limit apparatus placement and travel routes **(Figure 4.31)**.

Figure 4.31 This apparatus had to maneuver around parked cars on a narrow street. *Courtesy of Ron Jeffers, Union City, NJ.*

Obtaining a Three-Sided View of a Structure
The first-in company officer should consider pulling past the address to get a three-sided view of the structure to assist with both the size-up and the initial arrival report.

Upon arrival, engine officers should consider aerial apparatus access and spotting locations on all streets, especially narrow streets. Aerial apparatus cannot be used if access is unavailable. Loss of this asset can significantly change your attack plan. Aerial apparatus placement may limit street access or require using sidewalks for outrigger placement.

Structure Access

On commercial occupancies, multiple locks, security grilles, and roll-up doors delay entry, but are a good indication that the space is unoccupied. Residential occupancies may also have multiple locks and security grilles that will delay entry. A multiple family apartment dwelling may have numerous interior doors that need to be forced, requiring use of a hydraulic ram/rabbit tool and/or assignment of additional members to a forcible entry team. Upon arrival, begin by determining how many access doors there are. In townhouse, garden-style, or apartment type buildings, there may be only one entrance **(Figure 4.32)**. In row houses or attached structures, members will be delayed from accessing the rear of the structure because of the distance they will have to travel around multiple buildings.

Figure 4.32 Firefighters could have only one entranceway into an apartment building. *Courtesy of Ron Jeffers, Union City, NJ.*

> **Means of Egress** — (1) Safe, continuous path of travel from any point in a structure to a public way. Composed of three parts: exit access, exit, and exit discharge. (2) Continuous and unobstructed way of exit travel from any point in a building or structure to a public way, consisting of three separate and distinct parts: exit access, exit, and exit discharge. (**Source:** NFPA 101, *Life Safety Code®*).

Egress

Firefighter safety depends on the ability to exit a structure rapidly if conditions change. **Means of egress** that you should consider are:

— **Doors** — Note location of all doors and direction of swing. Determine if doors can be opened from the exterior. Note construction of doors and door frames in the event forcible entry must be used.

— **Windows** — Determine if windows can be opened. Find out the size of the window and determine if it is large enough for a person to fit through. Notice if windows are covered with security grilles or bars that will make egress difficult or impossible. Determine how high the windows are and if ladders will be needed to reach them **(Figure 4.33)**. Solid non-opening

Figure 4.33 While determining means of egress, personnel must determine if a ladder is needed to reach a window. *Courtesy of Bob Esposito.*

windows are energy efficient, but prevent access or egress and may allow heat and smoke to build up in the compartment or structure. Small windows between floors usually indicate stair placement. Bathroom windows are usually obscured for privacy and smaller than other windows.

— **Stairs** — Determine if exit stairwells provide egress directly to the outside of the building. Ensure that the exit doors can be opened from the exterior. Notice if stair wells permit reentry onto all floors from the stair well. Find the location of all standpipe connections in the stairwells. Look for indications that the stairwells are pressurized or designed to remove smoke from the structure.

— **Fire escapes** — Consider the location, operation, and condition of any exterior fire escapes. Notice if the fire escape has been pulled down, indicating that it has been used by occupants to escape the structure.

— **Dead-end corridors** — Based on the preincident plan, consider the presence of dead-end corridors. These corridors may exist in altered multifamily structures and small office buildings that use modular office units (cubicles).

Metal Security Doors and Bars

In areas of high crime rates, security grilles, gates, roll-down doors, or fortified entrances will cover windows and doors on residences and front and rear doors of commercial buildings **(Figure 4.34, p. 190)**. These obstructions will not only slow access, they will create egress hazards for occupants and firefighters who are inside the structure. In some communities, thick iron plates have been installed on the exposed flat roofs of commercial buildings to prevent people from cutting holes for illegal entry. These plates will prevent ventilation and pose a serious collapse hazard.

Security on windows and doors may consist of metal bars, Plexiglas®, wire grilles, or plywood or metal sheets. **Metal-clad doors** or heavy metal grilles

Metal-Clad Door — Door with a metal exterior; may be flush type or panel type. *Also known as* Kalamein Door.

Chapter 4 • Size-Up: Evaluation and Assessment **189**

Figure 4.34 Firefighters had to cut through a metal roll-down door to respond to a fire inside a commercial structure. *Courtesy of Chris Mickel, New Orleans (LA) F.D. Photo Unit.*

may be found on residences and commercial structures. Security doors, drop bars, and additional security are often found on the interior or exterior of rear doors to commercial occupancies. The IFSTA manual, **Building Construction for the Fire Service**, discusses a range of security measures that may be included in buildings.

Always consider calling additional resources for heavily secured buildings because forcible entry can be labor intensive and rapidly exhaust crews. If the security bars are installed to code in residential occupancies, there should be an interior release on the bars for emergency egress.

Special Hazards

Special hazardous conditions are rarely visible during your size-up. Your preincident plan should contain any information regarding these hazards, which can tell you:

- **Alterations** — Changes to the interior can create multiple rooms with dead-end corridors, multiple doors, and no designated exit ways. Alterations can also create concealed spaces and disrupt the original coverage of sprinkler systems.

- **Change in occupancy or use** — Can create situations that the existing fire protection system cannot control. Corridor and stairwell walls, compartment doors, and means of egress may be insufficient for the new use. Occupancy load limits may exceed the ability of exit passages to carry the new load.

- **Storage** — Illegal storage, compressed gases, flammable liquids and explosive materials, or drug labs may also exist that will increase the life safety risk to firefighters **(Figure 4.35)**.

Figure 4.35 The highly combustible chemicals found at an illegal drug lab pose a serious safety threat to firefighters.

Fire at a Building Built on a Hillside

Five companies and two command chiefs responded to an electrical fire at a multi-level residential building that had been built on a hillside. As an engine company made entry through the front door, firefighters noticed light smoke coming inside the structure. Several minutes passed, and when the Incident Commander (IC) did not get a response from the engine company, a battalion chief entered the building and spoke with the firefighters after finding them on the street level floor.

The lieutenant with the engine company believed the fire must be on the floor below them. However, when the battalion chief stated that the fire would be attacked from a different side of the building, the firefighters remained inside and did not follow him outside the structure. A few minutes later, the IC did not get a response when he attempted to contact the engine company.

Firefighters backing up the engine company tried to enter the building through a garage door, but they were unable to because of the extreme heat. When the battalion chief and another engine company forced a door open on the floor below street level, they encountered a fully involved basement fire. They knocked down the fire, allowing a rescue crew to enter the building to search for the first engine company that had not responded to the IC.

The rescue crew found two firefighters, including the lieutenant. Both firefighters had suffered injuries and were transported to a nearby hospital. A report determined several factors contributed to the incident, including the house's construction on a hillside, ineffective size-up and Command communications, and a lack of a personnel accountability system.

Life Hazard via Occupancy Condition

The life hazard to occupants will be largely determined by the occupancy type and the time of day. To be on the safe side, all structures should be considered occupied until proven otherwise. However, you must balance the risk to yourself and your personnel against the possibility that there is someone in the structure.

> **WARNING!**
> Once you enter a structure, it is occupied and you are at risk.

General Considerations

General considerations that you must remember when determining the life hazard of various types of occupancies:

- In occupancies with large open areas, such as car dealerships, warehouses, factories, or places of assembly, assume that there are exposed trusses supporting the roof structure. These trusses may degrade rapidly when exposed to heat and fire, creating a collapse hazard **(Figure 4.36)**.

Figure 4.36 Large open spaces allow fire to travel unchecked.

- Occupancy type will help with the rescue profile and estimating the potential fuel load. Assembly occupancies, such as nightclubs and theaters, have significant life safety dangers. Studies of human behavioral patterns indicate that most people will exit through the door they entered. However, building codes only require it to handle 50 percent of the occupant load to exit through that opening, providing there are other exits spaced evenly around the structure.

- Older commercial occupancies, such as warehouses or factories, may have been converted to residential occupancies, creating additional problems of access, egress, and fire spread. At the same time, residential structures on main streets may have been converted to light office use, such as small medical clinics, dental offices, and law offices. These buildings may have multiple doors with deadbolt locks inside.

- Type of occupancy use can indicate the mobility of the occupants and their ability to self-evacuate. For example, daycare centers for young children and babies, group homes for the mentally and physically challenged, nursing homes, and senior care facilities will contain occupants who will require assistance in evacuating the facility.

Clues to Occupancy Hazards

The name or type of business can indicate the possible contents and potential fire hazards to expect. For example, paint, hardware, pool supply, and automotive supply stores may contain flammable and combustible liquids, high thermal output materials such as hydrocarbons and polymers, and caustic or hazardous materials. When one of these occupancies is affected by fire, the rapid fire growth and high thermal output will cause you to consider exposures and evacuations earlier than normal.

Knowing firefighters' life safety hazards is your primary concern. This knowledge includes:

- **Air management** — The IC and personnel entering a structure must constantly evaluate the size of a structure, the contents in that structure, and the complexity of the structure's interior layout when making decisions to commit personnel to an interior offensive attack. Crews must be cognizant of air management on large or complex structures to ensure that sufficient air with reserve is available when exiting. The IC and divisions should keep interior crews apprised of additional exits or accesses to safe environments. Traditionally, one half the service pressure was considered the point to exit the structure in the means entered. This guide does not allow any reserve in the event the path is obstructed, conditions change, or disorientation occurs. Anytime a crew enters an IDLH environment, air management should be followed to ensure that all personnel have sufficient air for time to exit prior to the activation of the **End-of-Service-Time Indicator (ESTI)** **(Figure 4.37)**. If crews are not progressing in an IDLH environment or find by air consumption they need to exit, the chain of command should be notified. The work time in the IDLH will be determined by the personnel on the crew that consumes air the fastest and the team work time will be based on that worst case air consumption.

End-of-Service-Time Indicator (ESTI) — Warning device that alerts the user that the respiratory protection equipment is about to reach its limit and that it is time to exit the contaminated atmosphere; its alarm may be audible, tactile, visual, or any combination thereof. *Also known as* Low Air Alarm.

Figure 4.37 Air management is critical for firefighters working in an IDLH environment. *Courtesy of Chris Mickel, New Orleans (LA) F.D. Photo Unit.*

Figure 4.38 Personnel should be monitored during rehabilitation to ensure that they are feeling well and not in need of medical attention. *Courtesy of Ron Jeffers, Union City, NJ.*

- **Rehabilitation** — Although rehabilitation is not an initial concern, you will need to watch and monitor your personnel **(Figure 4.38)**. You should apply "the two cylinder rule" to determine when to send personnel to rehab. A firefighter should rehab after using two air cylinders. You must refer to your department's SOP/SOG for local rehab requirements.

- **Crew resources** — Hopefully a good size-up will determine your staffing needs early. Keep extra crews available, ensure that enough crews are on scene to clean up, and do not release resources too early. The IC should keep enough resources in staging to complete the incident. If you need to request additional resources, call for enough to mitigate the incident.

> **CAUTION**
> When conducting air supply checks ahead of entering a response environment, ascertain each responder's air supply.

Residential Occupancy Type

There are many types of residential occupancies. Single-family dwellings and apartments may be occupied 24 hours a day or just at night and on weekends. Hotels, motels, dormitories, and boarding houses are likely to be occupied 24 hours a day with staff on hand as well as guests or residents. Hotels, motels, and guest accommodations are likely to be occupied by transient residents who are unfamiliar with the structure, the exits, or the evacuation plan. Assistance in evacuating the structure may be necessary.

Multifamily dwellings may have unit access directly from the exterior or from a central interior hallway. These arrangements can affect your ability to advance hoselines and may increase reflex time **(Figure 4.39)**.

Figure 4.39 Advancing hoseline into a multifamily dwelling can take some maneuvering from the exterior. *Courtesy of Chris Mickel, New Orleans (LA) F.D. Photo Unit.*

Chapter 4 • Size-Up: Evaluation and Assessment

The number of doors, door-bells, mailboxes, and utility meters indicate the number of living units. Utility meters may indicate the number of units plus one for a building. Some apartment complexes, however, do not have separate utility service for every apartment. Exterior stairs with access to various levels can indicate multiple living areas. Window air conditioners, blinds, or curtains in windows may identify that the space is used as a living area.

Factors that will affect strategy and tactics with fires in these types of occupancies include:

- Occupied or unoccupied
- Multiple stories and/or a basement
- Construction (lightweight or legacy)

Business Occupancy Type

A business occupancy is a building or structure, or any portion thereof, that is used for office, professional, or service-type business. Business occupancies are typically only occupied during business hours; however, many will have cleanup or stocking crews inside well after being closed to the public **(Figure 4.40)**. Some business occupancies may be staffed around the clock if they are involved in global enterprises or computer services. In some areas, small businesses may have sleeping quarters in the structure, whether or not their existence is code compliant.

Figure 4.40 Carpet, fabric panels, and decorative wood and plastic add to the fuel load in an open floor plan office.

Some of the factors that will affect strategy and tactics with fires in these types of occupancies include:

- The fuel load may closely resemble residential occupancies. Modern furnishings and ordinary combustible materials throughout, which can contribute to rapid fire growth.
- Business office and cubicle arrangements can make access difficult and may require extended hose lays. The configuration may also contribute to disorienting the firefighters.
- The occupancy may be occupied at the time of the incident.
- May contain unique or non-replaceable business information that require aggressive salvage operations.
- The size and configuration of business occupancies may make it difficult to conduct effective ventilation operations.

Mercantile Occupancy Type

A mercantile occupancy is generally defined as any building that is used to display or sell merchandise, including retail locations such as:

- Department stores
- Pharmacies
- Supermarkets
- Shopping centers
- Malls

Mercantile occupancies contain both large quantities of combustible materials and the potential for high life loss. The arrangement of the merchandise, both on display and in storage, can result in high fire loads of combustible materials and at the same time restrict egress for customers **(Figure 4.41)**. Product displays are rarely fixed to the floor and can be moved to create new access patterns in the showrooms.

Some of the factors that will affect strategy and tactics with fires in these types of occupancies include:

- Type of business and associated hazard (equipment/materials, chemicals, manufacturing process)
- Size of the structure
- May be difficult to gain entry

Figure 4.41 This paint store has shelves filled with cans of paint and art supplies that would both increase the fire load and complicate egress for customers if a fire occurred.

Mixed Occupancy Type

A building used for two or more occupancy types, classified within different occupancy groups. A mixed occupancy building has the potential for creating hazardous situations if incompatible occupancy types are located near one another. Oftentimes, different occupancy classifications are separated by fire resistive partitions.

Mixed occupancy buildings are common in every community. For example, a strip mall that contains a tax-preparation business, a chiropractor, a liquor store, and a sandwich shop is a mixed-occupancy building **(Figure 4.42)**. Similarly, a light-industrial building contains an auto repair shop, a drop-off location for a thrift store, and a showroom for a cabinet company is also a mixed-occupancy, but the inherent hazards are much greater than the previous example.

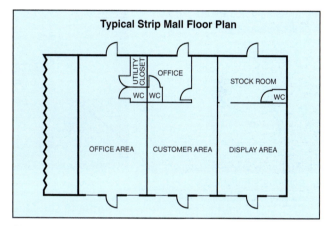

Figure 4.42 Typical floor plan for strip malls.

Some of the factors that will affect strategy and tactics with fires in these types of occupancies include:

- Preincident planning and a working knowledge of the occupancy are essential for safe and effective operations.
- Adjacent exposures to fire compartment may be more or less hazardous than the area involved.
- Depending on the frequency of fire prevention inspections, it is possible to have operations within a building that exceed the design and safety parameters of the building.
- Breaches in the fire resistive partitions can result in extension to adjacent and distant occupancies.
- Some mixed structures contain residential properties, so life safety concerns are ongoing.
- Type of individual businesses within a mixed structure.

Industrial and Storage Occupancies

Depending on the type of manufacturing, industrial sites may be staffed around the clock. Generally, you will find security staff on duty at the site. Personnel should be aware of the type of content and its combustibility/flammability present at storage occupancies, such as warehouses, storage units, and grain elevators **(Figure 4.43)**.

Figure 4.43 Containers of lacquer thinner, marked as a flammable liquid, are stacked on a pallet for storage.

Shelter in Place — Having occupants remain in a structure or vehicle in order to provide protection from a rapidly approaching hazard, such as a fire or hazardous gas cloud. Opposite of evacuation. *Also known as* Protection-in-Place, Sheltering, *and* Taking Refuge.

Some of the factors that will affect strategy and tactics with fires in these types of occupancies include:

- Amount and type of fuel load
- Size and configuration of the structure
- Racking system
- Any live processes (conveyer system, heavy machinery, chemical reactions)

Institutional Occupancy Type

Institutional occupancies include hospitals, nursing homes, daycare centers, and detention/correctional institutions. Occupants may be classified as either ambulatory or nonambulatory **(Figure 4.44)**. Ambulatory occupants are those who are capable of moving on their own, and nonambulatory persons are those who require assistance. In detention, correctional, and psychiatric institutions, such as prisons, jails, and drug rehabilitation facilities, occupants may be confined or restrained. Institutional occupancies operate on a 24-hour basis. In many of these situations, **sheltering** or protecting occupants where they are is more practical and effective than evacuation.

Figure 4.44 When daycare centers are in operation, staff should be present to assist in the evacuation of occupants.

Some of the factors that will affect strategy and tactics with fires in these types of occupancies include:

- Ambulatory vs. non-ambulatory
- Limited access
- Vulnerable population (under detention, non-ambulatory)

Assembly Occupancy Type

An assembly occupancy is defined as a building or structure, or any portion thereof, that is used for the gathering of 50 or more persons. Places of assembly vary in size and type, and are often used for civic, social, or religious functions, or for recreation, for food or drink consumption, or for entertainment **(Figure 4.45)**. Typically, exits are only required to be unlocked during business hours; however, many assembly occupancies will have cleanup or stocking crews inside well after being closed to the public. The owner/occupant is required to post the maximum number of people permitted in the room or structure. Many types of assembly occupancies are required to have a valid permit, regardless of the number of persons allowed.

Figure 4.45 Churches, synagogues, and other places of worship classified as assembly occupancies frequently host gatherings of at least fifty people.

Some of the factors that will affect strategy and tactics with fires in these types of occupancies include:

- The largest life-loss fires occurred in assembly occupancies. Individuals and families that are temporarily housed may be found at these occupancies, particularly after significant disasters.
- Modern and/or permitted assembly occupancies are typically protected with an automatic fire sprinkler system.
- Typically consist of large, open areas, but can also have complicated floor plans. Preincident planning and a working knowledge of the occupancy are essential for safe and effective operations.
- Extended hose lengths and/or a standpipe system may be necessary to effectively apply fire streams to the seat of the fire.
- Based on the assembly use and the construction type, roof systems may have extensive spans, which must be taken into account when determining the feasibility of employing roof operations.

Educational Occupancy Type

Educational occupancies, such as public and private schools and technical schools, are rarely occupied between midnight and 6 a.m. by students or faculty **(Figure 4.46)**. It is not unusual, however, for college

Figure 4.46 High-ceiling atriums have few features to contain smoke from a fire.

and university libraries to be open after midnight. Security and cleaning staff may be present after the facility is closed. Like transient residents, students may be unfamiliar with the structure or exits unless the school has provided emergency training to them.

Some of the factors that will affect strategy and tactics with fires in these types of occupancies include:

- Area of refuge location
- Fuel load
- Occupancy by individuals unable to self-evacuate
- Large building
- Limited fire suppression systems (in older facilities)

Unoccupied, Vacant, or Abandoned Structures

According to the NFPA, more than 11,000 fires occur annually in vacant structures. At the same time, the NFPA estimates that more than twenty civilians die and 6,000 firefighters are injured every year in these types of fires. Every type of structure may be classified as unoccupied, vacant, or abandoned at some time during its existence.

- *Unoccupied* — An existing residential or commercial property that is for sale may be considered unoccupied when an occupant has moved out of it. The utilities may still be turned on, and some contents may remain in the structure. The term may also be applied to a structure that is unoccupied when the business located there has closed. Fire protection systems may be operational.

- *Vacant* — A residential or commercial property that is empty and may have all entrances secured or boarded up. Vacant structures may still be on the local property tax rolls. The owners may be contributing to the structure's upkeep, and it may still have value. The utilities may be turned on and some contents may still remain in the structure. Fire protection systems may or may not be operational.

- *Abandoned* — A property that has been vacant for some time. It may be structurally unsound, or it may contain conditions that are in violation of the local building code **(Figure 4.47)**. There may also be some question as

Figure 4.47 Firefighters should be cautious when responding to an abandoned property because it may be structurally unsound and include other unsafe conditions. *Courtesy of Chris Mickel, New Orleans (LA) F.D. Photo Unit.*

to legal ownership of the structure. The utilities will have been terminated due to lack of payment. Fire protection systems may not be operational. Although the terms *vacant* and *abandoned* may be used to describe the same conditions, the key difference is the length of time the structure has been unused or uninhabitable.

Some jurisdictions, such as New York City, have the authority to inspect and designate abandoned or vacant structures as *vacant*. According to local ordinances, the owners must secure the structure to prevent access and post signs indicating the building is vacant. In those cases, fire department SOP/G prohibits interior fire attacks. However, the situation would change if an individual at the scene informs firefighters upon arrival that people are known to be inside the building.

Vacant or Abandoned Structures

Signs that a vacant or abandoned structure may be occupied include:

- Electrical extension cords leading into the building
- Signs of candles or campfires
- Evidence of forced entry
- Secure doors or windows that have been opened
- Make-shift curtains/shades or decorations
- Signs of construction or demolition
- Presence of Dumpsters, construction material, parked vehicles, and temporary electric or utility meters

Risk-Benefit Analysis — Comparison between the known hazards and potential benefits of any operation; used to determine the feasibility and parameters of the operation.

When making your size-up, implement a **risk-benefit analysis** for all buildings that are suspected to be unoccupied, vacant, or abandoned. Too many firefighters are injured or killed in the growing number of vacant buildings. Hazards that may be found in vacant or abandoned buildings include:

- Delay in discovery and notification of fire
- Fires located in multiple locations
- Increased exposure hazard
- Open shafts, pits, or holes in floors
- Structural instability caused by weather, vandalism, or lack of maintenance **(Figure 4.48)**
- Exposed structural members
- Penetrations in fire barriers that would permit rapid fire extension
- Maze-like configuration in interior floor plan
- Blocked or damaged stairs
- Combustible contents

Figure 4.48 This abandoned building shows apparent signs of being structurally unstable. *Courtesy of Chris Mickel, New Orleans (LA) F.D. Photo Unit.*

Chapter 4 • Size-Up: Evaluation and Assessment **199**

Some of the factors that will affect strategy and tactics in these types of occupancies include:

- These occupancies often do not have any type of utility connections. Many fires in these structures are human caused, so life safety is a concern.
- Personnel must determine if the structure is actually unoccupied or is illegally occupied by individuals.
- Structural members might be damaged or missing, often occurring when people strip the copper and wire from the structure.

Arrival Condition Indicators

Often initial decisions must be made with incomplete information. You may have to depend on probabilities and indicators as well. Then, you make your initial fire attack or rescue if you have the necessary resources to do so safely. The following sections explain how simple factors can provide a wealth of information.

Time of Day

The time of day can indicate whether the structure is occupied and if the occupants are awake or asleep. The structure still needs to be searched to ensure that all occupants are out. Time of day will also give you an idea of the amount of traffic congestion adjacent to the structure and the possibility that on-street parking will obstruct direct access. For example, around the time school is dismissed, the street in front of most public schools is crowded with waiting school busses and personal vehicles. During evening and nighttime hours, visibility will be naturally limited, independent of incident factors, requiring the use of artificial lighting and reflective barrier to protect the street side of apparatus **(Figure 4.49)**. Fires in occupancies that are closed may have a delayed transmission of the fire alarm, resulting in an advanced fire by the time true notifications are sent and received.

Figure 4.49 Artificial lights and reflective barriers may be needed during incidents at night when visibility is limited. *Courtesy of Chris Mickel, New Orleans (LA) F.D. Photo Unit.*

Weather

Weather conditions can have a negative effect on structural fire fighting operations by reducing the effectiveness and efficiency of your resources. Besides slowing your response to the scene, weather can make it difficult for personnel to work. In addition, occupants who are removed from the structure will have to be moved to a place of shelter.

- *Ice* — Ice is not always visible. Regardless of its visibility, it creates a slipping hazard, increases the load on equipment, and impairs function **(Figure 4.50)**. It complicates the process of advancing attack or supply hoselines, performing vertical ventilation, and doing forcible entry difficult. Hoses, ladders, and apparatus can become ice covered, causing slipping/sliding hazards.

Figure 4.50 The temperature was so cold during an incident that ice formed on a firefighter's helmet, bunker gear, and even his eyelashes. *Courtesy of Ron Jeffers, Union City, NJ.*

- *Snow* — Snow creates some of the same difficulties as ice. It also obscures tripping hazards.
- *Rain* — Rain and fog can reduce your ability to see the entire scene. It can make metal surfaces slippery. As the temperature drops, it can freeze on equipment and apparatus.
- *Humidity* — High humidity can cause smoke to remain close to the ground obscuring visibility of the building. It can also affect personnel, causing them to fatigue quickly and become dehydrated through perspiration loss.
- *Temperature extremes* — In extremely hot climates, personnel may succumb to heat stress rapidly and require **rehabilitation** earlier than normal. They may also become dehydrated, requiring additional fluids and medical care **(Figure 4.51, p. 202)**. Extreme cold temperatures can cause skin and clothing to stick to metal tools and equipment, cause frostbite injuries, and reduce stamina. Hoselines, pumps, and water supplies can freeze causing a loss in water supply or pressure. Hose, tools, and equipment can be damaged or become inoperable.

Rehabilitation— (1) Activities necessary to repair environmental damage or disturbance caused by wildland fire or the fire-suppression activity. (2) Allowing firefighters or rescuers to rest, rehydrate, and recover during an incident; also refers to a station at an incident where personnel can rest, rehydrate, and recover. *Also known as* Rehab.

Figure 4.51 Firefighters may need additional fluids and a chance to cool down after working in extreme heat. *Courtesy of Ron Jeffers, Union City, NJ.*

- *Wind* — In windy conditions, flame spread can become dangerously rapid if windows fail or are opened improperly, or if doors are left open. Strong winds can increase the dangers of working on the peak of a roof. A fire that appears to have started inside a structure may be the result of a wind-driven fire that began on the outside and extended into the structure. High velocity winds can spread fire brands onto exposures and can affect the accuracy and effectiveness of defensive hoseline streams. The IC should determine the source of the fire and not call for an interior attack that is not needed initially.

Visual Indicators

Visual indicators provide you with insight on conditions upon which to provide the initial report and base the initial decisions. Visual condition indicators vary widely depending on the type of emergency incident you encounter. Various indicators, including smoke, heat, and flame will assist in determining the size, location, and extent of the fire.

A large plume of black smoke may indicate a hydrocarbon fire with accelerants or a structure containing a large quantity of plastic materials. A 360-degree survey would allow you to see the large plume of black smoke that may be an outside fire in the rear of the structure. You should never rely on only one indicator to make a decision. Once a decision is made, you must continually size up the situation looking for indicators to confirm that your decision was correct. If indicators are contradictory to your decision, you must reevaluate and change the plan.

Chapter Review

1. What factors can be determined while responding that will affect pre-fire planning?
2. How do facts, perceptions, projections and probabilities influence decision-making?

3. What can observing smoke tell you about probable fire conditions?
4. What conditions may be observed during size-up that will influence pre-fire planning and response decisions?
5. What factors may influence firefighter and occupant survivability?
6. How will firefighter and occupant survivability factors influence prefire planning?
7. List the building characteristics that you can observe during a 360-degree survey. What do these characteristics tell you about the building and possible hazards?
8. What hazards might be associated with each building occupancy classification?
9. How do different construction types and occupancy classifications affect prefire planning?

Strategy

Photo courtesy of Chris Mickel, New Orleans (LA) F.D. Photo Unit.

Chapter Contents

Incident Priorities 207
    Life Safety... 207
    Incident Stabilization 210
    Property Conservation 210

Risk versus Benefit 211

Operational Strategies 212
    Offensive Strategy 212

    Defensive Strategy 214
    Transitioning Strategies 216

Incident Action Plan 217
    Developing the IAP 217
    Implementing the IAP 218

Chapter Review 222

chapter 5

Key Terms

Defensive Strategy 214	Mayday 215
Fireground 208	Offensive Strategy 212
Life Safety 207	

Strategy

FESHE Learning Outcomes

After reading this chapter, students will be able to:

4. Describe the steps taken during size-up.
5. Examine the significance of fire ground communications.
6. Identify the roles of the National Incident Management System (NIMS) and Incident Management System (ICS) as it relates to strategy and tactics.
7. Demonstrate the various roles and responsibilities in ICS/NIMS.

Chapter 5
Strategy

This chapter provides a foundation for the in-depth discussion of tactics in Chapter 6:

- Incident priorities
- Risk versus benefit
- Operational strategies
- Incident Action Plan

Incident Priorities

Life safety, incident stabilization, and property conservation (LIP) are the incident priorities that apply to all types of incidents and are listed in order of priority **(Figure 5.1)**. Placing them in this order makes it easier to make strategic and tactical decisions.

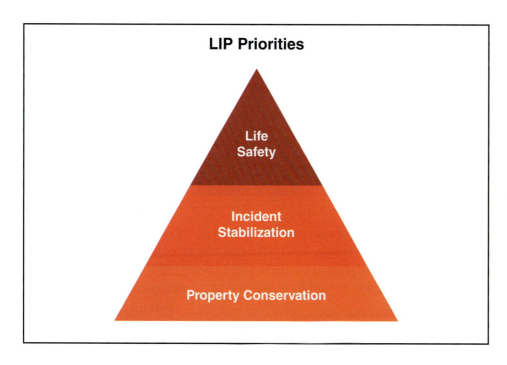

Figure 5.1 When firefighters respond to an incident, they should know that life safety is always their first priority, followed by incident stabilization, then property conservation.

Life Safety

Life safety is always the first priority. When a structure is known or suspected to be occupied, the concern is for both the occupants and the fire fighting personnel. The IC has direct effect on personnel safety based on a number of decisions, including:

Life Safety — Refers to the joint consideration of the life and physical well-being of individuals, both civilians and firefighters.

- Making the initial size-up
- Having accountability in place beyond passport and key cards
- Having a sound IAP
- Assigning division and group supervisors where appropriate
- Having a rapid intervention crew or team (RIC/RIT) team in place **(Figure 5.2)**
- Having a safety officer and medical/rehabilitation resources

Figure 5.2 A RIC/RIT is tasked with standing by at an incident to provide assistance to firefighters in need.

While it is true that responders will risk a lot for a savable life, they must also be concerned with their own safety. Attempting to save a vacant building or to attempt to rescue victims who are beyond saving is not worth risking a firefighter's life. Firefighter life safety is affected by a number of criteria, including:

- Using the correct personal protective equipment (PPE) properly
- Assigning tasks based on the abilities of the personnel and having sufficient resources
- Applying a risk-benefit analysis model
- Following the local standard operating procedures or guidelines (SOP/Gs) for fireground operations
- Maintaining discipline on the **fireground**
- Using tools and equipment correctly **(Figure 5.3)**

Fireground — Area around a fire and occupied by fire fighting forces.

Figure 5.3 Firefighters who are responsible for ventilating a roof could be injured if they improperly use an ax or chainsaw. *Courtesy of Bob Esposito.*

NIOSH Recommendations

The National Institute for Occupational Safety and Health (NIOSH) conducted a study in which it investigated 335 incidents involving 372 firefighter fatalities from 1998-2005. The leading recommendations, grouped into ten categories, are:

1. Medical Screening
2. Fitness and Wellness
3. Standard Operating Procedures and Guidelines
4. Communications
5. Incident Command
6. Motor Vehicle Operations
7. Personal Protective Equipment
8. Strategies and Tactics
9. Rapid Intervention Team
10. Staffing

Source: *NIOSH.*

The second concern of life safety is locating, stabilizing, protecting, and removing living victims from the hazardous area. If it is believed that people are alive and still savable in the structure, use an offensive strategy to rescue them. If it is known that there are viable victims inside the structure, then the two-in, two-out rule does not have to be followed. If conditions within the structure are found to be unsurvivable for any occupants, then adopt the strategy appropriate to stabilize the incident and conserve property with every consideration for firefighter safety.

Incident Stabilization

Incident stabilization is the process of controlling the fire. Although incident stabilization is listed as a lower priority than life safety, sometimes the best method to achieve life safety is through incident stabilization. For example, personnel might need to place attack lines into operation and/or initiate ventilation in order to halt fire progress, providing a safer entry for search teams **(Figure 5.4)**. Incident stabilization can also result in property conservation through a rapid fire attack that prevents the extension of a fire into unburned compartments or areas of the structure(s). Rapid fire attack may also lead to a reduction in water damage due to the use of fewer gallons (liters) for extinguishment. Incident stabilization is not achieved until the fire is under control. For structure fires, the fire is considered under control when it is incapable of growing larger due to the control measures in place. It also means that no added fuel, oxygen, or heat will result in fire growth.

Figure 5.4 Personnel could use attack lines to stabilize an incident.

Property Conservation

Property must take a lower priority than lives. Traditionally, property conservation is thought of as salvage or loss control activities. Like incident stabilization, some property conservation activities may begin early in the incident and be performed simultaneously with the other two priorities **(Figure 5.5)**. For instance, exposure protection should always be paramount once life safety has been addressed.

While the conservation of property has to be viewed as a necessary activity, at no time should property conservation take precedence over life safety or incident stabilization. If a defensive strategy is implemented, then strict adherence to collapse zones must be followed for firefighter safety.

Figure 5.5 The use of waterproof covers to protect property has long been an essential element of property conservation.

Risk versus Benefit

As a fire and emergency services responder, you are expected to take calculated risks to provide for life safety, incident stabilization, and property conservation. Calculated risks mean that you gather information through size-up and determine what level of risk is acceptable for the given situation. The fire service has continued to have line-of-duty deaths (LODD) every year. As a result, in 2001, the International Association of Fire Chiefs (IAFC) developed a model policy called *The 10 Rules of Engagement for Structural Fire Fighting*. The model policy was developed to help ensure that everyone can return home safely.

Line of Duty Deaths Over Time
Rate of Deaths for Firefighters Due to Cardiac Arrest

1970s – 2.6 per 100,000 fires

1990s – 1.9 per 100,000 fires

Increase of Firefighter Deaths Due to Traumatic Injuries

1970s – 1.8 per 100,000 fires

1990s – Almost 3 per 100,000 fires

Decrease of Fires

- 1978 – about 1.1 million
- 2008 – about 300,000
- Firefighter deaths per 100,000 fires went from 6 to 4.5
- Non-sudden death cardiac inside structures went from 1.5 to 3 per hundred thousand structures

Source: NFPA

The Ten Rules of Engagement for Structural Fire Fighting
Acceptability of Risk

1. No building or property is worth the life of a firefighter.
2. All interior fire fighting involves an inherent risk.
3. Some risk is acceptable in a measured and controlled manner.
4. No level of risk is acceptable where there is no potential to save lives or savable property.
5. Firefighters shall not be committed to interior offensive fire fighting operations in abandoned or derelict buildings that are known to be or reasonably believed to be unoccupied.

Risk Assessment

1. All feasible measures shall be taken to limit or avoid risks through risk assessment by a qualified officer.
2. It is the responsibility of the Incident Commander to evaluate the level of risk in every situation.
3. Risk assessment is a continuous process for the entire duration of each incident.
4. If conditions change, and risk increases, change strategy and tactics.
5. No building or property is worth the life of a firefighter.

Your decisions are made on acceptable risk based on the initial assessment of the incident scene. The model is a departmental SOP/G that is used to help Phoenix (AZ) Fire Department officers make reliable emergency response decisions. The essence of the model is stated as follows:

- Each emergency response is begun with the assumption that *responders can protect lives and property.*
- Responders will *risk their lives a lot, if necessary, to save savable lives.*
- Responders will *risk their lives a little, and in a calculated manner, to save savable property.*
- Responders will *NOT risk their lives at all to save lives and property that have already been lost* **(Figure 5.6)**.

Figure 5.6 During training or incident evolutions, responders may need to be reminded of the rules of response: Take no risk for no gain.

The safety of both firefighters and occupants is your primary concern. Your decisions must be based on this model that should be applied to your initial size-up. When committing to an interior offensive attack, you must balance the risk to your personnel if there is a doubt that a savable life is at risk.

Operational Strategies

Once you have determined the appropriate level of risk, you must decide on the strategy to implement. Traditionally, there are two strategies that the fire service uses: *offensive or defensive.*

> **Offensive Strategy** — (1) In wildland fire fighting, a direct attack on the fire perimeter by crews, engines, or aircraft, or an aggressive indirect attack such as backfiring. (2) Overall plan for incident control established by the incident commander (IC) in which responders take aggressive, direct action on the material, container, or process equipment involved in an incident.

Offensive Strategy

The **offensive strategy** used at a structure fire usually means that resources are deployed for interior operations to accomplish incident priorities. Countless possible variations exist for this scenario, depending upon life hazards and structural stability.

The Incident Commander may need to talk to building occupants who have escaped the structure and neighbors or other witnesses. These witnesses may be able to help determine whether there are any occupants still inside and, if so, whether there is a reasonable chance that they are still alive **(Figure 5.7)**.

Figure 5.7 Building occupants might be able to provide information to firefighters about other residents who may be trapped inside the structure. *Courtesy of Ron Jeffers, Union City, NJ.*

When confirmation is made that lives are threatened but not lost AND the structure is stable enough for operations, an offensive strategy may be justified. When adequate resources are available, fire attack should start simultaneously with a search and rescue operation. If not, the search might have to be delayed until the fire is confined (stabilized) or a hoseline is placed between the occupants and the fire.

NOTE: Search and rescue operations are discussed extensively in the IFSTA manual, **Fire Service Technical Search and Rescue**.

The offensive strategy may involve rescue, extinguishment, or both. In some fire incidents, rescue and extinguishment will occur simultaneously with engine crews attacking the seat of the fire while other personnel search for victims. In extreme cases, where a victim is known to be trapped, rescue will become the primary activity and fire attack will be performed only to protect the rescuers and the victim.

Based on the latest research from the National Institute of Standards and Technology (NIST), traditional offensive strategy should now include the option to apply water from the exterior of the structure as crews advance into the interior. This is known as the *transitional attack*. As crews deploy and charge handlines while preparing for interior attack, water can be directed into the structure for initial knockdown at the discretion of the unit officer. Historically, a transitional attack was considered moving from an offensive to defensive mode; however, this tactic focuses on crews moving in the direction from exterior to interior operations. Chapter 6 will go into more detail about the transitional attack.

Defensive Strategy — Overall plan for incident control established by the incident commander that involves protection of exposures, as opposed to aggressive, offensive intervention.

Defensive Strategy

The decision to operate in a **defensive strategy** is indicated when no threat to occupant life exists, occupants are not savable, or when the property is not salvageable. The defensive strategy is intended to isolate or stabilize an incident and keep it from getting any worse or larger. In the case of a structure fire, a defensive strategy may mean sacrificing a building that is on fire to save adjacent buildings that are not burning. A defensive strategy is usually an exterior operation that is chosen because of insufficient resources or an interior attack is unsafe.

WARNING!
Exposure protection can be accomplished as a property protection strategy and does not always mean defensive operations are the strategy.

Defensive strategic operations involve personnel and apparatus that are kept at a safe distance from the incident hazards. This strategy may be employed if the incident is too large or hazardous to safely resolve with an offensive strategy, such as large structure fires or potential structural collapses **(Figure 5.8)**.

Figure 5.8 An Incident Commander may call for a defensive strategy at an incident where the conditions are too unsafe and large in scope to attack with an offensive strategy. *Courtesy of Chris Mickel, New Orleans (LA) F.D. Photo Unit.*

For example, this strategy is used when protecting exposures from the spread of a large fire rather than extinguishment. Defensive strategic operations are justified in the following conditions involving a structure fire:

- **Volume of fire** — Amount of fire exceeds the resources available to confine or extinguish. Examples include:
  - Lack of personnel or trained personnel

- Inability to provide adequate fire flow; that is, gallons per minute (gpm) (liters per minute [L/min]) because of insufficient pumping capacity or availability of water supply
- Lack of appropriate apparatus or equipment to implement the required tactics

• **Structural deterioration** — Structure is unsafe for interior entry.

• **Risk outweighs benefit** — If the risk to emergency responders is greater than the benefit.

• **Occupancy** — Structure is known to be vacant or abandoned.

If the initial IC determines that a defensive strategy is justified, resources may be assigned to apply water to protect adjacent exposures and/or other action to prevent the fire from spreading to these exposures. A defensive strategy may also be employed if resources are limited to the point where offensive operations cannot be implemented safely.

When firefighter fatalities and injuries occur at structural fires, they are sometimes the result of firefighters being in offensive positions on defensive fires or rapid changes in the situation, such as the following:

• One structural member weakening can cause a ceiling, floor, wall, or other structural member to collapse.

• Introduction of fresh air into a superheated compartment can cause a rapid fire development condition (such as backdraft or flashover) **(Figure 5.9)**.

Figure 5.9 Firefighters need to be aware of how fires can rapidly change with the introduction of fresh air, such as when the glass doors at a storefront fail. *Courtesy of Chris Mickel, New Orleans (LA) F.D. Photo Unit.*

• Heavy or dense smoke can obscure vision, causing firefighters to become disoriented and run out of breathing air.

• When a firefighter is trapped or unaccounted for and a **Mayday** has been sounded.

• Lack of a proper Incident Command System in place and failure to change strategy when the situation changes.

Mayday — Internationally recognized distress signal.

Figure 5.10 Tracking and accountability systems include a procedure for personnel to check in before entering the incident scene.

- IC not gathering all available information to complete a proper size-up (360-degree view of the structure).
- Wind-driven fire changes conditions quickly.

These changes may occur without the IC being aware of them. As a result, ICs must conduct personnel accountability reports (PAR) to check on the welfare of all firefighters at predictable intervals, such as:

- On the scene every 10 to 20 minutes **(Figure 5.10)**
- In accordance with your department's SOP/G
- When any one of the above conditions has occurred

Transitioning Strategies

When a situation rapidly changes, the IC needs to be prepared to transition the operation from an offensive to a defensive strategy. The IC must efficiently communicate the change to all personnel and units operating at the incident **(Figure 5.11)**. The IC is responsible for the decisions made and personnel must not freelance. This type of situation may require an operational retreat where all interior crew are made aware of the need to evacuate the building. Sounding the apparatus horn several times in uniform manner can be an effective retreat signal. A PAR should follow to ensure that all personnel have been advised and have withdrawn from offensive positions. This change in transition must be communicated definitively, and must be followed explicitly. Although crew members may feel that they can extinguish the fire or complete vertical ventilation, the crew does not have the same information and exterior view of changes that the IC can see. Effective decision-making should include communications from all crews, divisions, and groups at the scene and relaying their information to the IC. The IC must also effectively relay information (such as change in smoke condition or structural integrity) as well as his or her goals and objectives to all personnel.

Figure 5.11 An Incident Commander must effectively communicate a change in strategy to personnel at the scene to avoid any confusion that could put firefighters in danger. *Courtesy of Ron Jeffers, Union City, NJ.*

WARNING! Do not freelance on the fireground.

For a transition to be efficient and effective, it must occur as soon as the need is recognized. The IC must maintain current and accurate knowledge of the situation. The IC must be aware of changing conditions throughout the structure. Officers must give situation or status reports to the IC or their immediate supervisor regularly.

When the transition between offensive and defensive strategies is necessary, the steps needed in the transition may depend on the speed at which the situation changes. Regardless of the circumstances, all personnel must be made aware of the transition. ICs must always know the location of personnel assigned under their command and must conduct personnel accountability checks when withdrawal is complete.

During an orderly withdrawal, hoselines should not be abandoned unless absolutely necessary since they cannot provide protection if they are left behind. Some units may need to remain in place to protect the withdrawal of other units. In contrast, in some conditions, all personnel will be directed to abandon equipment and evacuate with haste. RIT/RIC personnel must be ready to assist any units that require assistance during the transition.

Incident Action Plan

Depending on local policy and procedures, the first-arriving fire officer or acting fire officer is responsible for managing the incident scene. When you are in that position, you must act in accordance with the locally adopted ICS. Besides determining the incident priorities and strategy previously mentioned, you must also develop and implement the Incident Action Plan (IAP). This plan will allow you to assign resources based on the desired results, track your resources, and accomplish the tactics required to fulfill the incident priorities.

Developing the IAP

The IAP is based on the information you gathered in the incident size-up. The IAP may result in a written plan. If the incident is small and can be handled by the first-alarm assignments, the plan does not have to be written down **(Figure 5.12)**. As the incident grows in size, the use of a tactical worksheet can be of great value to the IC. If the incident is large or has the potential for involving multiple units or agencies for a prolonged period of time, it must be in writing. Standardized ICS forms are available to record the various elements of the plan.

Figure 5.12 A written Incident Action Plan is not needed for small incidents. *Courtesy of Chris Mickel, New Orleans (LA) F.D. Photo Unit.*

The plan must be communicated to all units and individuals operating at the scene before they are given a work assignment. This communication is done in person or over designated radio frequencies. All incident personnel must function within the scope of the IAP; actions taken outside the scope of the IAP are called *freelancing* and may place responders in jeopardy and reduce operational effectiveness. Fire officers should follow SOP/Gs that the agency and/or the IC identifies. Incident personnel should direct their actions toward achieving the incident objectives, strategies, and tactics specified in the plan. When all members understand their positions, roles, and functions in the ICS, the system can safely, effectively, and efficiently use resources to accomplish the plan.

Implementing the IAP

Putting an IAP into action is the next step in the process of controlling the emergency incident. At this point, the IC must select a command option and allocate your resources. The IC's choice of command options will be based on:

- How severe is the incident **(Figure 5.13)**

Figure 5.13 A firefighter watches a structure fully engulfed in flames burn from atop an apparatus. *Courtesy of Chris Mickel, New Orleans (LA) F.D. Photo Unit.*

- How rapidly it may develop
- How many resources are available

In some departments, the IC may have all the personnel, apparatus, and water supply within minutes of his or her arrival. In other departments, the IC may have to wait a significant amount of time before resources can be deployed.

Command Options

With the priorities established and the IAP ready for use, the IC must be ready to take Command of the situation. As the first-arriving officer, you have several options for Command, including the following three Command options available:

- Investigation option (nothing showing)
- Fast-attack or mobile command option
- Stationary command (Incident Command Post) option

Investigation Option (Nothing-Showing). When the problem generating the response is not obvious to the IC, he or she should establish Command and announce that *nothing is showing*. Then direct the other responding units to stage at a location that would allow maximum flexibility for their deployment of their tactical assignment at the incident based on the local SOP/G. The IC then accompanies unit personnel on an investigation of the situation and maintains Command using a portable radio. This approach applies to all types of most emergencies.

Fast-Attack or Mobile Command Option. The first-arriving unit has the responsibility of assuming command of the incident. At most incidents, the company officer will be the initial IC. Mobile Command is one of the Command options available to him or her. If the incident is one in which the direct involvement of the company officer will impact the outcome, a Mobile Command should be utilized.

A Mobile Command is not intended to be used throughout the incident, but so the company officer may remain with his or her crew. After giving the on-scene report and announcing a Mobile Command, the company officer should take his or her portable radio and accompany the crew. The use of the portable radio will allow the company officer to be involved in the attack without neglecting Command responsibilities. Mobile Command should be transferred at the earliest opportunity or a transition to a Stationary Command should take place. Command should never be transferred to anyone who is not on the scene. The transfer of Command procedures outlined by your agency should be followed.

Situations in which a Mobile Command may be used are:

- Life safety situations, such as a victim rescue
- Offensive fire attacks, which are in the marginal mode **(Figure 5.14)**
- Any incident where the safety of firefighters is a major concern
- Incidents that require further investigation by the company officer
- Crew members need closer supervision due to being new and inexperienced

Figure 5.14 A Mobile Command could be utilized during an offensive attack on a fire. *Courtesy of Ron Jeffers, Union City, NJ.*

Stationary Command (Incident Command Post) Option. Because of the nature, scope, or potential for rapid expansion of some incidents, immediate and strong overall Command is needed. In these incidents, the company officer should assume Command using the following steps:

- Naming the incident

- Designating the location of the Incident Command Post
- Giving an initial report on conditions
- Requesting the additional resources needed
- Initiating the use of a tactical worksheet

With this option, the company officer remains at the mobile radio in the apparatus, assigning tasks to other unit personnel, communicating with other responding units, and expanding the NIMS-ICS as the complexity of the incident requires **(Figure 5.15)**. The IC may assign one of the other members of his or her crew as the acting officer or direct the other crew members to work under the supervision of another company officer or assign them to other ICS positions. Before doing so, remember that crews assigned to another commander are no longer available to the IC, and the IC must ensure that their absence will not create a hardship.

When Command of an incident needs to be transferred to another officer, the transfer must be done correctly to prevent confusion about who is responsible for the incident response. The officer assuming Command must communicate with IC via face-to-face communication, which is preferred, or by radio. Command should only be transferred to improve it, such as to a higher ranking officer, a more experienced team member, or to get the initial IC back to his or her crew for close supervision. When transferring Command, the IC should brief the relieving officer on the following items:

Figure 5.15 The IC may establish a Command Post in the cab of an apparatus located at the scene.

- Incident status (such as fire conditions or number of victims)
- Safety considerations
- Goals and objectives listed in the IAP or tactical work sheet
- Progress toward completion of tactical objectives
- Deployment of assigned resources and their specific location in the hazard zone
- Assessment of the need for additional resources

Resource Allocation

The IC must determine if the available resources can support the initial or sustained attack. If the available fire flow and resource capabilities are equal to or greater than the estimated need, an initial offensive attack can be made. If the fire flow requirement exceeds the available water supply and/or resource capabilities, a defensive attack must be implemented **(Figure 5.16)**. In situations where there is no life hazard and the initial offensive attack is unsuccessful in controlling the fire or if the fire increases in intensity, the IC must increase the fire flow applied to the fire with additional larger hoselines. If this task is not possible, then you must shift to a defensive strategy until additional resources become available or the incident is terminated. All personnel and units as well as the communication center must be notified of the change of

Figure 5.16 The availability of resources will impact whether an IC calls for an offensive or defensive strategy. *Courtesy of Chris Mickel, New Orleans (LA) F.D. Photo Unit.*

operational strategy, and attacking units must be given time to disengage in an orderly manner. Personnel accountability is essential for a safe and efficient transition to the defensive strategy.

The IC's next action is to assign resources, which can be divided into two categories: assigning members of his or her unit and assigning other units that have been or will be dispatched to the incident.

Personnel. This task may be the more difficult of the two actions, especially if the IC has minimal staffing. Dividing up personnel to perform the various tasks that make up the most critical tactics and still adhere to local SOP/Gs and legal guidelines can be difficult. Generally, the driver/operator will be assigned to remain with the apparatus. Limited staffing may not permit the IC to remain at the Command Post when he or she is needed to investigate the source of the fire.

The IC's greatest challenge will be deciding the level of risk at which to place personnel when a life safety situation is obvious or perceived. The two-in, two-out rule can effectively prohibit the IC from making entry into the IDLH environment.

Other Units. During the IC's initial size-up, he or she must consider the number and type of units that are responding and the length of time until they are ready to deploy at the scene. If a preincident plan (operational plan) exists for the site, then the IC's job of assigning subsequent units will be somewhat easier **(Figure 5.17)**. However, the IC must not depend completely on it. The type of situation encountered may be different from the one that the plan is intended to address.

If no preincident plan exists or the situation is unique, assign arriving units based on the IC's tactical needs to meet the strategic goals and department SOP/G. A key to a successful operation is to assign units so that apparatus do not have to be repositioned during the incident. Assigning units to a staging area allows you time to determine the best location and use of personnel and equipment.

Figure 5.17 If available, a preincident plan and other documents on a structure could assist in the decision-making process while assigning personnel. *Courtesy of Ron Jeffers, Union City, NJ.*

Chapter Review

1. What are the priorities at an incident, in order from highest to lowest priority?
2. What factors should you consider when evaluating risk at an incident?
3. What conditions warrant an offensive strategy versus a defensive strategy?
4. Describe the three command options.
5. What must the IC consider when allocating resources, both personnel and other units?

Tactics

Photo courtesy of Chris Mickel, New Orleans (LA) F.D. Photo Unit.

Chapter Contents

Search and Rescue 227	Building Construction and Modern Furnishings 256
Search Safety Guidelines for the IC and Search Teams 231	Basement Fires 258
Conducting a Search 231	Fire Behavior Indicators 258
Exposures 234	Location and Extent of the Fire 259
Confinement 235	Type of Ventilation 259
Extinguishment 239	Location for Ventilation 260
Fire Attack 241	Weather Conditions 261
Transitional Attack 243	Exposures 262
Positive-Pressure Attack 245	Staffing and Available Resources 263
Overhaul 248	**Types of Tactical Ventilation** 263
Tactical Ventilation 253	Horizontal Ventilation 263
Considerations Affecting the Decision to Ventilate 255	Vertical Ventilation 267
	Other Types of Ventilation Situations 268
Risks to Occupants and Firefighters 256	**Salvage/Property Conservation** 271
	Chapter Review 273

224 Chapter 6 • Tactics

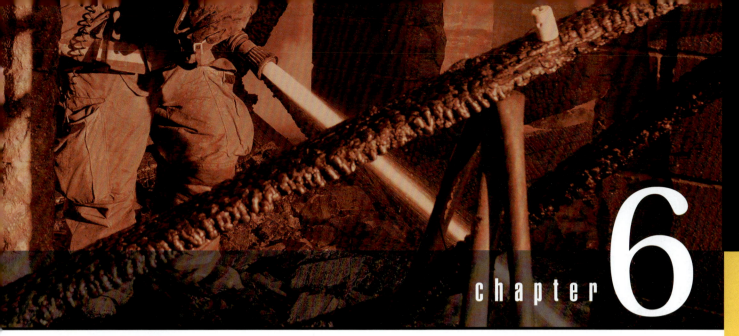

chapter 6

Key Terms

Confinement 229	Positive-Pressure Ventilation 264
Direct Attack 241	Primary Search 231
Exhaust 246	Rescue 227
Exposure 234	Salvage 265
Fire Watch 250	Search 227
Horizontal Smoke Spread 270	Search and Rescue Operation 228
Hydraulic Ventilation 263	Secondary Search 231
Intake 246	Shielded Fire 242
Mechanical Ventilation 263	Soffit 268
Medical Surveillance 248	Stack Effect 269
Natural Ventilation 263	Tactical Ventilation 253
Negative-Pressure Ventilation 264	Thermal Imager 232
Positive-Pressure Attack 245	

Tactics

FESHE Learning Outcomes

After reading this chapter, students will be able to:

1. Discuss fire behavior as it relates to strategies and tactics.
4. Describe the steps taken during size-up.
5. Examine the significance of fire ground communications.
6. Identify the roles of the National Incident Management System (NIMS) and Incident Management System (ICS) as it relates to strategy and tactics.
7. Demonstrate the various roles and responsibilities in ICS/NIMS.

Chapter 6
Tactics

This chapter discusses the tactics utilized on the fireground:

- Search and rescue
- Exposures
- Confinement
- Extinguishment
- Overhaul
- Tactical ventilation
- Considerations affecting the decision to ventilate
- Types of tactical ventilation
- Salvage/property conservation

Search and Rescue

Search involves the techniques that allow a rescuer to identify the location of victims and determine how to access those victims in order to remove them to a safe area. **Rescue** is the removal of victims from an untenable or unhealthy atmosphere. When Command decides that enough evidence exists to send firefighters into a structure to conduct a search, that operation is separate from a rescue operation. If there is a possibility that people may be in the structure, a primary search is conducted. The tactics in supporting a primary search can be as important as the primary search itself.

The search component requires that you as IC, your crew, or other units locate any victims who are in the hazard area. The rescue component requires that once you locate the victims, you separate them from the hazard. Separating the victims from the hazard involves moving them outside the involved structure or moving them to an area of safety located inside the structure, referred to as sheltering in place **(Figure 6.1)**.

> **Search** — Techniques that allow the rescuer to identify the location of victims and to determine access to those victims in order to remove them to a safe area.
>
> **Rescue** — Saving a life from fire or accident; removing a victim from an untenable or unhealthy atmosphere.

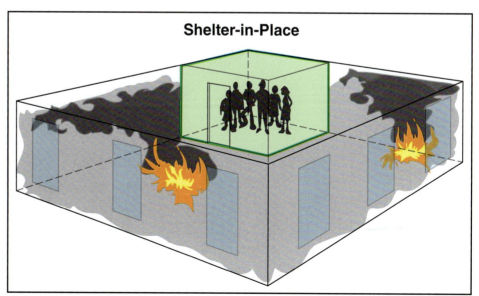

Figure 6.1 Shelter in place exists as an option where the inhabitants are unable to leave the confines of the structure and a safe place is defensible inside.

Search and Rescue Operation — Emergency incident operation consisting of an organized search for the occupants of a structure or for those lost in the outdoors, and the rescue of those in need.

Making the decision to perform **search and rescue operations** can be a subjective process. It is based on facts that you know, perceptions that you believe to be true, and projections of what you think may occur in the future. Your ability to complete a risk assessment, assign tasks, and determine a mode of attack is dependent on your understanding of the facts, perceptions, and projections — your situational awareness (**Figure 6.2**).

Figure 6.2 Incident Commanders should learn to recognize fire behavior indicators and what they may reveal about fire hazards.

You may take the approach that a life will be in danger and rescue will be required in all cases. This scenario will generate a thought pattern that includes an interior attack with search teams and ventilation. However, this scenario may also be beyond the ability of your resources, placing you and your crew at greater risk than necessary. While making the decision to commit to search and rescue, you must consider the facts available in developing your perceptions and projections. The plan should take into consideration where to start the search, where victims are most likely to be found, and the dangers involved. Fire attack and ventilation must be coordinated with the search effort.

Separated Firefighters at Three-Story Apartment Building

A tenant on the first floor of a three-story apartment building called 9-1-1 after a kitchen fire spread, sending flames and smoke up a common stairway that led to the upper floors. When personnel arrived at the Type III building built in the 1950s, the fire was coming out of the apartment's open sliding glass door and reaching a wooden balcony on the second floor. The fire compromised the sliding glass door of a second-floor apartment, igniting a sofa just inside the doorway.

The captain in command had responded to a kitchen fire at this apartment complex prior to this incident. As a crew was preparing to enter the building through the front door, they heard an occupant on the third floor calling for help and that he was about to jump to escape the fire. The captain and a driver reached the occupant with a ground ladder, but when a crew coming to assist in the rescue opened the building's front door, the intense heat knocked them to the ground. As crews fought the fire, two firefighters made their way to the third floor to search for any victims.

However, the two firefighters got separated inside an apartment when one of them went to check the bed, believing that a pile of clothes could be a victim. Within seconds, conditions unexpectedly worsened with high heat, zero visibility, and the hallway that they had used to enter the apartment filled with flames. Command called for the firefighters to evacuate the building. The firefighter who had checked on the clothes found a window inside the apartment, and unable to wait for a ground ladder to reach him, he broke the window and jumped out of it. Though injured, he informed personnel that the second firefighter was still inside the apartment.

Unable to find the window, the firefighter radioed several MAYDAY calls and gave a LUNAR (location, united number, assignment, and resources needed). However, crews took longer to reach him after he mistakenly gave his position as the Bravo-Charlie corner of the building. A firefighter standing on a ladder outside the building jumped onto the third-story balcony in an attempt to rescue the fallen firefighter, who was still conscious and breathing. However, his facepiece had melted from the intense heat from the rapid fire progression and he died before crews could drag him to safety.

In some cases, **confinement** of the fire must be initiated prior to the search and rescue effort for the sake of the firefighters' and occupants' safety. Confinement of the fire may be necessary before committing personnel to rescue. Confinement options can include:

- Creating a barrier between occupants and the fire (closing a door)
- Beginning suppression operations from the exterior
- Strategic placement of a hoseline between the occupants and the fire

Confinement — Fire fighting operations required to prevent fire from extending from the area of origin to uninvolved areas or structures.

Ideally, an entire crew conducts searches and is supplemented by an attack and ventilation crew. In some cases, however, a minimum search crew may include only two firefighters. The officer in charge of the search team will determine the number of personnel required to perform the search **(Figure 6.3)**. Size and complexity of the structure as well as the time of day will drive the decision for the number of personnel needed for search operations. For example, the report of smoke in a large hotel at 2 a.m. will require more search teams than a similar situation in a warehouse at 2 p.m. on a weekday.

Figure 6.3 The IC will make the determination on the number of personnel needed to perform a search, basing his or her decision on factors such as the size and complexity of the burning structure. *Courtesy of Chris Mickel, New Orleans (LA) F.D. Photo Unit.*

Tactics are achieved through the completion of tasks that units and individuals perform. Multiple tasks may be required to accomplish the tactic of rescue. Some of these tasks may be associated with other tactics, such as ventilation or fire extinguishment. When the rescue is complete, reassess your strategy and risk management plan.

Rescue Considerations for the Initial Incident Commander

- Performing size-up to include a 360-degree check and occupant survivability profile
- Performing ventilation in coordination with search and rescue as well as fire attack
- Gaining access
- Performing primary/secondary search
- Following ventilate, enter, isolate, and search (VEIS)
- Deploying charged hoselines
- Following the OSHA standard for the two-in, two-out rule
- Confining the fire
- Protecting and/or removing victims

Search Safety Guidelines for the IC and Search Teams

Some of the safety guidelines that the initial Incident Commander should use are:

- Forming a survivability profile
- Assigning tactical objective
- Maintaining communications and accountability **(Figure 6.4)**
- Monitoring conditions

Figure 6.4 Before responding to a large incident, firefighters use a tag system to maintain accountability. *Courtesy of Ron Jeffers, Union City, NJ.*

Some of the safety guidelines that the search teams should use are:

- Monitoring, evaluating, and communicating fire conditions and victim viability as appropriate
- Searching systematically
- Reporting status and PAR to Incident Command

Conducting a Search

In most structure fires, the search for life requires two types of searches: primary and secondary. A **primary search** is a rapid but thorough search that is performed either before or during fire suppression operations. The search should advance towards where occupants would be in the most danger. Entry should also be made as close to that point as possible. This area is typically above the fire and the rooms adjacent to the fire. During the primary search, the search team or teams can confirm that the fire conditions are as they appear from the outside or report any changes that personnel may encounter.

A **secondary search** is conducted *after* the fire is under control and the greatest hazards have been controlled. Personnel other than those firefighters who conducted the primary search should conduct the secondary search. The secondary search is a slower, more thorough search that attempts to ensure that no occupants were overlooked during the primary search. The secondary search team should approach the task as though the space or compartment has not been searched before.

Primary Search — Rapid but thorough search to determine the location of victims; performed either before or during fire suppression operations. May be conducted with or without a charged hoseline, depending on local policy.

Secondary Search — Slow, thorough search to ensure that no occupants were overlooked during the primary search; conducted after the fire is under control by personnel who did not conduct the primary search.

Primary Search

In single-family dwellings and small commercial occupancies, it is possible to make quick searches without a charged hoseline. Once resources are available, additional personnel must advance a hoseline into the structure to support and protect the search team **(Figure 6.5)**. After the IC assesses the occupant survival profile, he/she may then assign resources to complete a search using one of the methods included below.

Figure 6.5 While firefighters can perform a quick search of a single-family house without a charged hoseline, one must be advanced into the structure as soon as the resources are available. *Courtesy of Bob Esposito.*

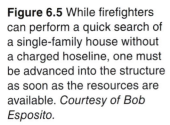

Thermal Imager — Electronic device that forms images using infrared radiation. *Also known as* Thermal Imaging Camera.

General/Traditional Search. Systematic pattern, such as left- or right-handed search. This type of search method is effective when entering a structure from a normal ingress such as a door and the structure is compartmentalized into small rooms such as a residential structures or offices. These searches require the least amount of training and supervision to be performed.

Oriented Search. Systematic approach in which the team can quickly and effectively search specific rooms or clusters of rooms. The team leader will position himself or herself in a fixed location to help keep the group's orientation within the structure while the other members of the search team conduct the physical searching. The team leader utilizing the thermal imager increases the efficiency of the search method and keeps track of the search progress **(Figure 6.6)**. This method can be utilized when entering the structure from any location, and will allow the search team to move directly to the areas of highest risk more quickly. These searches require advanced training in these methods and equipment utilized and a higher level of supervision to be performed.

Figure 6.6 During an oriented search, the team leader should use a thermal imager to identify where the fire is inside the structure and aid the search team in quickly getting to the area of highest risk. *Courtesy of Ron Jeffers, Union City, NJ.*

Wide-Area Search. Used to conduct a search of a large or complex area that is filled with smoke. These searches utilize specific search equipment and rope systems or techniques designed for use in large areas to prevent lost firefighters and provide for a thorough search of the open areas. These searches require advanced training in these methods and equipment utilized and a higher level of supervision to be performed.

VEIS (Vent, Enter, Isolate, Search). Utilized when it is suspected or probable that a victim is present in a specific room or compartment within the structure and there is a direct access point to enter the room, typically a window. This method may allow quicker access to the victim due to the fire's progress within the structure or the physical location of the room itself **(Figure 6.7)**. These searches require advanced training in these methods and should only be utilized when credible information warrants their use.

Figure 6.7 VEIS (Ventilation, Enter, Isolate, Search) is a method that allows personnel to quickly reach a victim who is known to be in a specific room inside a burning structure. *Courtesy of Jeremy Potter, Idaho Falls Fire Department.*

Vent, Enter, Isolate, and Search (VEIS)

VEIS should only be initiated following a 360-degree size-up of a structure, and the IC should consider risk assessment when making the VEIS assignment. Rooms that are involved with the fire or show indications of possible backdraft or flashover should not be searched using the VEIS technique. Only areas that appear to be survivable and unlikely to have fire extension when ventilated should be searched using the technique.

- V - Ventilate the opening to allow access to the compartment
- E - Enter the compartment to be searched
- I - Isolate the compartment by closing doors and other openings
- S - Search the compartment quickly

This ventilation opening is independent of fire attack operations and has the possibility to create a new unintended flow path. Remember, ventilating the structure introduces new oxygen. Fires need oxygen and will move towards new sources in ventilation-limited environments.

During the primary search, the IC should always assign a crew of two or more. Working together, two rescuers can quickly conduct a search while maintaining their own safety. During the initial operation, the IC should ensure that hoselines are being deployed within the structure for fire suppression as the primary search occurs.

Secondary Search

As the resources are available and conditions permit, personnel other than those who conducted the primary search are assigned to conduct a secondary search of the building. During the secondary search, thoroughness is more important than speed. The IC should be informed when these assignments are completed.

Victim Removal

When notified that a victim has been found, the IC should consider the following:

- Communicate with the company officer to determine the need for additional resources to assist with victim removal.
  — Do conditions warrant immediate victim removal or consider shelter-in-place?
  — Are personnel able to remove the victim from the same egress path?
- Determine location of the crew and victim inside the structure.
- Assign additional crews to back-fill assignments being vacated.
  — Was the attack crew or any other crew used to remove the victim?
  — Are more EMS units needed?
- Assign a medical group to treat victims (**Figure 6.8**).

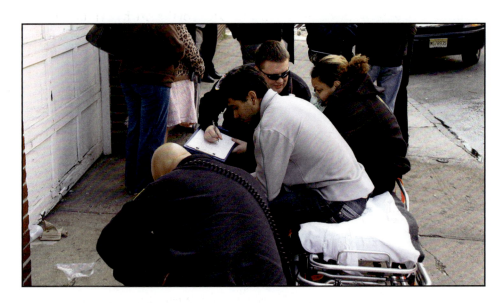

Figure 6.8 A medical group should be on the scene to treat any victims at an incident. *Courtesy of Ron Jeffers, Union City, NJ.*

Exposure — Structure surfaces or separate parts of the fireground to which a fire or products of combustion could spread.

Exposures

An **exposure** is any area to which fire could spread. Depending on the growth of a structure fire, it may have external or internal exposures. External exposures are other structures, equipment, vehicles, people, or natural features that the fire would endanger **(Figure 6.9)**. Internal exposures are unburned or unaffected portions of the building interior that the fire growth threatens. In some instances, protection may be a strategic goal. In other instances (particularly interior exposures), it may be a tactical objective supporting the strategic goal of containing the fire to the room/floor of origin. Protecting exposures involves activities that will prevent the spread of fire or products of combustion to them. Implementing the exposure tactic may also help meet the rescue and confinement objectives.

A thorough preincident survey will provide information on both interior and exterior exposures. Add to this information your knowledge of fire behavior, building construction, and information gathered on scene and you will enhance your ability to estimate the location and extent of the fire and project how it can spread.

Figure 6.9 Not all exposures are buildings. Vehicles, vegetation, and propane storage tanks may also be endangered.

Interior exposures to a fire are determined by the configuration of compartments and spaces and the location and arrangement of contents. As possible, initiate the following actions to provide support at a structural fire incident:

- Establish water supply
- Activate master stream appliances as needed
- Relocate exposed equipment, vehicles, or materials (if possible)
- Support fire protection systems in fire building and exposures as needed

Confinement

Personnel can determine the location and extent of the fire by evaluating the smoke and fire conditions. Smoke exiting the structure lazily typically indicates the fire is on a floor below. Smoke exiting under pressure with a defined neutral plane is a sign of a fire on the same structural level where the smoke is exiting but not necessarily the fire's origin or the only level that may have fire. Fire exiting an opening (most often a window) with a defined neutral plane indicates that the opening is supporting a bidirectional flow of products of combustion away from the fire and oxygen toward the fire. Knowing where the fire is receiving oxygen and what floor the fire is burning on allows personnel to make informed decisions on where to enter and how to confine the fire best using door control and extinguishment techniques.

Confinement includes fire fighting operations required to prevent fire from extending from the area of involvement to uninvolved areas or structures. Confining the fire to the involved area may be used to assist with rescue, separating trapped victims from the fire or flow paths.

The facts that you will need to consider when implementing confinement tactics are:

- Type of construction (according to building code)
- Type of fire protection systems or lack of systems
- Building occupancy type

- Contents within a compartment
- Flammability and combustibility of contents **(Figure 6.10)**

Figure 6.10 The Incident Commander should know that this room is filled with flammable chemicals while considering the use of confinement tactics on the structure. *Courtesy of Rich Mahaney.*

- Distance between involved and uninvolved contents
- Water supply vs. demand (fire flow)
- Number of stories
- Barriers to fire spread
- Heating, ventilation, and air conditioning systems **(Figure 6.11)**
- Current location and stage of fire
- Resources immediately available
- Rescue or exposure tactics implemented
- Additional fuel or ignition sources
- Explosion potential (internal and external)
- Fire behavior
- Closing fire doors

Size-up of the incident will give you an idea of how successful you will be in confining the fire to the compartment or structure of origin. For example, knowledge of how the building is configured, the type of construction and contents, and the methods of heat transfer, you will be better able to determine how and where to stop the spread of fire. Even with a thorough size-up, locating the origin of the fire and paths that it has taken or will take if

Figure 6.11 The type of heating, ventilation, and air conditioning system that a structure has could impact confinement operations. *Courtesy of Ron Jeffers, Union City, NJ.*

it is allowed to expand may be difficult. Examples of misleading clues include:

- The appearance of heavy smoke or flames in a cockloft may indicate the direction of spread but not the source.
- Smoke showing from the roof may lead you to believe that the fire is there while, in fact, it is a basement fire.
- You should also be aware of excessive heat buildup in areas that do not have visible fire or smoke.

Confinement Objectives

The proper application of ventilation techniques can help to release the heat and fire gasses from the structure, preventing the fire from spreading into other areas. When ventilation is not a viable option, confining the fire to a ventilation-controlled area may be a safer option. The following points will help the IC evaluate conditions for that tactic:

- Evaluate best access point for confinement
- Locate and isolate fire area
- Type of construction
- Type of fire protection systems or lack of systems
- Floor plan (open, compartmentalized, attic, and basement)
- Building occupancy type
- Number of stories
- Barriers to fire spread
- Heating, ventilation, and air conditioning systems (HVAC)
- Weather conditions (wind speed and direction)

The following factors determine whether fire extension will affect nearby external exposures:

- Type of construction of the building on fire and exposed structure
  - Indicates how quickly the exposure will become involved in fire, the area where exposure protection is needed, where fire will most likely spread from and to, and whether the fire is burning hot enough to spread.
- Type of fire protection systems or lack of systems in fire and exposed structure
  - Determines what types of protection are available, whether it can be controlled, and how it is supported.
- Building separation distances (often determined by building, fire, and sometimes zoning codes) **(Figure 6.12)**

Figure 6.12 In some communities, the space between structures may be so small that fire can easily move between buildings. *Courtesy of Bob Esposito.*

— Indicates how rapidly the fire can extend from the fire building to the exposure.

- Active and passive barriers to fire spread
  — Determines how long it will take for fire to extend into the exposure.
- Nonstructural exposures
  — Indicates additional hazards adjacent to the fire building. These hazards may include fuel storage and vegetation.
- Weather conditions (wind speed and direction, relative humidity, and air temperature)
  — Indicates direction, growth, and speed that fire will spread, and what exposures are in danger. Flying brands can cause fires several blocks away if not controlled.

If you are unable to perform a size-up of the structure's exterior, your perceptions of the fire situation may be incomplete. For instance, a fire reported in the early hours of the morning at a multistory warehouse in an older industrial district may create an orange-red glow and column of smoke from the back of the building. Does this mean that the fire is in the rear of the building, outside the building, or in an adjacent building across the alley? Even during daylight hours, you can mistake a column of smoke from a garbage fire behind a structure for a fire in the rear of the structure **(Figure 6.13)**. Or you may not realize that the garbage fire has spread into the building through a window or door. To the extent possible, you must verify your observations, based on fact, before committing your resources. You must also be ready to alter your perceptions after you get reports from interior and roof groups or divisions.

Figure 6.13 At times, the smoke from a garbage fire could lead firefighters into believing there is a fire in the rear of a building. *Courtesy of Chris Mickel, New Orleans (LA) F.D. Photo Unit.*

Your knowledge of the facts will guide you accurately in projecting how the fire will spread internally and externally and what exposures the fire will threaten. Some of the exposures, such as firewalls within the structure, will act as barriers to fire spread. Other exposures, such as high stack piles of combustible contents, will add to the fuel load and increase the magnitude of the fire.

The tasks you assign to your resources to implement the exposure protection tactical objective will depend on the stage of the fire and whether the exposures are internal or external. The type of exposure protection will mostly be application of water, foam, or gel, and largely dependent on what is available and in what amount.

Exposure Control

Your initial tactics are to determine the location and extent of the fire. This task can be accomplished through an investigation of the interior and a 360-degree walk around of the site when possible.

Internal — Resources may be assigned to:

- Investigation
- Fire attack
- Water supply
- Evacuation
- Fire suppression systems (activation/support)
- Ventilation
- Removal/protection of exposed contents (salvage)
- Supporting the fire protection system in the fire building as needed

External — Assigned where external exposures exist include:

- Perform a 360-degree walk-around
- Evacuate fire building and exposed structures
- Locate and isolate avenues of fire spread
- Coordinated appropriate ventilation techniques
- Place hoselines into use in appropriate attack strategy

Extinguishment

Extinguishing the fire is a natural extension of the confinement tactic. To accomplish this tactic, the extinguishing agent is placed on the seat of the fire or in the compartment containing the fire. Personnel may also use extinguishment to assist in rescue by removing the fire hazard altogether. Regardless of whether the structure is occupied, the IC should place high priority on fire extinguishment to remove the hazard.

The facts that the IC will need to consider include:

- Fire behavior
- Building construction
- Smoke and fire indicators
- Fuel load, including contents and structure
- Primary fuel type
- Location of the main body of fire

- Water supply
- Available personnel and equipment
- Attack mode in use **(Figure 6.14)**
- Response time

Offensive to Defensive Attack

Figure 6.14 Transitioning from an offensive attack to a defensive attack at the same structure fire. *Courtesy of Bob Esposito.*

An IC's perceptions may include:

- Effectiveness of fire confinement efforts
- Smoke color, density, velocity
- Neutral plane within ventilation openings
- Building stability indicators

The size-up must provide the necessary facts to determine the best extinguishment method to apply. If the fire is contained in a small compartment surrounded by fire barrier walls, then a rapid application of water may produce a rapid extinguishment. However, if the IC is faced with a fire that is growing quickly with sufficient fuel and oxygen available, he or she may need large quantities of water that large volume nozzles supply **(Figure 6.15)**.

Figure 6.15 A rapidly developing fire will require large quantities of water supplied by master streams or large-caliber nozzles. *Courtesy of Bob Esposito.*

In extinguishment as in rescue, the IC must determine if the benefit outweighs the risk that advancing hoselines into the structure poses. If rescue is unnecessary and the fire has expanded to consume much of the property or weaken its structural integrity, then using resources to protect exposures and confine the fire to the barrier walls of the structure may be the least risky approach. Continual evaluation of the risk management plan is a must.

In addition to the tasks needed to confine the fire, extinguishment will require that the IC's resources be committed to in no particular order:

- Selecting the appropriate size and type of nozzle
- Advancing hoselines
- Deploying master stream appliances, if applicable
- Selecting the most effective extinguishing agent
- Selecting the appropriate type of ventilation method
- Locating the seat of the fire

In the majority of cases, water will be the primary type of extinguishing agent that the IC will have at his or her disposal. However, if a sufficient supply of Class A foam is available for initial attack, this agent may extinguish the fire rapidly and as a result utilize less water on the supply side **(Figure 6.16)**. Remember that additional water will add to weight stress to the structure increasing the collapse hazard.

Figure 6.16 When available, Class A and B foam extinguishing agents may be required to extinguish a fire.

Fire Attack

Depending on the nature and size of the fire, firefighters may use a direct or indirect method of attacking the fire. Hoseline selection and stream selection are also made when the fire attack is conducted.

Direct Attack

A **direct attack** on the fire using a solid or straight stream uses water or foam most efficiently on fuel-controlled fires **(Figure 6.17)**. The primary fuels burning in these fires are the combustibles located within the compartment.

Considerations for the IC when choosing a direct attack:

- Arrival conditions
- Location, size, and extent of the fire
- Available resources (quantity and experience)
- Size of the structure
- Structural features that may affect interior suppression (lack of compartmentation)

Figure 6.17 A direct attack involves applying water or foam directly on a fire.

Direct Attack — (1) In structural fire fighting, an attack method that involves the discharge of water or a foam stream directly onto the burning fuel.

Chapter 6 • Tactics **241**

- Ease of access to the burning fuels
- Hose stream type and application
- Potential flow paths within the structure

Indirect Attack

An indirect attack on the fire using a straight or solid stream uses water most efficiently on ventilation-controlled fires. The primary fuel within these types of fires are the fire gases that the combustion process inside a compartment produces. An indirect attack can be considered gas cooling if performed on the interior of the structure. Additionally, an indirect attack can be considered transitional if performed beginning on the exterior.

Gas Cooling

Gas cooling is not a fire extinguishment method but is a way of reducing heat release from the hot gas layer. This technique is effective when faced with a **shielded fire** **(Figure 6.18)**. In these situations, you cannot apply water directly onto the burning material without entering the room and working under the hot gas layer.

> **Shielded Fire** — Fire that is located in a remote part of the structure or hidden from view by objects in the compartment.

Figure 6.18 A combination attack adds the cooling effect of an indirect attack to the suppression effect of a direct attack.

Considerations for the IC when choosing gas cooling:

- High heat, rollover, flashover conditions
- Arrival conditions
- Location, size, and extent of the fire
- Available resources (quantity and experience)
- Size of the structure
- Structural features that may affect interior suppression (lack of compartmentation)
- Hose stream type and application

- Potential flow paths within the structure
- Door control to increase efficiency
- Smoke color, volume, density, and velocity

Transitional Attack

The transitional attack is an offensive tactic used in an offensive strategy that the IC or initial company officer employs in which a hoseline is deployed from the exterior of the structure where smoke and/or flames are visible to gain control of the fire prior to the deployment of interior hoselines. The objective of the transitional attack is to improve conditions on the interior of the structure and maximize the safety of both potential victims and firefighters alike.

The transitional attack serves to interrupt the fire growth and limit further heat and smoke (fuel) production. In many instances, deploying a line to the exterior of the structure for initial knockdown can occur quicker than a traditional interior attack. Gaining control of the fire more quickly is essential to improve the occupant survivability profile. Additionally, once the fire is initially knocked down from the exterior of the structure, the danger to firefighters operating within the flow path is reduced because there is less of a chance for rapid fire development.

Forcing entry into the structure should be seen as opening an additional ventilation channel and creating a new flow path **(Figure 6.19)**. Firefighters operating within this flow path should be aware of any changes in fire conditions that could lead to rapid fire development. The transitional attack limits the potential for rapid fire behavior changes within the flow path as the crews advance from the exterior to the interior of the structure.

To achieve these objectives, initial arriving companies should conduct a comprehensive 360-degree size-up to attempt to determine the fire location, size, and extent within the structure.

Figure 6.19 Firefighters should be aware that forcing entry into a building, such as breaking a window, creates a new flow path for the fire.

Using a Thermal Imager
A thermal imager should be used during every size-up to assist in locating the fire within the structure. If fire is not visible from the structure upon arrival, a thermal imager can be used to locate heat signatures via doors, windows, walls, and other ventilation openings.

Once the fire is located within the building via arrival conditions or the 360-degree size-up, hoselines should be deployed to these locations on the exterior of the structure. If you arrive with fire showing and the ventilation opening into the compartment has already been made (whether self-vented or open prior to arrival), the hoseline should be directed into the opening. The crews should use a straight or solid stream bounced off the ceiling inside the compartment for maximum efficiency. Bouncing the stream off of the ceiling creates a sprinkler effect in which the water cools the gas layer, exposed surfaces, and fuels. The hose stream should be directed into the compartment until all visible fire is knocked down **(Figure 6.20)**.

Figure 6.20 A fog stream compared to a straight/solid stream.

Hose Stream For A Transitional Attack

The firefighters conducting the transitional attack should ensure that the nozzle is not placed on a fog pattern and must avoid movement of the straight or solid water stream within the opening. Firefighters must ensure that the ventilation opening does not block the hose stream and that it can still function as an exhaust for fire gases. If the opening is obstructed, the hose stream can entrain air, increase pressure within the compartment, and redirect the fire flow to another low pressure vent within the structure.

Considerations with Ventilation Openings

Creating a ventilation opening (window) to complete a transitional attack is a viable option. However, it should be known that this opening creates additional flow paths within in the structure. These openings should be controlled if at all possible. The hoseline must be charged, tested, and ready to flow prior to creating the opening.

If you arrive with no visible signs of fire and determine its location upon completing your 360-degree size-up, you may need to make a ventilation opening to complete a transitional attack. Once the ventilation opening has been made, the hose stream should be directed into the compartment, similar to the description above. Water application should continue until after the smoke production from the ventilation opening has decreased and the firefighters determine that the exterior hose stream has affected the fire.

At this point, the IC should determine whether firefighters are able to continue with an interior attack. The IC can either redirect the hoseline used for the transitional attack or deploy another hoseline for interior suppression. Firefighters that continue with interior suppression operations can employ either a direct or indirect attack.

Considerations when choosing transitional attack:

- Arrival conditions
- Location, size, and extent of the fire
- Available resources (quantity and experience)
- Size of the structure
- Structural features that may affect exterior suppression (such as deck and balcony)

Positive-Pressure Attack

Positive-pressure fans are most effective on fires confined to a compartment. The intent is to use high-volume fans to create a slightly higher pressure in adjacent compartments and force the products of combustion (smoke) to the exterior of the structure through exhaust opening(s) that either already exist or have been created **(Figure 6.21)**. This task is a challenge because a growing

Figure 6.21 Positive-pressure attack relies on personnel using high-volume fans to force smoke out of a structure through exhaust openings.

fire creates pressure. For the tactic to be effective, the fan must create enough pressure to force smoke and heat to the intended exhaust openings. Controlling the flow within the structure in this manner during the initial stages of the fire, prior to suppression is known as **positive-pressure attack (PPA)**. When done correctly PPA reduces the thermal effect on firefighters as they perform an interior operation. If PPA is not applied correctly, there is a greater potential for rapid fire development and spread.

Positive-Pressure Attack (PPA) — The use and application of high volume ventilation fans before fire suppression which are intended to force heat and smoke toward desired exhaust openings.

> **CAUTION**
> Smoke will exhaust toward the positive-pressure attack intake if the pressure is not sufficient.

Intake — In terms of ventilation, the location where air is being entrained toward a fire.

Exhaust — In terms of ventilation, the location where hot gases and the products of construction are leaving a structure.

The opening where the fan is set up and air flow is introduced is known as the **intake**. The location where the products of combustion will be removed from the structure is known as the **exhaust** (**Figure 6.22**). The size and type of fan used may vary greatly from one department to the next. Therefore, firefighters should be familiar with the fan utilized in their department. Follow the manufacturer's recommendations along with the standard operation procedures of your department for the location of the fan.

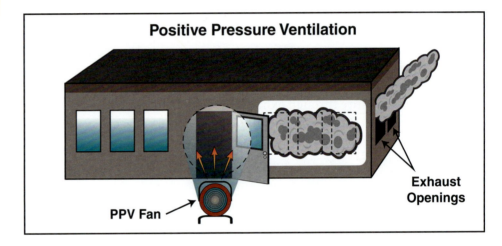

Figure 6.22 For positive-pressure attack, personnel set up a fan in an opening, known as the intake, to expel gas and smoke through exhaust openings.

The two main considerations for PPA are fire location and exhaust-to-intake-size ratio. The ratio is the comparison of the area of ALL exhaust openings compared with the surface area of all compartment intake openings. PPA is only an effective tactic if the location of the fire is known and the appropriate exhaust-to-intake-ratio (greater than 1) can be achieved. Note in **Figure 6.23** that the relevant intake area is the bedroom doorway.

Figure 6.23 In this example, the relevant intake area is the bedroom doorway.

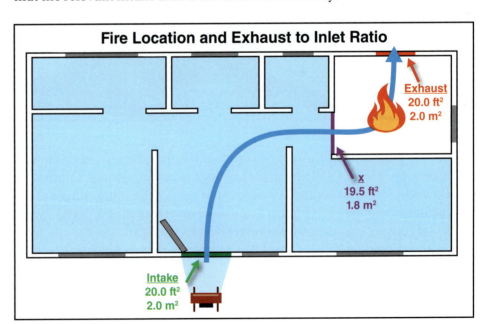

246   Chapter 6 • Tactics

In order for the fan to create the needed pressure to force the heat and toxic gases out of the structure, the exhaust openings must be located in the fire compartment. Additional exhaust openings remote from the fire compartment can draw air and pressure out of the established flow path and weaken the fan's ability to force smoke and heat toward the fire compartment exhaust openings. This can result in the pressure in the fire compartment being higher than the pressure in the rest of the structure. As flow always moves from high pressure to low pressure, this will draw the heat, smoke, and eventually fire towards the new exhaust point (**Figure 6.24**).

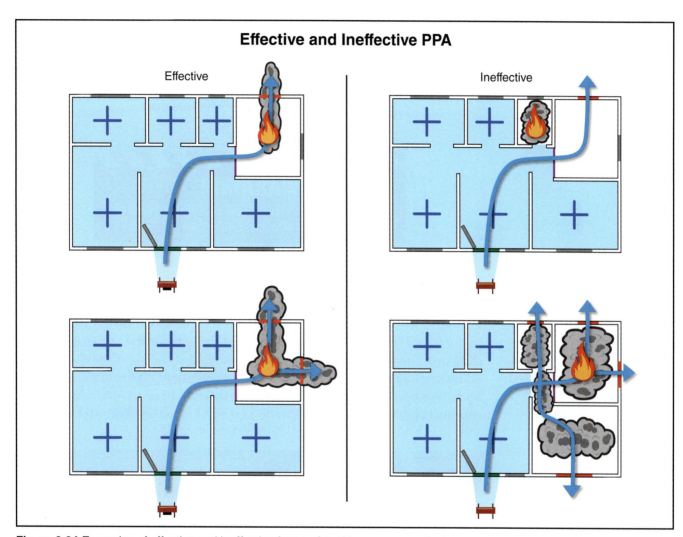

Figure 6.24 Examples of effective and ineffective forms of positive-pressure attack.

Creating the necessary pressure in the adjacent compartments is difficult because the fan is not the only component of the system creating pressure. The fire also creates pressure as it grows, which when combined with the pressure of the fan can result in flow in unintended directions. Reducing the pressure that the fire creates is not possible, so firefighters must control the pressure that the fan creates or apply water to cool the fire. Creating a large enough exhaust opening to ensure the fire room pressure remains lower than the remainder of the structure will allow firefighters to control the flow path and the fire. The size of that exhaust must be greater than the intake to the fire compartment.

Like all ventilation tactics, PPA requires coordination between crews. Changes made to the flow path have the potential to cause rapid fire development. Crews operating inside the structure can be cut off from their escape route, trapping them in the structure. Search and rescue operations have a high potential to change the flow path change the flow path if crews open and close doors during PPA. Crews need to enter and exit the structure from the intake only taking care not to block the intake opening. Once inside the structure crews should only search the open compartments as opening interior doors will change the flow path.

When handled correctly, the application of water to the fire compartment will reduce the pressure differences that the fire creates, making PPA more effective. The decreased temperatures along the path of approach allows this to happen faster. Although temperatures are reduced while approaching the fire compartment, crews should apply water to the fire compartment as soon as their streams will reach. As soon as the water enters the compartment, it will improve the effectiveness of the PPA.

> **WARNING!**
> Seemingly unrelated activities within a building can affect fire behavior.

Overhaul

Overhaul operations can begin once the main body of the fire has been extinguished and a safety assessment conducted at the direction of the IC deems that it is safe to begin. Prior to beginning overhaul, an attempt to determine the cause and point of origin should be made. This is an appropriate time for crews to rehab and go through **medical surveillance** (**Figure 6.25**). Overhaul operations can destroy or alter evidence of incendiary or accidental cause. Use caution to protect both evidence and personnel during this phase.

Medical Surveillance — Rehabilitation function during an incident intended to monitor responders' vital signs and incident-stress levels.

Figure 6.25 Firefighters should have their health monitored prior to beginning overhaul operations to ensure that their vital signs and stress levels are regular. *Courtesy of Bob Esposito.*

NOTE: Your AHJ will determine who is responsible for determining fire cause and origin and when an investigator is required at the incident.

Overhaul consists of:

- Searching for and extinguishing hidden or remaining fire, including contents and concealed spaces
- Placing the building and its contents in a safe condition, including the barricading of unsafe structures
- Determining the origin and/or cause of the fire
- Recognizing and preserving evidence of arson **(Figure 6.26)**
- Identifying the area and point of origin

> **WARNING!**
> Always verify that the structure is safe for responders to enter before overhaul begins.

Overhaul is a dangerous period on the fireground because:

- Building structural members may have been weakened
- Water weight has been added to the structure
- Dangerous fire gases such as carbon monoxide and hydrogen still exist, forcing personnel to use SCBA **(Figure 6.27)**

Figure 6.26 Personnel should preserve any evidence of arson during overhaul.

Figure 6.27 With dangerous fire gases still present, firefighters should not remove their SCBAs while performing overhaul. *Courtesy of Ron Jeffers, Union City, NJ.*

- Firefighters may be fatigued
- Debris can cause tripping hazards, making it important that personnel work in proper lighting
- Natural gas or electric lines may be present under debris or in enclosed spaces

During overhaul operations, firefighters can be exposed to a variety of hazards, including numerous toxic gases. Because the health effects of some of these gases are delayed – sometimes for years – it is easy for firefighters to be lulled into a false sense of security.

Air Monitoring Research

The International Agency for Research on Cancer (IARC) identified more than thirty carcinogens in fire smoke. It is impossible to measure how many such carcinogens are present in the fire fighting environment and in what concentrations. As a result, the Firefighter Cancer Support Network (FCSN) strongly recommends firefighters use SCBAs from the beginning of fire fighting operations until the end of overhaul. Fire departments should not tie SCBA use solely to air monitoring results. Policies that permit firefighters to remove their SCBAs when monitored levels of carbon monoxide (CO) and hydrogen cyanide (HCN) reach certain levels are no longer appropriate.

Source: FCSN.

Fire Watch — Usually refers to someone who has the responsibility to tour a building or facility on at least an hourly basis, look for actual or potential fire emergency conditions, and send an appropriate warning if such conditions are found.

Some of the tasks that must be performed to complete overhaul may be accomplished during extinguishment or confinement. The use of fresh or well-rested personnel or crews can help start this process. A **fire watch** or drive-by checks of the fire scene may be needed.

From a liability standpoint, personnel should make the scene as safe as possible before returning the property to the owner. Legally, in some jurisdictions, the fire department takes possession of the site throughout fire suppression activities. As a result, the site must be as safe as possible before the owner is allowed to enter or take possession. If the site is unsafe, the structure should be barricaded to prevent unauthorized entry.

If it is suspected during overhaul that the fire was intentionally set, the structure must remain in the possession of the fire department or law enforcement agency. Once possession is returned to the owner, it may not be legally possible for the fire department to return to gather evidence, take photographs, or search for evidence or clues. The chain of custody has been broken. Fire officers who are certified to NFPA 1021 should be able to determine the cause and origin of most fires.

Because overhaul generally occurs after the fire is mostly extinguished, you may have more time to decide where and how the overhaul tactic will be performed. Because overhaul must be thorough, your decision must be based on solid facts. Remember that when performing overhaul, you must be able to justify your actions to the owner.

Figure 6.28 Personnel should be aware of a structure's stability before they enter it to begin overhaul operations. *Courtesy of Chris Mickel, New Orleans (LA) F.D. Photo Unit.*

The facts that you need to know before performing overhaul are:

- Building stability **(Figure 6.28)**
- Fire behavior and extent of damage
- Building construction
  — Location of voids or concealed spaces
  — Construction materials used
  — Type of construction
- Estimated length of time the fire burned
- Indicators of fire ignition and spread (burn patterns)
- Indicators of intentional fire ignition
- Evidence of criminal activity
- Need for preserving evidence
- Type and combustibility of contents

Your senses may help you to determine where overhaul needs to be performed. You should:

- Look for signs of burning around openings, such as electrical outlets or air-conditioning vents in walls, floors, and ceilings
- Look for smoke coming from the wall, ceiling, or floor cavities
- Feel wall surfaces with the back of the hand to determine if areas conceal fire **(Figure 6.29)**
- Listen for the sound of burning materials in voids or concealed spaces
- Look for discoloration of walls in unburned compartments

Figure 6.29 A firefighter can place the back of his or her hand on a wall to feel for the possibility of heat coming from a concealed fire.

Chapter 6 • Tactics   **251**

- Look for temperature variations using a thermal imager
- Look for smoke coming from ventilation systems
- Look for trailers or paths on flooring that may indicate the use of accelerants
- Look for deep-seated fires in contents such as bales of cotton or rolls of paper

Breathing contaminated air can result in permanent injury or death. Respiratory protection must be worn during all overhaul operations unless the safety officer determines otherwise.

WARNING!
Do not use your sense of smell to determine the presence of fire or a combustible liquid.

Given your knowledge of fire behavior and the dangers of fire spread in concealed spaces and the potential for rekindle, you should be able to project how a fire might spread unseen. The damage that opening walls, floors, ceilings, and concealed spaces does is far less than the damage from a rekindle of the fire **(Figure 6.30)**. In addition, the rekindle would further endanger civilians working to clean up the area and firefighters who would be called to the new fire.

Figure 6.30 Firefighters attacking a structure fire used a chainsaw and then an ax to create an opening in the wall next to a second-floor door. *Courtesy of Bob Esposito.*

The hazards associated with overhaul extend beyond the potential for a rekindle. Added hazards include:

- Exposure to toxic or contaminated atmospheres
- Injury from sharp objects
- Energized electrical wiring

- Presence of gas lines
- Potential of structural collapse
- Injury or fatalities related to overexertion
- Injury or fatalities due to strokes or cardiac arrest
- Effects of heat stress
- Hazardous materials

    Taking the following steps can mitigate these hazards:

- Wearing complete PPE, including appropriate respiratory protection unless or until a qualified person with appropriate equipment declares the area free of toxic gases
- Ensuring the structural integrity of the building before committing resources to overhaul
- Providing rehabilitation facilities during the operation
- Providing on-site medical support
- Requesting additional personnel to replace initial assignment units, also known as a *Fire Detail*
- Securing utilities at all times
- Contracting with an outside company to use heavy equipment to remove debris (this activity may need to be coordinated with the owner)
- Shoring or securing structural members that may collapse
- Removing excess water that may contribute to collapse
- Using proper lifting, cutting, and prying techniques to reduce injuries
- Requesting assistance from utility companies

> **Tactical Ventilation** — Planned, systematic, and coordinated removal of heated air, smoke, gases or other airborne contaminants from a structure, replacing them with cooler and/or fresher air to meet the incident priorities of life safety, incident stabilization, and property conservation.

Tactical Ventilation

While the term *ventilation* is the traditional word used in the fire service, the activity is more accurately referred to as **tactical ventilation**. Tactical ventilation is the planned, systematic, and coordinated removal of heated air, smoke, gases, or other airborne contaminants from a structure, replacing them with cooler and/or fresher air to meet the incident priorities of life safety, incident stabilization, and property conservation **(Figure 6.31)**. It must be coordinated with fire-suppression operations to prevent unwanted consequences for crews or victims.

Figure 6.31 A firefighter cuts into a roof as part of ventilation operations on a house fire. *Courtesy of Bob Esposito.*

Figure 6.32 When a window fails, a new source of oxygen is suddenly available to a fire. The fire can rapidly move toward and consume this oxygen, creating a "blowtorch" effect. *Courtesy of Mike Wieder.*

Ventilation may occur before, during, or after fire suppression operations start. Some examples of unplanned ventilation include failure of windows, doors, and structural members as a result of heat/fire exposure **(Figure 6.32)**. To prevent rapid fire development from occurring, the initial IC must know how and when to direct the proper release of the heated toxic gases using safe and efficient forms of ventilation. Tactical ventilation should only be performed when the fire attack hoselines and teams are in place and ready to advance toward the fire.

The Effect of Fresh Air on an Oxygen-Deprived Environment

In an oxygen-limited environment, the introduction of fresh air without a coordinated fire attack will result in rapid fire development, the production of more heat, and an increased threat to life safety for occupants and firefighters.

Improper Ventilation LODD, 2010

On March 30, 2010, a fire in a one-story, single-family dwelling resulted in the death of a firefighter and an occupant as well as the injury of a second firefighter. When emergency units arrived, heavy fire conditions were visible at the rear of the house and moderate smoke conditions existed in the uninvolved interior areas. A search and rescue crew entered the house to search for a civilian who was reported trapped at the rear of the house. Three other firefighters advanced a charged 2½-inch (64 mm) hoseline into the house. Thick, black, rolling smoke banked down to knee level as the hoseline was advanced 12 feet (3.7 m) into the kitchen area.

Additional units were assigned to the roof to perform vertical ventilation and to ground-level windows on the D and B sides to perform horizontal ventilation. When the windows were opened, the fire intensified, spreading into the area between the attack hoseline crew and the A-side door. When this occurred, the search and rescue crew saw fire rolling across the ceiling within the smoke. They immediately yelled to the hoseline crew to "get out." The search and rescue crew were able to exit the structure safely and then returned to rescue the injured firefighter and then the victim. The victim received medical care at the scene and was transported to a local hospital where he was pronounced dead.

Lessons Learned:

The subsequent NIOSH investigation into the fatality determined that a contributing factor was the lack of coordination between the ventilation and fire suppression crews. During the incident, uncoordinated ventilation occurred while the hoseline and search and rescue crews were inside the house. The victim and the other firefighters within the small 950 square foot (88 square meters) house were caught between the fire and the ventilation source. One firefighter reported heavy, turbulent, black smoke pushing from a window on the B-side after it was broken. Shortly after, the house sustained an apparent ventilation-induced flashover.

Source: CDC NIOSH LODD Report F-2010-10.

Considerations Affecting the Decision to Ventilate

The Incident Commander (IC) should make the decision whether to ventilate the structure. The decision to ventilate a structure is based on a number of factors, including:

- Risks to occupants and firefighters **(Figure 6.33)**

Figure 6.33 As with every decision made, the IC should consider the safety of firefighters and occupants while determining whether to ventilate a building. *Courtesy of Bob Esposito.*

- Building construction and modern furnishings
- Fire behavior indicators
- Location and extent of the fire
- Type of ventilation
- Location for ventilation
- Weather conditions
- Exposures
- Staffing and available resources

Risks to Occupants and Firefighters

Timely and effective tactical ventilation can assist in both fire attack and search and rescue operations. Life hazards in a structure fire are generally lower if the occupants are awake and able bodied. On the other hand, if the occupants were asleep when the fire developed and are still in the building, a number of possibilities must be considered:

- Occupants may have been overcome by smoke and fire gases
- Occupants might have become lost in the structure
- Occupants may be taking refuge in their rooms because the doors were closed

In addition to the hazards that endanger occupants, there are potential hazards to firefighters. Smoke is unburned fuel and has the potential to aid in rapid fire development as it builds up within the structure. As firefighters enter the structure to complete fire attack and search the building, they need to be aware of any open door or window as these openings are considered ventilation and may introduce oxygen into a ventilation-limited compartment. This can cause the fire to rapidly develop within the flow path placing firefighters and occupants in danger.

Unknowingly Creating Ventilation Openings

All openings within the structure are considered ventilation. These openings include the point of entry that the firefighters make in addition to any open doorway or window within the structure. These openings can contribute to fire development and spread.

Building Construction and Modern Furnishings

Over the past 50 years, building construction in North America has changed significantly. In single-family residential structures, the footprint of houses has increased over 150 percent between 1973 and 2008. At the same time, lot sizes have shrunk approximately 25 percent, reducing firefighter access and increasing potential exposure risks.

Residential interior layouts and construction materials have also changed. Older structures were composed of smaller compartments, windows that could be opened for ventilation, and empty wall cavities that used air pockets to provide insulation.

Modern single-family structures are often built with an open floor plan to maximize the feeling of space and utility **(Figure 6.34)**. In this style of construction, features that may be filled with synthetic insulation include:

- High ceilings and atrium
- Lightweight manufactured structural components
- Sealed windows
- Wall cavities

Figure 6.34 Open floor plan houses have common areas for living, dining, and cooking.

Construction materials and interior finish consisting of synthetic materials and light composite wood components add to the fuel load of the structure and contribute to the creation of toxic gases during a fire. Because of energy-efficient designs, the structures also tend to contain fires for a longer period of time creating fuel-rich (ventilation-limited) environments. These problems are magnified in large-area residential structures.

Commercial, institutional, educational, and multifamily residential structures also rely on energy conservation measures that increase the intensity of a fire and make the use of tactical ventilation difficult. Open plan commercial structures, such as "Big Box" stores, have high fuel loads in the contents and no physical barriers to prevent the spread of fire and smoke in the space **(Figure 6.35)**.

Figure 6.35 Big box stores have a high fuel load, and the normally safe contents presents a wide variety of hazards during a fire or other emergency. *Courtesy of Ron Moore, McKinney (TX) Fire Department.*

In addition, the use of plastics and other synthetic materials in modern furnishings has dramatically increased the fuel load in all types of occupancies. These synthetic materials produce larger quantities of toxic and combustible gases at a quicker rate when compared to legacy fuels. The heat generated from these fuels can reach its peak in a much shorter time. The increased production of both heat and smoke (fuel) can lead to much shorter time to flashover within the compartment.

Knowledge of the building involved is a great asset when making decisions concerning tactical ventilation. This information can be obtained from the pre-incident plan, inspection reports, or observation of similar types of structures.

Basement Fires

Basement fires can be among the most challenging situations firefighters will face. Ventilating basement fires must be coordinated with fire attack. Personnel must determine what stage the fire is in before ventilating a basement. Unless the basement has vents installed, heat and smoke will quickly spread upward into the building. Without effective ventilation, access into the basement is difficult because firefighters would have to descend through the intense rising heat and smoke to get to the seat of the fire. Access to the basement may be through interior or exterior stairs, cellar doors, exterior windows, or hoistways **(Figure 6.36)**. Barriers to these external entrances include iron gratings, steel shutters, or wooden doors, whether alone or in combination. All of these features may impede attempts at ventilation.

Figure 6.36 Basement fires could be difficult to access, requiring personnel to reach them through an exterior window. *Courtesy of Ron Jeffers, Union City, NJ.*

Fire Behavior Indicators

Reading the fire and understanding the effect of changes in ventilation on the fire behavior is essential to effective tactical operations and firefighter safety. The following observations about smoke, taken together, can help firefighters obtain a clear picture of the actual, interior fire conditions:

- Volume of smoke discharge
- Location of smoke discharge
- Smoke color, density, and velocity
- Type of flow at ventilation opening (bi-directional, uni-directional)
- Location of neutral plane at ventilation opening

Flow path is volume between an inlet and an outlet that allows the movement of heat and smoke from the higher pressure within the fire area towards the lower pressure areas accessible via doors and window openings. Varying building configurations and multiple compartments yield more potential flow paths. An evaluation of all ventilation openings (type of smoke flow, location of neutral plane) within the structure can indicate the potential flow paths present. In the case of a single open doorway to a structure fire, the air inlet is the lower portion of the doorway (low pressure below the neutral plane) and the smoke exhaust is the upper portion of the doorway (high-pressure area above the neutral plane). In other cases, the entire open doorway may be the air inlet and a window might serve as the smoke/hot gas exhaust vent. The flow path is the connection between the inlet and the outlet.

> **CAUTION**
> A thermal imager should be used during the size-up to assist in locating the fire and determining the location of where ventilation openings should be made or controlled.

Visible flames may provide an indication of the size and location of the fire, for example, fire showing from one window versus fire showing from all windows on the fire floor. The effect (or lack of effect) of fire streams on flaming combustion may also indicate the size and extent of the fire.

Location and Extent of the Fire

A fire may have traveled some distance throughout a structure before firefighters arrive. Therefore, first-arriving units must quickly determine the size and extent of the fire as well as its location.

Tests and experience indicate that creating tactical ventilation openings in an uncoordinated manner can spread the fire to uninvolved areas of the building and cut off escape routes for building occupants. The severity and extent of the fire depend upon a number of factors, including the type of fuel and the amount of time it has been burning, activation of fire detection and suppression systems, and the degree of confinement. The stage of the fire and whether it is fuel or ventilation controlled are a primary consideration in determining tactical ventilation procedures.

Type of Ventilation

To be safe and effective, tactical ventilation operations must be coordinated with other tactical operations including fire suppression and search and rescue. Before orders are given to ventilate a structure, the IC must consider the effects that ventilation will have on the fire's behavior. Fire attack crews with charged hoselines, search and rescue teams, and exposure protection must be in place before tactical ventilation begins. The IC first determines if ventilation is necessary and when, where, and in what form it should be initiated **(Figure 6.37, p. 260)**. Conditions present upon arrival will influence ventilation decisions. Some incidents may simply require locating and extin-

Figure 6.37 The Incident Commander makes the determination when to ventilate a structure and how it should be done, if needed. *Courtesy of Ron Jeffers, Union City, NJ.*

guishing the fire and then ventilating afterward to clear residual smoke from the structure. Other incidents will require immediate ventilation to enable firefighters to apply water from the exterior of the building to improve interior conditions and allow for entry. The type of ventilation (vertical vs. horizontal) and the means of ventilation (natural vs. mechanical) used must be the most appropriate for the situation.

Location for Ventilation

Before selecting a place to ventilate, firefighters should gather as much information as possible about the fire, the building, and the occupancy. Factors that have a bearing on where to ventilate include the following:

Figure 6.38 Skylights mark a penetration in a roof that is covered by a different kind of material than the rest of the roofing components.

- Location of occupants
- Availability of existing roof openings, such as skylights, ventilator shafts, monitors, and hatches, which access the fire area **(Figure 6.38)**
- Location of the fire
- Desired air flow path
- Type of building construction
- Wind direction
- Extent of progress of the fire
- Condition of the building and its contents
- Indications of potential structural collapse
- Effect that ventilation will have on the fire
- Effect that ventilation will have on exposures
- State of readiness of fire attack crews
- Ability to protect exposures prior to ventilating the structure
- Protecting means of egress and access

Weather Conditions

Any opening in a building, whether part of the building design or that the fire caused, allows the surrounding atmosphere to affect what is happening inside the building. The following weather conditions can affect tactical ventilation:

- Wind
- Temperature
- Atmospheric pressure
- Precipitation
- Relative humidity

The most important weather-related influence on ventilation is wind. Wind conditions must always be considered when determining the proper means and location of tactical ventilation in all types of structures. Wind can blow the fire toward an external exposure, supply oxygen to the fire, or blow the fire into uninvolved areas of the structure **(Figure 6.39)**. The means of tactical ventilation selected should work with the prevailing wind and not against it.

Figure 6.39 An Incident Commander should consider the wind conditions at a scene before calling for a specific type of ventilation. Wind could blow the fire toward external exposures and supply more oxygen to the fire. *Courtesy of the Los Angeles Fire Department – ISTS.*

> **CAUTION**
> Wind has the potential to overpower the natural convective effect of a fire and drive the smoke and hot gases back into the building. Additionally, wind can overpower the pressure of a hose stream and/or a positive pressure fan.

Exposures

When beginning tactical ventilation operations, firefighters must consider both internal and external exposures **(Figure 6.40)**. Internal exposures include the building occupants, contents, and any uninvolved rooms or portions of the building. When ventilation does not release heat and smoke directly above the fire, some routing of the smoke is necessary. The routes the smoke and heated fire gases would naturally travel to exit the building may be the same corridors and passageways that occupants need for evacuation and firefighters need for working.

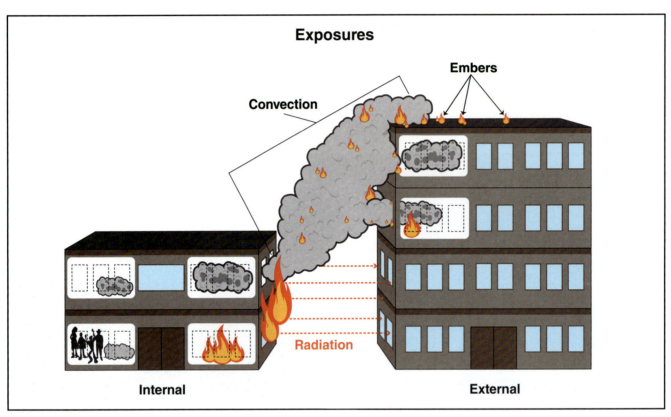

Figure 6.40 Firefighters must consider both internal and external exposures that tactical ventilation could affect.

Ventilation below the highest point of the building creates the danger that the rising fire gases will ignite portions of the building above the exhaust point. Heat and fire gases may be drawn into open windows or attic vents above a ventilation opening, and they may also ignite the eaves of the building or adjacent structures. Ventilation above the area of the building involved in fire can be dangerous and has the potential to create additional flow paths within the structure, drawing fire to uninvolved areas, if the fire compartment is not isolated. The tactic of vent-enter-isolate-search (VEIS) is an example.

Radiation and/or direct flame contact can affect external exposures, such as structures located adjacent to the fire building. Window-mounted air conditioning units or HVAC intake vents may draw smoke into adjacent buildings as well. Nearby structures and vegetation can be ignited if convection carries hot fire brands or embers aloft. Fire may be drawn into exterior windows or openings of the adjacent exposures.

Staffing and Available Resources

All tactical ventilation operations require personnel and resources. Staffing requirements range from two firefighters to multiple companies. In a small structure, ventilation may only require two firefighters. Additional personnel and companies are required when it is necessary to make a roof opening or activate fans. Staffing and available resources dictate the order in which tactics are accomplished. In some incidents, ventilation is a secondary priority when compared to water application.

Types of Tactical Ventilation

There are generally two types of tactical ventilation used for structure fires: horizontal and vertical. The means used for horizontal and vertical ventilation are **natural**, **mechanical**, and **hydraulic ventilation**. Natural horizontal ventilation involves opening doors and windows to allow natural air currents and pressure differences to move smoke and heat out of the building. Natural vertical ventilation uses the buoyancy of heated smoke and gases to draw them out of the structure through the roof openings while entraining (pulling or drawing) fresh air into the structure **(Figure 6.41)**.

In general, the need to use mechanical or hydraulic ventilation is indicated when:

- The location and size of the fire have been determined.
- The layout of the building is not conducive to natural ventilation.
- Natural ventilation slows, becomes ineffective, and needs support.
- The fire is burning below ground in the structure.
- The involved area within a compartment is so large that natural ventilation is inefficient.
- The type of building or the fire situation dictates its use.

Horizontal Ventilation

Structures that lend themselves to the application of horizontal ventilation include:

- Buildings in which the fire has not involved the attic or cockloft area
- Involved floors of multistoried structures below the top floor or the top floor if the attic is uninvolved
- Buildings so weakened by the fire that vertical ventilation is unsafe
- Buildings with daylight basements
- Buildings in which vertical ventilation is ineffective

Natural Horizontal Ventilation

When conditions are appropriate, natural horizontal ventilation operations should work with existing atmospheric conditions, taking advantage of natural air

> **Natural Ventilation** — Techniques that use the wind, convection currents, and other natural phenomena to ventilate a structure without the use of fans, blowers, smoke ejectors, or other mechanical devices.

> **Mechanical Ventilation** — Any means other than natural ventilation; may involve the use of fans, blowers, smoke ejectors, and fire streams.

> **Hydraulic Ventilation** — Ventilation accomplished by using a spray stream to draw the smoke from a compartment through an exterior opening.

Figure 6.41 Natural vertical ventilation works with the buoyancy of heated smoke to clear a structure of smoke and gases.

Negative-Pressure Ventilation (NPV) — Technique using smoke ejectors to develop artificial air flow and to pull smoke out of a structure. Smoke ejectors are placed in windows, doors, or roof vent holes to pull the smoke, heat, and gases from inside the building and eject them to the exterior.

Positive-Pressure Ventilation (PPV) — Method of ventilating a room or structure by mechanically blowing fresh air through an inlet opening into the space in sufficient volume to create a slight positive pressure within and thereby forcing the contaminated atmosphere out the exit opening.

flow. Natural ventilation requires no additional personnel or equipment to set up and maintain.

Considerations for natural horizontal ventilation:

- Wind speed and direction
- Fire location and proximity to an external opening
- Size of opening(s) available

Building with limited openings or fires in confined spaces are not effective for natural horizontal ventilation. When the IC gives the order, windows and doors on the downwind side of the structure (low pressure side) should be opened first to create an exhaust point. Openings on the upwind side of the structure (high pressure side) are then opened to permit fresh air to enter forcing the smoke toward the exhaust openings.

Mechanical Horizontal Ventilation

When the natural flow of air currents and the currents that the fire creates are insufficient to remove smoke, heat, and fire gases, mechanical ventilation is necessary. Mechanical ventilation is accomplished through negative pressure or positive pressure.

Negative-pressure ventilation. **Negative-pressure ventilation (NPV)** is the oldest type of mechanical ventilation. Personnel use smoke ejectors to expel (pull) smoke from a structure by developing artificial air flow or enhancing natural ventilation **(Figure 6.42)**. Smoke and fire gases are drawn out of the structure. Fans can be placed in windows, doors, or roof vent openings to exhaust the smoke, heat, and gases from inside the building to the exterior.

Considerations for negative-pressure ventilation:

- Typically not used to support fire attack
- Requires smaller windows or openings that can be sealed around the fan
- Not to be used in a flammable environment

WARNING!
Negative-pressure ventilation should not be used if the fire is not under control.

Figure 6.42 Mechanical negative-pressure ventilation pulls smoke and fire gases out of a room.

Positive-pressure ventilation. The same high volume fans utilized in positive-pressure attack (PPA) can be utilized to exhaust the smoke, heat, and toxic gases to the exterior after fire suppression **(Figure 6.43)**. This tactic is known as **positive-pressure ventilation (PPV)**. PPV is one means of accelerating the natural ventilation of the structure. When the pressure is higher inside the building, the smoke inside the building moves toward a selected, lower-pressure exhaust opening. Unlike with PPA, the application of PPV requires less emphasis on the exhaust-to-intake ratio. When using PPV, the intent is to increase the pressure higher than the exterior of the structure, not higher than the fire can produce. This allows for more exhaust openings, creating more flow. The more air the fan can move through the structure, the faster the interior environment will approach ambient conditions.

When PPV is chosen as a tactic to remove smoke after fire suppression, the sooner the PPV is begun, the more effective the ventilation will be. The increased ventilation effectiveness from the fan will decrease temperatures, decrease toxic gas concentration and increase visibility. All of these lead to safer and more effective interior operations such as search and rescue, overhaul and **salvage**.

Although PPV has the potential to expedite the ventilation of the structure, it also has the potential to hide fire extension while fans are running. Smoldering fires can be intensified with the additional airflow. When using PPV, personnel must identify and control fire extension. Once conditions improve on the interior, re-evaluate interior conditions. Shutting off the fan or closing the intake for a brief period of time allows interior crews to evaluate smoke conditions and identify possible extension before it becomes a larger hazard.

Figure 6.43 Positive-pressure ventilation calls for the same high volume fans used in positive-pressure attack to push smoke and gases out of a building. *Courtesy of Iowa State Training Bureau.*

To ensure an effective PPV operation, take the following actions:

- Ensure that your exhaust surface area is greater than the intake area.
- Monitor the operation of the PPV fan.
- Maintain communications between the IC, the interior attack crews, and the PPV operator.
- Take advantage of existing wind conditions **(Figure 6.44, p. 266)**.

The advantages of PPV compared to NPV include the following:

- Firefighters can set up PPV blowers without entering the smoke-filled environment.
- PPV is equally effective with either horizontal ventilation or vertical ventilation because it supplements natural air currents.
- Removal of smoke and heat from a structure is more efficient.
- Exposed buildings or adjacent compartments can be pressurized to reduce fire spread into them.

The disadvantages of PPV are:

- The structure must have some remaining compartmentation.
- Exhaust from gas-powered fans can add carbon monoxide to a structure.
- Hidden fires may be accelerated and spread throughout the building.

Salvage — Methods and operating procedures by which firefighters attempt to save property and reduce further damage from water, smoke, heat, and exposure during or immediately after a fire; may be accomplished by removing property from a fire area, by covering it, or by other means.

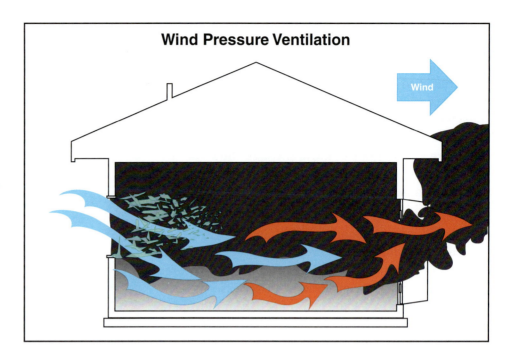

Figure 6.44 Wind can affect ventilation operations and may create a pressure differential between the interior and exterior pressure, which causes the windows to fail.

> **WARNING!**
> Positive-pressure ventilation requires coordination with ALL other fireground tactics. Positive-pressure ventilation has the potential to create rapid fire development if the fire is not yet extinguished.

Hydraulic ventilation. Hydraulic ventilation may be used in situations where other types of forced ventilation are unavailable. Hydraulic ventilation is used to clear a room or building of smoke, heat, steam, and gases after a fire has been controlled.

Hydraulic ventilation uses a spray stream from a fog nozzle to entrain smoke and gases and carry them out of the structure through a door or window **(Figure 6.45)**. To perform hydraulic ventilation, a fog nozzle is set on a wide fog pattern. The larger the opening, the faster ventilation will occur. In fact, a master stream device can be set up in an open commercial or industrial doorway such as those on loading docks.

Considerations for hydraulic ventilation:

- Limited use immediately after fire attack or during overhaul operations for clearing individual rooms
- Not as effective in large areas with small handlines
- If improperly done, more potential damage can be created
- Weather conditions, such as freezing temperatures

Figure 6.45 With hydraulic ventilation, personnel should use a fog nozzle set on a wide fog pattern to push smoke and gases out of a structure.

Vertical Ventilation

Vertical ventilation is the opening of the upper part of the structure to allow smoke, gases, and heat to escape.

Considerations for vertical ventilation:

- Determine that vertical ventilation is needed and can be done safely and effectively
- Consider the age and type of construction involved
- Multi-roof construction (tongue-and-groove ceiling)
- Photovoltaic (PV) panels, commonly known as solar panels
- Consider the location, duration, and extent of the fire
- Observe safety precautions
- Identify escape routes
- Select the place to ventilate
- Move personnel and tools to the roof

Vertical ventilation presents the following increased risks:

- Placing personnel above ground level
- Working on both peaked and flat surfaces
- Working above the fire, such as an attic fire **(Figure 6.46)**
- Working on roofs that may have been weakened because of age or fire damage

Figure 6.46 Vertical ventilation puts firefighters at risk while working on a roof above the fire. *Courtesy of Bob Esposito.*

WARNING!
The IC must coordinate the vertical ventilation with the fire attack operation and ensure that suppression is controlled prior to creating an opening.

Soffit — Lower horizontal surface such as the undersurface of eaves or cornices.

Information gathered from the incident scene size-up and any preincident planning should identify buildings that have roofs supported by lightweight or engineered trusses or pulling soffits to identify the construction type **(Figure 6.47)**. These roofs may fail early in a fire and are extremely dangerous to work on or under.

Figure 6.47 Lightweight trusses can be used to support a roof. *Courtesy of McKinney (TX) Fire Department.*

Dependent on jurisdictional operating procedures, the IC may identify the need for vertical ventilation early in the incident. The roof ventilation team should be in constant communication with their supervisor or the IC. Some responsibilities of the roof ventilation team leader can include:

- Coordinating the team's efforts with those of firefighters inside the building
- Ensuring the safety of all personnel who are assisting with ventilation operations
- Ensuring that there are two means of egress from the roof
- Ensuring that the team leaves the roof as soon as their assignment is completed
- Ensure that a hoseline is present if the fire involves the compartment directly beneath the roof team

> **CAUTION**
> Work in groups of at least two, but with no more personnel than absolutely necessary to perform the assigned task.

Other Types of Ventilation Situations

Most ventilation operations will involve residential occupancies requiring horizontal or vertical ventilation tactics. However, you may be faced with incidents involving basements, windowless buildings, or high-rise buildings that will require a variation on these tactics.

Basement Ventilations

Basement ventilation can be accomplished in several ways. If the basement has ground-level windows or even below ground-level windows in wells, horizontal ventilation can be employed effectively **(Figure 6.48)**. Natural paths from the basement, such as stairwells and hoistway shafts, can be used to evacuate heat and smoke if there is a way to expel the heat and smoke to the atmosphere without placing other portions of the building in danger **(Figure 6.49)**. As a last resort, an opening may be cut in the floor near a ground-level door or window, and the heat and smoke can be forced from the opening through the exterior opening using fans. If a basement fire is found and there is no exterior access to deploy hoselines for fire attack, consider a transitional attack through ventilation of exterior basement windows prior to making entry.

Figure 6.48 Ventilation may be directed from an upper level toward a below-ground fire to horizontally vent the space.

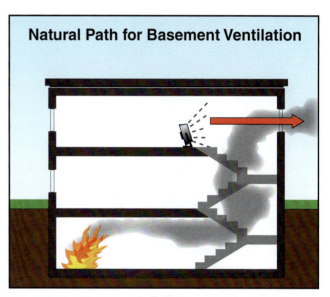

Figure 6.49 Natural ventilation paths may be utilized to mechanically direct smoke through parts of the building that will not spread the hazard.

High-Rise Fires

High-rises buildings may contain hospitals, hotels, apartments, or offices. Because there are more occupants in high-rise buildings than in other occupancies, life safety considerations are an even higher priority. Tactical ventilation in a high-rise building must be carefully coordinated to ensure the safest and most effective use of personnel, equipment, and extinguishing agents. The personnel required for search and rescue and firefighting operations in high-rise buildings is often four to six times as great as required for a fire in a typical low-rise building.

Openings within a high rise can contribute to a **stack effect** (natural movement of heat and smoke throughout a building), creating an upward draft and interfering with evacuation and ventilation **(Figure 6.50, p. 270)**. These openings include:

- Pipe shafts
- Stairways
- Elevator shafts

> **Stack Effect** — (1) Tendency of any vertical shaft within a tall building to act as a chimney or "smokestack", by channeling heat, smoke, and other products of combustion upward due to convection. *Also known as* Stack Action. (2) Phenomenon of a strong air draft moving from ground level to the roof level of a building; affected by building height, configuration, and temperature differences between inside and outside air. *Also known as* Chimney Effect.

- Unprotected ducts
- Other vertical and horizontal openings in high-rises

Heated smoke and fire gases travel upward until they reach the top of the building or they are cooled to the temperature of the surrounding air. When this equalization of temperature occurs, the smoke and fire gases stop rising, spread horizontally, and stratify (form layers). In some cases, such as high-rise buildings, these layers of smoke and fire gases will collect on floors below the top floor. Additional heat and smoke will eventually force these layers to expand and move upward to the top floor of the building. **Horizontal smoke spread** and hot gas layer development can also occur when a vertical exhaust opening is not large enough to exhaust the smoke and gases. Tactics involving mechanical ventilation, whether horizontal or vertical, must be developed to cope with the ventilation and life hazard problems inherent in stratified smoke. In many instances, ventilation must be accomplished horizontally with the use of mechanical ventilation devices and the building's HVAC systems.

Tactical vertical ventilation in high-rise buildings must be considered during preincident planning. In many buildings, only one stairwell penetrates the roof. This stairwell can be used much like a chimney to ventilate smoke, heat, and fire gases from various floors while another stairwell is used as the escape route for building occupants. However, during a fire, the doors on uninvolved floors must be controlled so occupants do not accidentally enter the ventilation stairwell as they

Figure 6.50 Evacuation and ventilation can be more difficult at a high-rise fire because of a phenomenon known as stack effect. *Courtesy of Matthew Daly/MMattyPhoto, Bronxville, NY.*

Horizontal Smoke Spread — Tendency of heat, smoke, and other products of combustion to rise until they encounter a horizontal obstruction. At this point they will spread laterally (ceiling jet) until they encounter vertical obstructions and begin to bank downward (hot gas layer development).

are evacuating. Before the doors on the fire floors are opened and the stairwell is ventilated, roof access door must be blocked open or removed from its hinges. Preventing the door at the top of the shaft from closing ensures that it cannot compromise established ventilation operations. Remember that when ventilating the top of a stairwell, you will be drawing the smoke and heat to you or anyone else in the stairwell between the fire floor and the roof. When an enclosed secondary stairwell is used for evacuating occupants, PPV fans should be located at the bottom floor to pressurize the stairwell and keep smoke from entering it.

In some high-rise structures, ventilation fans are built into the top of the stairwell to assist in ventilation. When activated, these fans draw smoke from the fire floor into the stairwell and out the top. This technique may make it difficult for the fire suppression team to make entry onto the fire floor from this stairwell. The safest and most effective technique may be to pressurize the stairwells with PPV fans to confine the smoke on the floors. Firefighters can advance to the fire floor in a safe atmosphere.

270 Chapter 6 • Tactics

Confining A Fire In A High Rise Building

The most important aspect of tactical ventilation in high rise buildings is the ability to confine the fire to its compartment or floor of origin. This task is accomplished through door control. The IC needs to be aware that if you open up the stairwell prior to opening the roof access door, you place those personnel in the position of the new flow path. A secondary means of egress will now be required for those individuals. Keeping the heat and fire gases within the affected areas allows for safe evacuation of building occupants and the most effective means of fire attack.

Salvage/Property Conservation

Salvage operations are performed during or immediately after a fire. This task involves removing property from a fire area, covering it, or using other means. Salvage tactical operations begin when resources are available to implement the tasks required to complete salvage. Although salvage operations are intended to meet the third strategic goal of property conservation, it is part of the initial operations and continues throughout the event. Salvage is defined as methods and procedures that firefighters use in an attempt to save property and reduce further damage from water, smoke, heat, and exposure **(Figure 6.51)**.

Figure 6.51 Salvage operations serve to protect property and minimize the fuel load available to a fire.

Your preincident plan and size-up will provide the information needed to decide when to begin salvage operations and the areas that must be addressed first. While these decisions must be fact-based, you may also be confronted with the pressure of owners or occupants who believe more should be done to save their property. In commercial properties, merchandise, business records, computers, and electronic equipment may need to be removed rather than covered in place. At the same time, once the items are removed, they must be protected from the weather and theft.

The preincident plan should provide an idea of the contents that will need to be removed or protected in place. In business and commercial properties, business records contained in filing cabinets, safes, and on computers are sometimes more valuable than a structure's merchandise or contents. The preincident plan should indicate whether records are backed up at an off-site location.

Other facts you should be aware of include:

- Type and value of contents
- Location and type (hard copy or computer) of business records
- Type of fire suppression system (water-based or special-agent systems)
- Type of contents that are susceptible to water damage
- Location of fire or smoke barriers
- Height and floor space of structure
- Location of fire
- Ability to protect removed items outside the structure

When a preincident plan does not exist and witnesses are not around, you must use your own observations to decide what should be protected and in what manner. Look for such things as:

- Type and value of the structure
- Reports from owners or occupants
- Smoke conditions in structure
- Weather conditions (potential damage to items removed from structure)
- Difficulty involved in removing contents
- Difficulty involved with protecting removed contents

Determining the potential damage property may get from fire, heat, smoke, or water can be difficult. You must know how the fire is spreading, whether confinement and extinguishment activities are successful, and where water will accumulate or travel. Structural collapse is a possibility when a large quantity of water has been applied to the fire. Removing the contents can require more resources than removing the water. You will need to know the most effective method for water removal that will take the least effort and cause the minimum of damage. Salvage covers may not provide complete protection in the event of ceiling collapse or deep water.

In the case of residential properties, no preplan will exist. As a result, your decision-making process should follow common sense. Items on lower floors should be protected from water damage or structural collapse, including ceiling tiles, gypsum board, or light fixtures falling in. Items in adjacent compartments should be protected from smoke damage through ventilation or removal. Items in the fire compartment should be removed or covered if possible.

Protect as much property as you are able to when conducting overhaul operations **(Figure 6.53)**. Some items that you may consider worthless may be irreplaceable

Figure 6.53 While conducting overhaul, firefighters should try to protect an individual's property as much as possible.

to their owners. This is especially true in residential properties and places of worship where the items can have sentimental or ceremonial value that exceeds their monetary value. Examples include:

- Religious artifacts and books
- Photographs and albums
- Children's artwork
- Furniture

You will have to decide how much effort must be put into saving these types of items when you have limited resources available. Do not remove or destroy evidence of arson while performing salvage operations, regardless of the occupancy type.

Salvage operations require specific tools and equipment that should be stored in a specially designated salvage toolbox or other containers to make them easier to carry. Salvage materials and supplies may be kept in a plastic tub and brought into the structure as soon as possible. The tub itself provides a useful water-resistant container to protect items such as computers, pictures, and other water-sensitive materials.

Typical tools and equipment used in salvage operations include but are not limited to:

- Salvage Covers
- Electrical tools
- Mechanical tools
- Plumbing tools
- General carpentry tools
- Mops, squeegees, and buckets

Chapter Review

1. Describe the different search techniques.
2. What are the reasons to protect exposures during a fire incident?
3. What factors may influence your decision to implement confinement tactics?
4. Describe the factors to consider when choosing a method of attack?
5. What are the objectives of a transitional attack?
6. What factors are needed for a successful positive pressure attack?
7. Describe signs to help you determine where overhaul needs to be performed.
8. Why must tactical ventilation be carefully coordinated with other fire-ground activities?
9. What factors will affect the ICs decisions to ventilate?
10. Describe the three means of tactical ventilation.
11. What factors should be considered when making decisions about overhaul priorities and procedures?

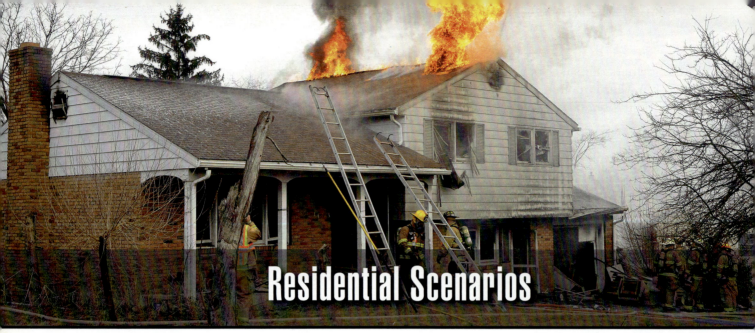

Residential Scenarios

Photo courtesy of Bob Esposito.

Chapter Contents

Scenario 1 277
Scenario 2 283
Scenario 3 289
Scenario 4 295
Scenario 5 301

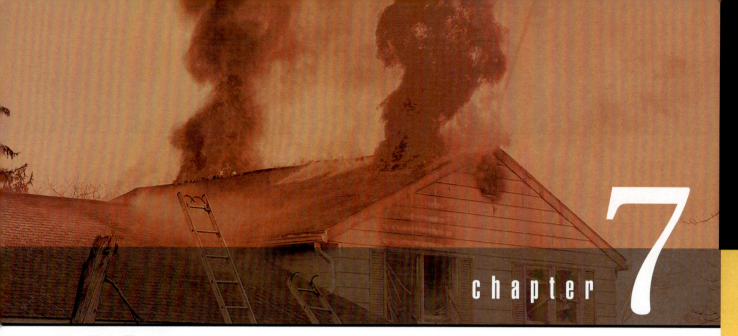

chapter 7

Key Terms

This chapter contains five hypothetical scenarios based on residential occupancy types that are found in most response areas. The scenarios contain best practices that can be applied to each type of situation. However, these are suggestions only and not hard and fast rules that must be applied in every incident.

Because these scenarios are examples to help you learn how to make decisions based on the information in the previous chapters, there are limitations placed on them. The resources available to you will arrive within 10 minutes. You are the Incident Commander (IC) and must allocate your resources to control/mitigate the incident to the best of your ability. The resources you have reflect those in your local department, including first-alarm assignments, standard operating procedures/guidelines (SOP/Gs), mutual aid, equipment, and staffing.

Residential Scenarios

FESHE Learning Outcomes

After reading this chapter, students will be able to:

1. Given information on a residential fire and resources available to your department, develop an Incident Action Plan (IAP) for achieving the incident objectives.

Note to Instructors

The following scenarios contain recommended practices based on the facts provided and on information contained in the text of this manual. These scenarios are intended to be used as teaching aids. It is suggested that you develop scenarios specific to your department's resources, procedures, and response area. Your scenarios should contain structures, hazards, water supplies, photographs, and site plans that exist in your jurisdiction. Do not include probabilities, incident priorities, strategies, command options or tactics. Students should supply this information based on their understanding of the text. Your scenarios can be varied based on changes to the water supply (hydrants out of service, peak water demand times), delayed resources, weather conditions, minimum staffing, time of day, or local standard operating procedures/guidelines (SOP/Gs).

The considerations listed are not absolute. The scenarios are designed for discussion purposes and training perspectives. The AHJ and their respected policies within will dictate each department's strategies, tactics, and considerations.

Chapter 7
Residentail Scenarios

Photo courtesy of Chris Mickel, New Orleans (LA) F.D. Photo Unit.

Scenario 1

On Wednesday, May 12, at 11:15 a.m., dispatch reports a residential structure fire at 4186 McLelland Court. The dispatcher states that a pedestrian reported the incident after noticing brown smoke coming from the house's gables and eves during a morning walk. Current weather conditions are dark skies with a strong chance of thunderstorms.

On Arrival

You are the first-arriving officer. You see a large two-story, single-family residence with brown smoke coming from the gables and eaves. A car is parked in the driveway and the garage doors are closed. But one occupant, who is holding a child in the driveway, says the family dog is still in the house.

Water Supply

A blue hydrant 1,500 gpm (6,000 L/min) is located every 800 feet (240 m) in the neighborhood.

Weather

- Temperature — 48° F (9°C)
- Wind direction — North
- Wind speed — 10 mph (16 km/h)
- Humidity — 100%

Alpha Side

Bravo Side

Charlie Side

Delta Side

1. Size-Up

Based on the dispatch information and your visual observation of the scene on arrival, what are the facts and probabilities for this scenario?

Facts

- Type V construction with brick veneer
- General floor arrangement typical of this type of structure
- High ceilings in part of the living area
- Steep pitch roof
- Smoke is visible coming from the eves on all sides
- Water supply is close to the structure with one 1,500 gpm hydrant [6,000 L/mn]
- No exposures threatened
- Charlie-side access is limited due to the lot arrangement and privacy fences
- All occupants are outside and accounted for by the homeowner

Probabilities

- Lightweight truss construction
- Rapid internal fire spread due to an open floor plan
- Sleeping rooms are on the upper floor, and the master bedroom is on main floor
- Structural weakness due to lightweight construction
- Increased fire load due to the size of the structure
- Possible attic fire due to smoke from the eves and clear windows on the second floor
- Brown smoke indicating involvement of structural elements

2. Incident Priorities

What are your considerations for addressing the incident priorities of life safety, incident stabilization, and property conservation?

Life Safety

- Establish safety of responders; occupants are outside and accounted for by the homeowner
- Conduct primary search
- Establish primary and backup lines for interior crews
- Secure utilities
- Establish evacuation zone from smoke and weather

Incident Stabilization

- Find the seat of the fire
- Ladder the structure
- Ventilate through existing vents in coordination with fire suppression
- Pull the ceiling

Property Conservation
- Remove valuables
- Salvage covers over the furniture
- Cover any holes in the roof or walls to withstand weather

3. Operational Strategy

What are your considerations when selecting your operational strategy?
- Risk-benefit analysis
- Incident appears to be limited to the attic
- No visible fire
- Fire extent
- Resource availability
- Offensive strategy to locate and extinguish the fire

4. Command Options

What are your considerations when selecting a Command option: Investigation, Fast Attack, or Command?
- Stationary Command Post on the Alpha side (address side) due to the structure size

5. Tactics

What are your considerations, if any, for the tactics used to accomplish your incident priorities?

Rescue
- Occupant declares that everyone is safe and out of the house
- No smoke inside the livable space
- Time of day

Exposure
- Internal exposures
- Limited external exposures

Confinement
- Locate the area of origin
- Confined to the roof and attic area
- Limit openings between the attic and the livable space below

Extinguishment
- Access from the interior, if possible
- Consider piercing nozzles or distributor nozzles
- Steep roof
- Multiple roof structures
- High ceilings

Overhaul
- Hidden fires in void spaces
- Air monitoring
- Conduct extensive overhaul to ensure no rekindling occurs from hidden fires

Ventilation
- Open windows and doors to allow natural ventilation (post-extinguishment)
- Consider positive-pressure ventilation (post-extinguishment)
- Make inspection holes in the roof to find the seat of the fire
- Inspection holes will also ventilate the attic

Salvage
- Ongoing as resources allow
- Use of salvage covers and floor runners

6. Unusual Hazards or Conditions

What are any unusual hazards or conditions that must be addressed?
- Lightweight truss construction
- Open floor plan with high ceilings in the entry and living area
- Steep roof
- Fuel load in attic
- Weather conditions (wind/rain/lightening)

Notes

Scenario 2

On Saturday, October 15, at 4:30 a.m., dispatch reports a house fire at 8005 Cornerford Court. Dispatch reports that the caller is a physically impaired resident who is still inside the structure. The caller reported he awoke to the sound of a smoke detector as well as smoke filling his second-floor bedroom. Current weather conditions are mostly cloudy with cold temperatures.

On Arrival

You are the first-arriving officer. While en route, you notice a column of smoke from down the street. Upon arrival, you find a two-story, single-family residence of approximately 2,600 square feet (234 m) with fire and smoke showing from the first floor on the Alpha/Bravo corner of the structure. There are no signs of the physically impaired resident outside of the house.

Water Supply

A blue hydrant 1,500 gpm (6,000 L/min) is located 200 feet (60 m) down the street at the corner of Cornerford Court and Halston Drive.

Weather

- Temperature — 20° F (-7°C)
- Wind direction — South
- Wind speed — 5 mph (8 km/h)
- Humidity — 40%

Alpha Side

Bravo Side

Charlie Side

Delta Side

Chapter 7 • Residential Scenarios

1. Size-Up

Based on the dispatch information and your visual observation of the scene on arrival, what are the facts and probabilities for this scenario?

Facts

- Two-story, single-family, wood-framed type construction
- Smoke and fire from the first-floor windows on the Alpha and Bravo sides of the building near the Alpha/Bravo corner
- Hydrant within 200 feet (60 m) of the residence
- No visible sign of occupant
- No other external exposures

Probabilities

- Possible victim
- Fire/smoke appears to be confined to the room(s) near the Alpha/Bravo corner of the structure
- Potential for lightweight truss
- Delay of water on the fire can result in fire progression/extension
- Potential for the fire to involve some extension both horizontally and vertically

2. Incident Priorities

What are your considerations for addressing the incident priorities of life safety, incident stabilization, and property conservation?

Life Safety

- Immediate deployment of a search team when resources are available for a possible occupant trapped due to given information
- Two-in, two-out team set up transitioning to a RIT/C team for personnel
- EMS group for a potential victim
- Rehab area designated with the medical group team for personnel

Incident Stabilization

- Rapid 360-degree size-up of the structure
- Knock down fire initially from exterior (transitional attack)
- Knock down remaining fire on the interior and check for extension
- Utility control
- Determine ventilation concerns to make residency tenable

Property Conservation

- Control extension into remaining portion of residence
- Salvage with tarps; removal of household items

3. Operational Strategy

What are your considerations when selecting your operational strategy?

- Risk-benefit analysis
- Reported victim trapped

- Fire appears to be limited to the first floor
- Fire extension
- Resource availability
- Offensive strategy to locate and extinguish the fire

4. Command Options

What are your considerations when selecting a Command option: Investigation, Fast Attack, or Command?

- Stationary Command post on the Alpha side of the structure
- Mobile Command option

5. Tactics

What are your considerations, if any, for the tactics used to accomplish your incident priorities?

Rescue

- Reported victim trapped
- Potential for VEIS operation
- Rapid entry for primary search
- Last known victim location is on the second floor

Exposure

- Interior spread to adjacent compartments on the first floor
- Potential interior vertical spread through pipe chases/void spaces
- Potential exterior vertical spread through open windows
- Potential for lateral extension to other structures

Confinement

- Door control on the interior of the structure to limit compartment-to-compartment spread
- Quick water application from the exterior to knock down the bulk of fire
- Interior hoseline(s) to ensure no further spread

Extinguishment

- Exterior knockdown of visible fire
- Interior deployment of hoseline for remaining suppression activity and extension

Overhaul

- Air monitoring
- Potential for fire in void spaces

Ventilation

- Open windows and doors to allow natural ventilation (post-extinguishment)
- Consider positive-pressure ventilation (post-extinguishment)

Salvage

- Ongoing as resources allow
- Use of salvage covers and floor runners

6. Unusual Hazards or Conditions

What are any unusual hazards or conditions that must be addressed?

- Potential for unknown structure contents
- Change in grade from the Alpha to the Charlie side
- Cold temperatures (water and firefighter exposure)
- Lightweight construction

Notes

Scenario 3

On Monday, August 16, at 6 p.m., dispatch reports smoke in an apartment on the second floor of a building at 1911 Kent Drive. The dispatcher states that a tenant on the second floor called 9-1-1 after smelling smoke and noticing that the wall of an adjacent apartment was warm to the touch. Current weather conditions are clear.

On Arrival

You are the first-arriving officer. You see light smoke coming from the second-floor windows on the Alpha/Bravo corner of a two-story, Type V, multi-family structure. You also learn that the building has not been evacuated. Cars are parked in the parking lot.

Water Supply

A blue hydrant 1,500 gpm (6,000 L/min) is located near the Alpha/Delta corner of the structure.

A fire department connection is located on the Alpha side near the Alpha/Delta corner. This is a sprinklered building.

Weather

- Temperature — 75° F (24°C)
- Wind direction — West/northwest
- Wind speed — 20 mph (32 km/h)
- Humidity — 80%

Alpha Side

Bravo Side

Charlie Side

Delta Side

Chapter 7 • Residential Scenarios **291**

1. Size-Up

Based on the dispatch information and your visual observation of the scene on arrival, what are the facts and probabilities for this scenario?

Facts

- Occupied multi-family dwelling
- Light smoke laminar flow from the windows
- Type V construction
- A two-story structure
- Sprinklered building
- Composite pitched roof

Probabilities

- Open attic space
- Lightweight truss construction
- Room and content fire with the possibility of extension
- Trapped occupants/unaware occupants
- Unsprinklered attic area
- Access may be required to multiple apartments

2. Incident Priorities

What are your considerations for addressing the incident priorities of life safety, incident stabilization, and property conservation?

Life Safety

- Establish two-in, two-out team
- Vehicles in the parking lot and the time of day indicate rescue potential, leading to the evacuation of residents
- Large amount of resources necessary to provide for this task

Incident Stabilization

- Determine the location and the extent of the fire
- Quick deployment of initial hand line
- Pull ceiling/open walls to provide for extinguishment
- Horizontal or vertical ventilation as necessary
- Establish water supply

Property Conservation

- Salvage covers in the first-floor apartments and areas where the ceiling is pulled
- Locate sprinkler control valve or secure activated sprinklers
- Amount of content based on multiple occupancies

3. Operational Strategy

What are your considerations when selecting your operational strategy?

- Risk-benefit analysis

- Survivability profile
- Availability of resources/ first due assignment and staffing size
- Extent of fire involvement
- Information derived from residents

4. Command Options

What are your considerations when selecting a Command option: Investigation, Fast Attack/Mobile, or Command?

- Depending on the resources available, the IC may select either fast attack/mobile or Command
- If rescue is imminent, fast attack/mobile Command may be the most appropriate
- If active fire is present and sufficient resources are available and on-scene, stationary Command is appropriate

5. Tactics

What are your considerations, if any, for the tactics used to accomplish your incident priorities?

Rescue

- Primary search of involved apartments
- Prioritize search areas
- Evacuation of nearby apartments
- Establish RIT/C team
- High occupancy load
- Exposures

Exposure

- Assign an exposure group to enter adjoining apartments to assess for extension
- Possibility of extension to adjoining units
- Possibility of vertical and horizontal spread

Confinement

- Initial attack lines to locate and confine the fire to area(s) of origin
- Check for extension
- Door control

Extinguishment

- Offensive deployment of hoseline(s)
- Backup line placement
- Consideration given for checking the attic space to check for fire spread in the attic
- Consideration given for checking the walls on the first and second floors to detect for fire/smoke in the wall areas
- Secure utilities

Overhaul
- Walls and attic thoroughly checked to ensure no fire and/or smoke is present
- Consideration must be given for weakened structures directly affected by fire/water

Ventilation
- Vertical ventilation, if needed, must be coordinated with the fire attack crew/group
- Horizontal ventilation, if used, can be achieved using gable vents, existing windows, or positive-pressure vents

Salvage
- Use salvage covers to protect contents in the fire room(s) and on lower floors
- Control the sprinkler system

6. Unusual Hazards or Conditions
What are any unusual hazards or conditions that must be addressed?
- Lightweight truss construction
- High fuel loads
- Pitched roof
- Common attic space

Notes

Scenario 4

On Friday, April 3, at 10:07 a.m., dispatch reports a garage fire at 4445 Hardaway Drive. The dispatcher states that a relative with mobility issues is still in a back bedroom of the split-level house. Current weather conditions are clear with little wind.

On Arrival

You are the first-arriving officer. You see flames and heavy black smoke coming from an open garage door, which also reveals that a blue sedan in the garage is also involved in the fire. The homeowner is outside, frantically pointing to the window on the Alpha/Delta corner of the house. The garage is also on the Alpha/Delta corner.

Water Supply

An orange hydrant 500-1,000 gpm (2,000-4,000 L/min) is located 400 feet (120 m) from the house.

Weather

- Temperature — 50°F (10°C)
- Wind direction — Southwest
- Wind speed — 2 mph (3 km/h)
- Humidity — 40%

Alpha Side

Bravo Side

Charlie Side

Delta Side

Chapter 7 • Residential Scenarios 297

1. Size-Up

Based on the dispatch information and your visual observation of the scene on arrival, what are the facts and probabilities for this scenario?

Facts

- Split-level Type V wood frame
- Visible fire and smoke
- Heavy fuel load with the vehicle involved
- Water supply with a hydrant 400 feet (120 m) away
- Initial resources available

Probabilities

- A person is trapped and needs rescue
- Started as a vehicle fire
- Confined to the garage if built to code
- Quick knockdown of the fire
- Access to the structure through front door

2. Incident Priorities

What are your considerations for addressing the incident priorities of life safety, incident stabilization, and property conservation?

Life Safety

- Report by the owner that indicates a need for search and rescue
- Safety for the personnel working on the fireground
- Risk-benefit analysis
- Survivability profile
- Secure the garage door
- Secure utilities

Incident Stabilization

- Aggressive attack with initial line to the garage to support search and rescue and prevent extension to the upper story and soffit
- Obtain sustained water supply

Property Conservation

- Check for extension on the upper level and attic
- Ensure the integrity of the doorway between the house and the garage
- Use salvage covers to protect contents inside the structure

3. Operational Strategy

What are your considerations when selecting your operational strategy?

- Offensive strategy based on the report of a person still in the structure
- Fire appears to be isolated to the garage with a high probability of a rescue and savable property
- Risk to firefighters is low

4. Command Options

What are your considerations when selecting a Command option: Investigation, Fast Attack, or Command?

- Fast attack/Mobile command would allow simultaneous attack and rescue activities
- Stationary Command based on resources

5. Tactics

What are your considerations, if any, for the tactics used to accomplish your incident priorities?

Rescue

- Aggressive entry; witness is providing detailed information on the victim location
- Rapid victim location due to assumed conditions on the interior

Exposure

- The floor of the structure is the highest concern
- External exposures do not appear to be an issue

Confinement

- Attempt to isolate to the garage
- Check door between the garage and house
- Monitor for extension

Extinguishment

- Rapid fire attack and extinguishment are required to protect the search and rescue team(s) and the victim believed to be above the fire
- Direct attack from the Alpha side

Overhaul

- Check for extension in the walls and ceiling in the garage
- Consideration of possible extension in the walls and floor above the garage and attic
- Utilization of thermal imaging camera
- Air monitoring

Ventilation

- Positive-pressure ventilation horizontally at the front door once the fire is out
- Natural ventilation could be an option

Salvage

- Prevent further damage with runners and tarps on the main level
- Check for excessive water in the house with same level of the garage
- Secure valuables at the owner's requests

6. Unusual Hazards or Conditions

What are any unusual hazards or conditions that must be addressed?

- Unknown contents and storage in the garage
- Possible confusion of the floor level due to multiple levels with stairs
- Possible weakened floor in the area above the fire
- Potential fire spread to the upper floors

Notes

Scenario 5

On Tuesday, January 3, at 7:45 a.m., dispatch reports a residential structure fire at 110 Mountain Ranches Road. The dispatcher states that a passerby reported the incident and was unsure if occupants were home. The current weather condition is snowy.

On Arrival

You are the first-arriving officer. You see a two-story, single-family residence with heavy smoke coming from the basement window of the structure. Neighbors report that they saw family members at a second-story window and they did not see anyone exit the structure.

Water Supply

One orange 800 gpm (3,200 L/min) hydrant is located at the corner of Sooner Road and W. McElroy Street.

Weather

- Temperature — 15° F (-9°C)
- Wind direction — Northeast
- Wind speed — 4 mph (6 km/h)
- Humidity — 15%

Alpha Side

Bravo Side

Charlie Side

Delta Side

Chapter 7 • Residential Scenarios **303**

1. Size-Up

Based on the dispatch information and your visual observation of the scene on arrival, what are the facts and probabilities for this scenario?

Facts

- Type V wood-frame construction
- Multistory structure with a basement and two floors aboveground
- Heavy smoke coming from the basement window and soffit
- Water supply is nearby
- Exposure on the Alpha/Bravo sides
- Heavy snow buildup on the ground and on a covered swimming pool on the Charlie side

Probabilities

- Radiant heat transfer to the exposure on the Bravo side
- Primary fire located in the basement
- Fire may spread up the wall cavities and into the attic
- Structure may be occupied and the occupants may be trapped on the second floor
- Basement fire may have weakened exposed floor joists, creating a collapse hazard
- Utility service in the house, including electricity and propane gas

2. Incident Priorities

What are your considerations for addressing the incident priorities of life safety, incident stabilization, and property conservation?

Life Safety

- Neighbors reported seeing residents in the second-floor windows
- Search and rescue are the primary priorities
- Swimming pool covered in snow
- Personnel safety

Incident Stabilization

- Attack lines should be directed through the basement windows for initial knockdown
- Advance hoselines through the front door and to the basement door after an exterior knockdown
- Consider the use of a piercing or drive-in nozzle
- Protecting exposures

Property Conservation

- Rapid fire extinguishment in the basement is the most effective way to protect the contents of the house above the fire
- Door control can be used to decrease smoke contamination and damage
- Exposure must also be considered at risk

3. Operational Strategy

What are your considerations when selecting your operational strategy?

- Consider direct routes to the interior basement door through the front and back doors for a straightforward offensive operation after an exterior knockdown
- Rapid confinement and extinguishment to protect the exposure on the Bravo side
- Use the nearby water supply, which should be adequate for an offensive or a defensive attack for exposure protection

4. Command Options

What are your considerations when selecting a Command option: Investigation, Fast Attack, or Command?

- Mobile/fast attack due to imminent rescue
- Stationary Command if resources allow

5. Tactics

What are your considerations, if any, for the tactics used to accomplish your incident priorities?

Rescue

- The integrity of the floor must be considered when conducting a search
- VEIS could be considered
- Resources committed to search and rescue may exceed those assigned to fire extinguishment
- If traditional methods used, consider stairwell protection with the hoseline

Exposure

- Interior exposures can be protected by the primary attack hoseline through the front or back door after exterior knockdown of the fire
- The unattached garage on the Bravo side can be protected by a hoseline

Confinement

- Rapid placement of a hoseline directed through the exterior basement windows
- Rapid placement of a hoseline onto the first floor to confine the fire to the basement
- Attack the fire down the stairs or supply a piercing or drive-in nozzle through the floor with a hose
- Opening the exterior walls on the Bravo side to prevent fire extension to the upper floors

Extinguishment

- Initial attack from the exterior through the basement windows
- Subsequent placement of a hoseline on the interior of the structure
- Consideration must be given to the structural condition of the first floor
- Consider using piercing or drive-in nozzles
- Consider the potential for floor collapse

Overhaul
- Consider structural integrity during overhaul
- Open and check walls for extension
- Monitor air

Ventilation
- Open windows on the first floor and natural ventilation will improve visibility
- Use negative ventilation in the basement windows once the fire is extinguished
- Use positive-pressure ventilation at the front door once the fire is extinguished
- Consider hydraulic ventilation

Salvage
- Properly applied ventilation techniques can reduce smoke damage
- Remove water from basement when fire is extinguished
- Use salvage covers and floor runners if the floor is stable

6. Unusual Hazards or Conditions

What are any unusual hazards or conditions which must be addressed?

- Potential floor collapse
- Personnel must work from above the fire
- Slippery conditions on the roof, creating hazards during vertical ventilation
- Snow and ice
- Unknown contents stored in basement

Notes

Commercial Scenarios

Photo courtesy of Ron Jeffers, Union City, NJ.

Chapter Contents

Scenario 1 311
Scenario 2 317
Scenario 3 323
Scenario 4 329
Scenario 5 335

Chapter 8

Key Terms

This chapter contains five hypothetical scenarios based on commercial occupancy types that are found in most response areas. The scenarios contain best practices that can be applied to each type of situation. However, these are suggestions only and not hard and fast rules that must be applied in every incident.

Because these scenarios are examples to help you learn how to make decisions based on the information in the previous chapters, there are limitations placed on them. The resources available to you will arrive within 10 minutes. You are the Incident Commander (IC) and must allocate your resources to control/mitigate the incident to the best of your ability. The resources you have reflect those in your local department, including first-alarm assignments, standard operating procedures/guidelines (SOP/Gs), mutual aid, equipment, and staffing.

Commercial Scenarios

FESHE Learning Outcomes

After reading this chapter, students will be able to:

1. Given information on a commercial fire and resources available to your department, develop an Incident Action Plan (IAP) for achieving the incident objectives.

> **Note to Instructors**
>
> The following scenarios contain recommended practices based on the facts provided and on information contained in the text of this manual. These scenarios are intended to be used as teaching aids. It is suggested that you develop scenarios specific to your department's resources, procedures, and response area. Your scenarios should contain structures, hazards, water supplies, photographs, and site plans that exist in your jurisdiction. Do not include probabilities, incident priorities, strategies, command options, or tactics. Students should supply this information based on their understanding of the text. Your scenarios can be varied based on changes to the water supply (hydrants out of service, peak water demand times), delayed resources, weather conditions, minimum staffing, time of day, or local standard operating procedures/guidelines (SOP/Gs).
>
> **The considerations listed are not absolute. The scenarios are designed for discussion purposes and training perspectives. The AHJ and their respected policies within will dictate each department's strategies, tactics, and considerations.**

Chapter 8
Commercial Scenarios

Photo courtesy of Ron Jeffers, Union City, NJ.

Scenario 1

On Saturday, August 1, at 5:45 a.m., dispatch reports a structure fire that is a 5,000-square-foot (450 m) restaurant located at 2136 E. Kaden Avenue. Dispatch reports that the building has been in the process of remodeling for the last three months. Current weather conditions are clear skies.

On Arrival

You are the first-arriving officer. You see a single-story commercial building with heavy smoke and fire showing from the roof of the structure.

Water Supply

One blue 1,500 gpm (6,000 L/min) hydrant is located at E. Kaden Avenue and the Landon Mall Access Road.

Weather
- Temperature — 80° F (27°C)
- Wind direction — South
- Wind speed — 10 mph (16 km/h)
- Humidity — 68%

Alpha Side

Bravo Side

Charlie Side

Delta Side

1. Size-Up

Based on the dispatch information and your visual observation of the scene on arrival, what are the facts and probabilities for this scenario?

Facts

- Unoccupied/vacant structure
- Fire is showing through the roof
- Water supply is within one block of the structure
- Type V wood frame
- The structure has a flat roof
- No access issues to the structure
- There are no exposures near the structure
- The building is under renovation

Probabilities

- Probably a large volume fire
- Assign a water supply officer
- Opening in the walls and ceiling may exist to permit rapid fire extension
- Lightweight wood trusses in the attic are being exposed to heat and flames due to extension; can lead to roof collapse
- The fire suppression system may be out of service

2. Incident Priorities

What are your considerations for addressing the incident priorities of life safety, incident stabilization, and property conservation?

Life Safety

- Firefighter life safety would take precedence, justifying a defensive strategy
- Establish collapse zone
- Secure utilities
- Chance of civilian life is minimal

Incident Stabilization

- Apparatus placement
- Large volume of water

Property Conservation

- Try to protect any ancillary exposures
- Property conservation may no longer be a priority because of the large volume of fire visible

3. Operational Strategy

What are your considerations when selecting your operational strategy?

- This is a defensive strategy fire
- All life safety concerns
- Advanced fire in a vacated/unoccupied structure

4. Command Options

What are your considerations when selecting a Command option: Investigation, Fast Attack, or Command?

- Stationary Command
- Establish groups and divisions

5. Tactics

What are your considerations, if any, for the tactics used to accomplish your incident priorities?

Rescue

- Survivability profile is low
- Firefighter life safety would take precedence justifying a defensive strategy
- The time of morning and the volume of fire reduces the possibility that a savable victim is in the structure
- Search and rescue should not be attempted

Exposure

- An ember patrol may be required if the embers start to spread because of the wind

Confinement

- Master streams should be used to confine the fire to the building of origin

Extinguishment

- Master stream appliances should be used to extinguish the majority of the fire
- Handlines can be used to extinguish the remaining fires
- Class A foam may be used if available in the correct quantity to complete fire extinguishment

Overhaul

- Air monitoring
- Look for collapse potential
- Fire watch detail

Ventilation

- The structure has self-ventilated
- Ventilation will not be required at this stage

Salvage

- Limited, if any

6. Unusual Hazards or Conditions

What are any unusual hazards or conditions that must be addressed? How will you address those hazards and/or conditions?

- Unknown quantity and type of hazardous materials; construction materials, including paints, thinners, and cleaners, may be present
- Lightweight wood truss construction; all personnel must monitor the structural integrity

- Construction activities may have created openings in the floor slab or walls
- Weakened roof and wall structures may collapse without warning
- Unknown fire time that could lead to the potential for a building collapse
- Resource intensive in both deployment and allocation

Notes

Scenario 2

On Thursday, January 9, at 11:50 a.m., dispatch reports a structure fire at the Colossal Savings Supercenter located at 1632 W. Frontage Road. The dispatcher states that the fire started in the tire and lube section of the building. Current weather conditions are clear skies.

On Arrival

You are the first-arriving officer. You see a single-story, large area commercial building with black smoke showing from the Bravo side of the structure.

Water Supply

One blue 1,500 gpm (6,000 L/min) hydrant is located at the corner of North Jardot and the North parking lot entrance.

One blue 1,500 gpm (6,000 L/min) hydrant is located at the corner of North Jardot and the South parking lot entrance.

Weather

- Temperature : 52° F (11°C)
- Wind direction: South
- Wind speed: 5 mph (8 km/h)
- Humidity: 71%

Alpha Side

Bravo Side

Charlie Side

Delta Side

Chapter 8 • Commercial Scenarios

1. Size-Up

Based on the dispatch information and your visual observation of the scene on arrival, what are the facts and probabilities for this scenario?

Facts

- Type II construction
- Fire suppression (sprinkler) and standpipe system
- Mercantile occupancy
- High fuel load
- Classified as a Target Hazard generating a preincident plan
- Smoke is visible coming from the overhead doors of the vehicle maintenance area on the Charlie side of the structure
- Water supply is close to the structure with two 1,500 gpm (6,000 L/min) hydrants available
- The structure is occupied and in the process of being evacuated

Probabilities

- The fire suppression system may have been activated and attempting to confine the fire
- Smoke color and volume indicate the fire suppression system is not controlling the fire but may be confining it to the maintenance area
- The area in which the fire is located contains flammable and combustible liquids, tires, compressed gases, plastics, and vehicles
- Limited customer access
- A barrier wall separates the maintenance bay from the remainder of the store
- Employees and customers may be trapped in or near the fire

2. Incident Priorities

What are your considerations for addressing the incident priorities of life safety, incident stabilization, and property conservation?

Life Safety

- Secure utilities
- Firefighter safety through defensive attack, large hose streams and large water supply
- Ensure that all occupants have been evacuated
- Consider hooking up to the fire department connection to augment the existing suppression system
- Accountability of employees and customers

Incident Stabilization

- Rapid fire extinguishment is necessary to prevent the fire from extending into the store
- The fire suppression system should be supported and supplemented with attack hoseline
- Water supply

Property Conservation

- Remove tarping
- Door control to limit the spread of smoke

3. Operational Strategy

What are your considerations when selecting your operational strategy?

- Consider transitional attack with a large volume of water due to the proximity to the exterior
- Extinguishing the fire is the best way to provide an evacuation route for occupants

4. Command Options

What are your considerations when selecting a Command option: Investigation, Fast Attack, or Command?

- Because of the structure's size, the initial IC should select the Command mode and manage the incident from a Command Post

5. Tactics

What are your considerations, if any, for the tactics used to accomplish your incident priorities?

Rescue

- Consider using the building's PA system to direct occupants to the best route of egress
- Search and rescue for victims near the fire should begin first
- Store employees may be involved in the evacuation of customers but should not be assigned this task
- The store manager can assist in accounting for employees

Exposure

- The primary exposures are internal, consisting of the sales area and the stockrooms or other spaces adjacent to the maintenance area

Confinement

- Consider a blitz attack using a 2 ½-inch hoseline or larger to make a rapid knockdown
- The fire suppression system performs initial confinement; resources must be assigned
- Initial fire attack hoselines can be directed into the maintenance area through the overhead doors
- Additional lines can be connected to standpipe outlets within the structure as resources become available

Extinguishment

- A coordinated attack by hoselines from inside the structure and at the overhead door openings along with the fire suppression system
- Class A or B foam may be required to extinguish some burning materials

Overhaul
- Attention to the adjoining walls of the structure and inspection of the roof
- Removal of debris may be necessary to help locate hidden fires

Ventilation
- Horizontal ventilation should be used to pull the smoke out of the maintenance area
- Mechanical horizontal ventilation can be used from the barrier wall side
- If necessary, vertical ventilation can be used over the fire
- The HVAC system needs to be secured to prevent it from spreading smoke and fire

Salvage
- Shut off the sprinkler system
- Salvage covers and the removal of merchandise
- Application of proper ventilation techniques can prevent smoke from entering the unburned areas
- Retracted overhead doors must be monitored for structural stability

6. Unusual Hazards or Conditions

What are any unusual hazards or conditions that must be addressed? How will you address those hazards and/or conditions?

- Collapse hazard due to the type of construction
- Potentially high occupant load
- Exposed structural members
- Hazardous materials
- Potential for rapid fire spread into a large, uncompartmentalized area
- Large open area
- High fuel load
- Vehicles present in the maintenance area
- Overhead doors may be subject to collapse or closing

Notes

Scenario 3

On Friday, February 9, at 6 a.m., dispatch reports a structure fire at the Lehane Strip Mall at 320 North Redbud Avenue. Current weather conditions are clear skies.

On Arrival

You are the first-arriving officer. You see a single-story small strip shopping center with heavy smoke and fire showing from a unit on the west end of the Alpha side of the structure.

Water Supply

One red 500 gpm (2,000 L/min) hydrant directly in front of the structure on E. Redbud Avenue.

One red 500 gpm (2,000 L/min) hydrant half a block West on E. Redbud Avenue.

Weather

- Temperature — 45° F (7°C)
- Wind direction — South
- Wind speed — 10 mph (16 km/h)
- Humidity — 48%

Alpha Side

Bravo Side

Charlie Side

Delta Side

1. Size-Up

Based on the dispatch information and your visual observation of the scene on arrival, what are the facts and probabilities for this scenario?

Facts

- Confirmed structure fire on the Alpha/Delta corner of the strip mall
- No vehicles in the parking lot
- Fire has self-vented through the roof
- Commercial occupancy
- Type III construction
- Internal exposures
- Flat roof with HVAC unit

Probabilities

- No civilian life safety hazard
- Truss roof system
- Possible roof collapse
- Facade collapse; Delta side exterior wall
- Possible firewall exists

2. Incident Priorities

What are your considerations for addressing the incident priorities of life safety, incident stabilization, and property conservation?

Life Safety

- No civilian hazard because of the vacant parking lot and time of day
- Personnel safety
- Establish a RIT/RIC
- Secure utilities
- Force the door on the Charlie side of the structure while keeping the door control

Incident Stabilization

- Extinguish the fire from the Alpha side of the structure
- Check units and the attic space in all Bravo side exposures

Property Conservation

- Protect contents with salvage covers in all Bravo side exposures
- Volume of fire in Unit 320 would indicate that the contents are beyond saving
- Salvage covers for contents in Units 312 through 318

3. Operational Strategy

What are your considerations when selecting your operational strategy?

- Offensive attack because the fire is being confined to one unit
- Consider a defensive attack in the main fire unit and offensive fire attack in exposure(s) if the fire has extended
- Establish a collapse zone

4. Command Options

What are your considerations when selecting a Command option: Investigation, Fast Attack, or Command?

- Fast attack because the fire appears to be confined in one unit only
- Consider Command option if fire has progressed to more than one unit

5. Tactics

What are your considerations, if any, for the tactics used to accomplish your incident priorities?

Rescue

- Perform a primary search in exposure units closest to the unit on fire

Exposure

- Assign resources to check units and the attic space in adjacent units
- Assign resources with a hoseline
- Monitor the condition of the roof for sagging and potential collapse

Confinement

- Deploy personnel with handlines to control and extinguish the fire in fire unit
- Monitor conditions in exposure units

Extinguishment

- Deploy handlines from the Alpha side of the structure
- Apply a Master stream

Overhaul

- Check walls and the attic space in the main fire unit and adjacent exposure units

Ventilation

- No tactical ventilation because of the fire is venting from the roof
- Consider horizontal ventilation to remove smoke once the fire has been extinguished

Salvage

- Protect contents in fire unit and adjacent exposure units

6. Unusual Hazards or Conditions

What are any unusual hazards or conditions that must be addressed? How will you address those hazards and/or conditions?

- High occupant load depending on the time of day
- Lightweight construction overhang on Alpha side
- Unknown condition of the interior Bravo-side wall, including possible penetrations
- Possible common attic and exposed metal roof trusses
- Retaining wall and narrow access on Charlie side
- Alternative water source for master stream

Notes

Scenario 4

On Wednesday, December 17, at 2:10 p.m., dispatch reports a structure fire at the Elderly Care Clinic located at 1311 Monaco Avenue. Dispatch reports that the fire alarm in the lobby area has been activated. Current weather conditions are clear skies.

On Arrival

You are the first-arriving officer. You see a single-story commercial building with smoke and fire showing from the Alpha side of the structure.

Water Supply

One red 500 gpm (5 680 L/min) hydrant on the southeast corner of the parking lot.

Weather

- Temperature — 25° F (4°C)
- Wind direction — North
- Wind speed — 5 mph (8 km/h)
- Humidity — 80%

Alpha Side

Bravo Side

Charlie Side

Delta Side

Chapter 8 • Commercial Scenarios   **331**

1. Size-Up

Based on the dispatch information and your visual observation of the scene on arrival, what are the facts and probabilities for this scenario?

Facts

- Type V wood frame construction with brick veneer
- Business/medical occupancy that is occupied
- Fire is visible in a front door on the Alpha side
- Water supply is close to the structure
- A ditch and an overhead power line obstruct access to the Delta side
- A second access/egress door is on the Bravo side
- There is a large-volume attic space above the main floor
- No occupants visible

Probabilities

- Time of day, vehicles, and occupancy type indicate that the structure may be occupied
- Some occupants may be nonambulatory and/or elderly
- Hazardous medical equipment and waste present

2. Incident Priorities

What are your considerations for addressing the incident priorities of life safety, incident stabilization, and property conservation?

Life Safety

- Lack of visible occupants outside the structure indicates that search and rescue are required initially
- Access through the door on Bravo side may be the most direct route
- Rapid fire attack will have greatest impact on survivability of occupants
- Medical transport of patients to another medical facility

Incident Stabilization

- Access through the door on the Alpha side will help in search and rescue, and place an attack hoseline between the occupants and the fire
- A positive pressure ventilation fan placed in the Bravo side after extinguishment

Property Conservation

- Rapid fire extinguishment will reduce structural and content damage
- Application of proper salvage and ventilation techniques will limit property damage

3. Operational Strategy

What are your considerations when selecting your operational strategy?

- An offensive strategy is required to locate and evacuate occupants and extinguish the fire
- Risk-benefit analysis

- Survivability profile
- Location and size of the fire

4. Command Options
What are your considerations when selecting a Command option: Investigation, Fast Attack, or Command?

- The need for rapid action justifies the use of the Fast Attack option
- Once additional resources arrive, the IC should transfer Command to Stationary Command mode. If resources are available, use Stationary command from the onset

5. Tactics
What are your considerations, if any, for the tactics used to accomplish your incident priorities?

Rescue

- Search, rescue, and evacuation of occupants should be accomplished initially
- The access door on the Bravo side should be used for evacuation
- If resources are limited, search, rescue, and evacuation should be accomplished secondary to fire attack

Exposure

- There is no external exposure requiring protection
- The primary internal exposures are the offices adjacent to the lobby and the attic above the fire area
- Placing a secondary hoseline between the lobby fire and the remainder of the interior will protect this exposure

Confinement

- Taking the hose through the access door can confine the fire to the lobby

Extinguishment

- If the fire is small in nature and limited in area, a possible attack from the lobby entrance door with a rapid application of water
- Watch for opposing hoselines

Overhaul

- Walls and voids should be opened to locate any remaining fire
- Air monitoring

Ventilation

- Mechanical horizontal ventilation can be used through the access door on Bravo side if an exit point is available for the smoke closer to the lobby

Salvage

- Limit access of personnel to uninvolved parts of the structure if possible
- Application of proper ventilation techniques will reduce smoke damage
- Use of salvage covers and floor runners

6. Unusual Hazards or Conditions

What are any unusual hazards or conditions that must be addressed? How will you address those hazards and/or conditions?

- Medical equipment (oxygen tank, biomedical waste, and radioactive sources)
- Potential roof collapse if fire gets into the attic area
- Lightweight construction
- Ambulatory/nonambulatory patients
- Limited paths of egress
- Water supply would be an issue if the fire extends past the area of origin

Notes

Scenario 5

On Sunday, October 8, at 1:15 p.m., dispatch reports a structure fire located at 5789 Gwinnett Road. Dispatch reports that a passerby saw the fire and reported heavy fire and smoke coming from the hayloft of the barn.

On Arrival
You are the first-arriving officer. You see a large burn structure with heavy smoke and fire showing from the hayloft of the structure.

Water Supply
No hydrant supply in the immediate area. A pond is located behind the structure.

Weather
- Temperature — 58° F (14°C)
- Wind direction — Northeast
- Wind speed — 10 mph (16 km/h)
- Humidity — 75%

Alpha Side

Bravo Side

336 Chapter 8 • Commercial Scenarios

Charlie Side

Delta Side

Chapter 8 • Commercial Scenarios **337**

1. Size-Up

Based on the dispatch information and your visual observation of the scene on arrival, what are the facts and probabilities for this scenario?

Facts

- Type V wood frame construction with metal siding and roof
- A storage building containing hay and possibly livestock
- Fire is visible from the loft door on the Alpha side
- Water supply consists of a pond on the Charlie side of the structure
- There are no exterior exposures
- Electric lines provide power to the structure

Probabilities

- There is a potential for structural collapse
- Initial assignment may not have enough water on hand to control the incident without establishing a shuttle operation or drafting from the pond
- Establishing either a drafting operation or water shuttle will be time-consuming
- Evacuation of livestock will require high level of staffing and skill
- Overhaul of the structure will require a high level of staffing
- Full fire extinguishment and overhaul will require copious amounts of water
- Structural stability is unknown due to the fact that it may be wood or exposed steel
- The barn may contain unknown quantities of farm chemicals including fertilizer, insecticides, fuel, machinery, and explosives
- Fence may be electrified

2. Incident Priorities

What are your considerations for addressing the incident priorities of life safety, incident stabilization, and property conservation?

Life Safety

- Cannot rule out potential occupants
- Possible livestock inside
- Collapse potential
- Ensure firefighters are wearing proper PPE/SCBA since there could be unknown chemicals inside the barn. Personnel may be tempted to remove SCBA due to exterior operation
- Secure utilities

Incident Stabilization

- Establish water supply operations; consider drafting and tender operations
- Assign water supply officer
- Master stream operations

Property Conservation

- Livestock
- Farm equipment if structure is stable

3. Operational Strategy

What are your considerations when selecting your operational strategy?

- Direct access to the fire from exterior
- Low likelihood of occupants
- Complicated water supply
- These factors suggest either defensive or transitional

4. Command Options

What are your considerations when selecting a Command option: Investigation, Fast Attack, or Command?

- Due to multiple complications (water supply and removal of livestock), consider Stationary Command post

5. Tactics

What are your considerations, if any, for the tactics used to accomplish your incident priorities?

Rescue

- If rescue of any employees or evacuation of livestock is required, access should be gained through the openings on the Delta side
- It will be easy to see into the structure during a 360 from multiple openings
- Time of day and several exits suggests self-evacuation

Exposure

- Exterior exposure is the dry grass surrounding the structure
- Internal

Confinement

- Rapid attack from the exterior and extinguishment should confine the fire to the loft area; however, the volume of fire and the potential for collapse may limit the ability to achieve this

Extinguishment

- Direct application of water into the hayloft from the exterior Alpha side should be used to extinguish the fire
- Consider foam
- Apply elevated Master streams

Overhaul

- To limit the collapse hazard, the hay must be removed from the loft as soon as it is extinguished
- The structural integrity of the loft must be ensured
- Move in with handlines if structure is stable

Ventilation

- Ventilation is not an issue due to the type and configuration of the structure
- Under no circumstances should vertical ventilation be attempted until the structural integrity of the roof is ensured
- Vertical ventilation is not an option due to the fire impinging on the metal roof

Salvage

- Remove farm equipment and other property if possible

6. Unusual Hazards or Conditions

What are any unusual hazards or conditions that must be addressed? How will you address those hazards and/or conditions?

- Lack of fire hydrants
- Removal of livestock
- Heavy fuel load
- Water-soaked hay increases potential for collapse of hay loft
- Potential for roof collapse
- Potential grass fire

Notes

Special Hazards Scenarios

Photo courtesy of Ron Jeffers, Union City, NJ.

Chapter Contents

Scenario 1 345
Scenario 2 351
Scenario 3 357
Scenario 4 363
Scenario 5 369

chapter 9

Key Terms

This chapter contains five hypothetical scenarios based on special occupancy types that are found in most response areas. The scenarios contain best practices that can be applied to each type of situation. However, these are suggestions only and not hard and fast rules that must be applied in every incident.

Because these scenarios are examples to help you learn how to make decisions based on the information in the previous chapters, there are limitations placed on them. The resources available to you will arrive within 10 minutes. You are the Incident Commander (IC) and must allocate your resources to control/mitigate the incident to the best of your ability. The resources you have reflect those in your local department, including first-alarm assignments, standard operating procedures/guidelines (SOP/Gs), mutual aid, equipment, and staffing.

Special Hazards Scenarios

FESHE Learning Outcomes

After reading this chapter, students will be able to:

1. Given information on a residential fire and resources available to your department, develop an Incident Action Plan (IAP) for achieving the incident objectives.

> **Note to Instructors**
> The following scenarios contain recommended practices based on the facts provided and on information contained in the text of this manual. These scenarios are intended to be used as teaching aids. It is suggested that you develop scenarios specific to your department's resources, procedures, and response area. Your scenarios should contain structures, hazards, water supplies, photographs, and site plans that exist in your jurisdiction. Do not include probabilities, incident priorities, strategies, command options, or tactics. Students should supply this information based on their understanding of the text. Your scenarios can be varied based on changes to the water supply (hydrants out of service, peak water demand times), delayed resources, weather conditions, minimum staffing, time of day, or local standard operating procedures/guidelines (SOP/Gs).

Chapter 9
Special Hazards Scenarios

Photo courtesy of Ron Jeffers, Union City, NJ.

Scenario 1

On Tuesday, December 19, at 10:20 p.m., dispatch reports a structure fire at the Allerdale Inn hotel located at 412 W. Frontage Road. The dispatcher states that the night manager called in the report. The night manager reported that the fire is located in a room on the fourth floor. The fourth-floor corridor is filling with smoke, and the building's fire suppression system has apparently not activated. Guests are evacuating the structure. The current weather condition is cold.

On Arrival

You are the first-arriving officer. You see a four-story hotel structure with fire showing from the Alpha/Bravo corner sides of the structure.

Water Supply

One blue 1,500 gpm (6,000 L/min) hydrant is located at the corner of S. Murphy and W. Frontage Road.

One blue 1,500 gpm (6,000 L/min) hydrant is located on the Northeast side of the hotel parking lot.

Weather
- Temperature – 28° F (-2°C)
- Wind direction – North
- Wind velocity – 5 mph (8 km/h)
- Humidity – 100%

Alpha Side

Bravo Side

Charlie Side

Delta Side

Chapter 9 • Special Hazards Scenarios **347**

1. Size-Up

Based on the dispatch information and your visual observation of the scene on arrival, what are the facts and probabilities for this scenario?

Facts

- The fire is visible on the Alpha-Bravo corner of the fourth floor
- Two fire hydrants are available for water supply
- Cold weather will play a role in relocation of evacuated guests
- The fire suppression system has apparently not activated
- Residential multifamily occupancy
- Evacuation is in progress

Probabilities

- High occupancy load due to the time of night; guests are likely not familiar with evacuation routes or procedures
- Some guests may be asleep
- Not all guests may have been alerted to the hazard
- Reduced visibility on the fourth-floor corridor due to smoke
- Limited hotel staffing
- The door to the room with the fire may be closed, possibly creating a vent-limited fire
- The exit doors on the Bravo and Delta sides may be locked from the inside
- Command could consider an evacuation group
- Consider calling an outside agency (such as law enforcement) for assistance in evacuation and relocation of guests
- Make a division for the fire-floor operations
- Have an engine feed the standpipe riser if available

2. Incident Priorities

What are your considerations for addressing the incident priorities of life safety, incident stabilization, and property conservation?

Life Safety

- Evacuation of the fire floor (Will the fire alarm system be active in all floors?)
- Shelter in place of other guests if feasible
- Treat injured if applicable
- Protect paths of egress
- Warm shelter needed for displaced guests and rehabbing firefighters

Incident Stabilization

- The IC should make groups/divisions for suppression activities
- Identify room(s) involved and fire extension
- Make the building tenable for operational suppression activities
- Establish lobby/accountability control

- Support the system riser if practical
- Protect exposures based on the conditions found on the fourth floor/the fire floor

Property Conservation

- Salvage floors below the fire for possible water damage
- Smoke/heated gas/fuel removal
- Aggressively confine/extinguish the fire in the room of origin
- Overhaul the room of origin

3. Operational Strategy

What are your considerations when selecting your operational strategy?

- Remove endangered occupants
- Property conservation (Salvage)
- Accountability
- Identifying the fire floor and the room involvement along with extension
- Water supply since there is an apparent nonoperational system
- The need for search, rescue, and evacuation and fire extinguishment requires an offensive strategy
- Identify the stairwell for attack and the stairwell for evacuation

4. Command Options

What are your considerations when selecting a Command option: Investigation, Fast Attack, or Command?

- Due to the high life safety hazard, the size of the structure, and the lack of confirmed fire protection, the initial IC should use the Command option and manage the incident from a Command post or from the fire control panel in the lobby

5. Tactics

What are your considerations, if any, for the tactics used to accomplish your incident priorities?

Rescue

- Evacuation of the structure occurs simultaneously with fire extinguishment
- If the structure is equipped with a voice notification system, guests should be directed to leave the building; they made also be directed to remain in place if they are not on the floor in which the alarm is activated
- Protect stairways and avenues of egress
- Elevators must be controlled to prevent guests from using them
- Aerial apparatus for rescue/egress
- Primary and secondary searches of fire floor
- Control flow paths
- Extensive search operations

Exposure
- Adjoining rooms
- The floor below the fire
- Attic space

Confinement
- Aggressive offensive attack to confine the fire to the room of origin
- Confinement could also include closing the door to the fire room

Extinguishment
- Offensive attack with handlines
- Depending on the issue with the suppression system, feeding the riser is another method of extinguishment that can be considered

Overhaul
- Check attic spaces and adjoining rooms for extension

Ventilation
- Control flow paths
- Ventilate the corridor and stairwells
- Ventilation should be coordinated with extinguishment operations

Salvage
- Protect from water damage on the floor below the fire
- Remove smoke and products of combustion from the corridor and adjoining rooms
- Cover furnishings

6. Unusual Hazards or Conditions

What are any unusual hazards or conditions that must be addressed? How will you address those hazards and/or conditions?

- Subfreezing temperatures
- Transportation and accommodations for displaced occupants
- Large area and controlled access for primary search
- Ventilation may not be possible if windows do not open or the system is not designed to remove smoke

Notes

_____

_____

_____

_____

_____

Scenario 2

On Monday, July 8, at 2:30 p.m., dispatch reports a passerby called 9-1-1 after spotting smoke coming from the roof of a nursing home located at 2124 Midshipmen Street. Current weather conditions are clear skies in the hot afternoon.

On Arrival

You are the first-arriving officer. You see smoke coming from a roof on the Delta side of the nursing home. The nursing home staff is in the process of evacuating residents from the structure when you arrive.

Water Supply

One blue 1,500 gpm (6,000 L/min) hydrant is located right in front of the nursing home on Midshipmen Street.

One blue 1,500 gpm (6,000 L/min) hydrant is located 200 feet to the north of the nursing home on Midshipmen Street.

One blue 1,500 gpm (6,000 L/min) hydrant is located 200 feet to the south of the nursing home on Midshipmen Street.

Weather

- Temperature — 101° F (38°C)
- Wind direction — East
- Wind speed — 2 mph (3 km/h)
- Humidity — 45%

Alpha Side

Bravo Side

Charlie Side

Delta Side

Chapter 9 • Special Hazards Scenarios

1. Size-Up

Based on the dispatch information and your visual observation of the scene on arrival, what are the facts and probabilities for this scenario?

Facts

- Black smoke showing
- Type I construction
- Evacuation in progress
- Ample water supply
- Confirmed fire
- Commercial structure
- Smoke showing from the Delta side

Probabilities

- Residents will have physical limitations inhibiting their egress
- Some residents have already been evacuated, and some residents may still be inside the structure
- Sufficient resources will be needed for the rescue
- Rescue assessment profile indicates a probability of survivability in certain areas of the structure

2. Incident Priorities

What are your considerations for addressing the incident priorities of life safety, incident stabilization, and property conservation?

Life Safety

- Remove occupants in immediate fire area
- Shelter in place for the rest of the occupants
- Close doors to isolate victims from the fire
- Determine priority search areas based on rescue survivability profile
- Establish resident accountability control
- Ensure additional transport units are on the scene

Incident Stabilization

- Isolate the fire first
- Aggressive fire attack. This can probably be achieved quicker with fewer firefighters than first evacuating every occupant
- The IC should make groups/divisions for suppression activities
- Identify room(s) involved and fire extension
- Make the building tenable for operational suppression activities
- Support fire suppression system if available and needed
- Protect exposures based on the conditions found on the fire floor

Property Conservation

- Shut down the sprinkler system if applicable and when the fire is out
- Salvage covers to protect furnishing and personal property

3. Operational Strategy

What are your considerations when selecting your operational strategy?

- Probable survivable victims
- Bulk of building is not involved
- Offensive

4. Command Options

What are your considerations when selecting a Command option: Investigation, Fast Attack, or Command?

- Stationary Command
- Extensive search and rescue operations
- Fire attack
- Number of personnel responding on initial arrival apparatus
- Due to the high life safety hazard and the size of the structure, the initial IC should use the Command option and manage the incident from a Command Post

5. Tactics

What are your considerations, if any, for the tactics used to accomplish your incident priorities?

Rescue

- Rescue takes place simultaneously as attack; remove the threat and make rescue more effective by extinguishment and vent
- Do not fully staff rescue assignments until extinguishment is addressed
- Coordinate ventilation and fire attack

Exposure

- Close interior doors to limit interior exposure and spread
- Consider assigning resources to protect external exposures
- Ensure enough resources are available to address internal/external exposure issues

Confinement

- Quick direct attack
- Close doors to isolate the residents and the fire

Extinguishment

- Fast aggressive fire attack
- Establish a continuous water supply early
- Deploy attack line and backup line

Overhaul

- Check all six sides of the fire compartment
- Crew resources management
- Minimize secondary damage

Ventilation
- Hydraulic ventilation from the fire area is possible immediately
- Hold off on positive-pressure ventilation until the fire is extinguished
- One story with ample windows; horizontal ventilation may be effective
- Ventilate in coordination with fire attack and the IC
- Control flow path

Salvage
- Remove medical equipment and resident belongings
- Consider salvage operations during fire attack based on building content(s)

6. Unusual Hazards or Conditions

What are any unusual hazards or conditions that must be addressed? How will you address those hazards and/or conditions?

- High density of disabled residents; physical limitations will inhibit egress. Call early for additional resources
- Must plan for alternate means of care for the patients, ambulances for those needing advanced care. Can call for public transportation for large numbers of stable patients and their caregivers
- Notify local hospitals of potential influx of disabled residents who may have medical complaints
- Notify local Red Cross for housing and food needs for displaced residents

Notes

Scenario 3

On Saturday, June 6, at 2 p.m., dispatch reports a passerby noticed a strong acrid smell coming from Fred's Pool Store located at 5121 Demond Avenue. Current weather conditions are light rain in the afternoon.

On Arrival

You are the first-arriving officer. You see a white cloud coming from the Charlie/Delta corner of a mercantile building of 3,000 square feet. Four employees and two customers are sitting on the curb and complain of difficulty breathing.

Water Supply

One blue 1,500 gpm (6,000 L/min) hydrant is located 300 feet from the store on Demond Avenue.

Weather

- Temperature — 65° F (18°C)
- Wind direction — West
- Wind speed — 15 mph (24 km/h)
- Humidity — 55%

Alpha Side

Bravo Side

358 Chapter 9 • Special Hazards Scenarios

Charlie Side

Delta Side

1. Size-Up

Based on the dispatch information and your visual observation of the scene on arrival, what are the facts and probabilities for this scenario?

Facts

- Type II construction
- Single-story structure
- Multiple occupancy strip mall
- Dead load on the roof
- Multiple patients
- Water supply available
- Easy access for a complete 360-degree check of the building

Probabilities

- Does not appear to be a busy area
- Will require a Hazardous Materials group
- Additional transport units

2. Incident Priorities

What are your considerations for addressing the incident priorities of life safety, incident stabilization, and property conservation?

Life Safety

- Building evacuation
- Area evacuation
- Hazmat identification/response
- EMS notification/response
- Maintain responder accountability and safety

Incident Stabilization

- Leak identification, isolation, and control
- Shutdown of HVAC to reduce release to the atmosphere and surrounding area
- Establishment of ICS and identify the possibility of multiple operational periods
- State notification
- Identify early if there are additional resources needed

Property Conservation

- Control of the leak to reduce exposure
- Control damage by Hazmat/fire crews during operations

3. Operational Strategy

What are your considerations when selecting your operational strategy?

- Hazmat response
- Hazmat decontamination
- Fire suppression (is it a fire or a chemical reaction?)

- Isolate and control the release
- Vapor suppression

4. Command Options

What are your considerations when selecting a Command option: Investigation, Fast Attack, or Command?

- Stationary/static Command post at a safe distance
- Address possibility of more than one operational period

5. Tactics

What are your considerations, if any, for the tactics used to accomplish your incident priorities?

Rescue

- Level B protection for responders
- Rapidly remove victims from hazardous atmosphere
- Ensure accountability of all individuals in the area
- Address hazmat decontamination
- Establish control zones
- Continually monitor weather and wind conditions

Exposure

- Control the spread of leak
- Identify and control any other exposures, such as waterways and groundwater

Confinement

- Find the leak and stop it if one exists
- Identify needs for stopping the leak

Extinguishment

- Identify what is burning and extinguish it

Overhaul

- Air monitoring
- Vapor suppression

Ventilation

- Control of the HVAC system

Salvage

- Consider all factors for the cleanup of the spill if one is found and who is responsible for that function

6. Unusual Hazards or Conditions

What are any unusual hazards or conditions that must be addressed? How will you address those hazards and/or conditions?

- Hazmat
- Utilize business owners for assistance on control

- Utilize state resources
- The spread of the leak addressed through control and stoppage of leak

Notes

Scenario 4

On Thursday, April 30, 4 p.m., dispatch reports a structure fire at 200 E. Main Street. There was a language barrier with the 9-1-1 caller and no other information was available. Current weather conditions are rapidly declining with a likely storm approaching.

On Arrival

You are the first-arriving officer. You see no fire showing on the front of the building that has mercantile on the first floor and residential on the second floor. While conducting a 360-degree survey, you see smoke showing from the first floor on the Charlie side. The structure on the Charlie side is three stories. Access to the Charlie side is available only through an alley.

Water Supply

One blue 1,500 gpm (6,000 L/min) hydrant is located 200 feet away on Main Street.

One blue 1,500 gpm (6,000 L/min) hydrant is located 500 feet away on Main Street.

One blue 1,500 gpm (6,000 L/min) hydrant is located 400 feet away on Pine Drive on the Charlie side.

Weather

- Temperature — 45° F (7°C)
- Wind direction — West
- Wind speed — 25 mph (40 km/h)
- Humidity — 40%

Alpha Side

Bravo Side

364 Chapter 9 • Special Hazards Scenarios

Charlie Side

Delta Side

1. Size-Up

Based on the dispatch information and your visual observation of the scene on arrival, what are the facts and probabilities for this scenario?

Facts

- Three stories
- Type III construction
- Mixed-use occupancy
- Masonry exterior wall construction
- Common wall with Bravo exposure
- Smoke showing from a door at the Charlie-Delta corner
- Exterior stairwell impacted by smoke
- Large antenna on roof
- Major potential
- Gas and electrical utilities are on the Charlie side

Probabilities

- Occupied at the time of the incident
- Multiple tenant improvements
- Wood frame (dimensional lumber) interior construction; major fire load
- Basement
- Not sprinklered
- Potential for large fire flow

2. Incident Priorities

What are your considerations for addressing the incident priorities of life safety, incident stabilization, and property conservation?

Life Safety

- Responder safety — primary and backup hoselines
- Primary search and rescue starting in the area of the fire
- Primary search and rescue above the fire
- Hoseline above the fire and on the fire floor
- Ventilation
- Extinguish fire

Incident Stabilization

- Locate the fire and identify what is burning
- Hoseline above the fire and the fire floor
- Coordinated ventilation and fire attack
- Extinguish the fire
- Perform overhaul

Property Conservation

- Contain and extinguish the fire

- Salvage operations
- Control flow paths
- Ventilation
- Perform Overhaul

3. Operational Strategy

What are your considerations when selecting your operational strategy?

- Offensive strategy
- Responder and occupant safety
- Size, construction, and age of the structure
- No sprinkler system
- Exposures within the fire building
- Bravo exposure
- Potential for a basement

4. Command Options

What are your considerations when selecting a Command option: Investigation, Fast Attack, or Command?

- Fast attack
- Smoke showing on the Charlie-Delta corner
- Potential for rescue
- Potential for a heavy fire load
- Resource needs

5. Tactics

What are your considerations, if any, for the tactics used to accomplish your incident priorities?

Rescue

- Probability of being occupied based on type of occupancy and time of day
- Location of the fire
- Building size and number of stories
- No sprinkler system
- Resources needs

Exposure

- Likely wood frame interior construction – internal flame spread
- Masonry exterior
- No sprinkler system
- Tenant improvements and remodels may have created void spaces and chases

Confinement

- Likely wood frame interior construction; internal flame spread

- No sprinkler system
- Tenant improvements and remodels may have created void spaces and chases
- Winds at 25 mph can potentially lead to a wind-driven fire
- Building access
- Hoseline deployment considerations

Extinguishment

- Based on building construction and occupancies, it should be an ordinary combustibles-fueled fire

Overhaul

- Tenant improvements and remodels may have created void spaces and chases
- May have tongue and groove and/or lath and plaster; resources

Ventilation

- Building size and internal configuration
- Winds at 25 mph can potentially lead to a wind-driven fire

Salvage

- Extinguish the fire to eliminate the cause of property damage
- Close doors to compartmentalize the smoke/fire

6. Unusual Hazards or Conditions

What are any unusual hazards or conditions that must be addressed? How will you address those hazards and/or conditions?

- Size, construction, and age of the structure
- No sprinkler system
- Masonry exterior walls, creating a possible collapse hazard
- Antenna on the roof

Notes

Scenario 5

On Sunday, April 15, at 4 p.m., dispatch reports an automatic fire alarm was activated and there have been multiple calls reporting smoke on the 14th floor of a 16-story residential building located at 2178 Keith Court. Current weather conditions are clear skies.

On Arrival

You are the first-arriving officer. You see light smoke coming from a closed window on the Alpha side on the 14th floor of the structure. You hear fire alarms that were activated. The parking lot is full of cars.

Water Supply

One blue 1,500 gpm (6,000 L/min) hydrant is located 300 feet away on Keith Court.

One blue 1,500 gpm (6,000 L/min) hydrant is located 600 feet away on Keith Court.

One blue 1,500 gpm (6,000 L/min) hydrant is located 900 feet away on Keith Court.

Weather

- Temperature — 75° F (23°C)
- Wind direction — (Blowing toward A side)
- Wind speed — 25 mph (40 km/h)
- Humidity — 60%

Alpha Side

Bravo Side

1. Size-Up
Based on the dispatch information and your visual observation of the scene on arrival, what are the facts and probabilities for this scenario?

Facts

- Type I or II construction
- Smoke showing from an upper floor
- Building occupied due to time of day and cars parked in the garage/parking lot
- Sufficient water supply in the area
- Building supplied with a standpipe system

Probabilities

- Sprinkler system activation in fire area
- Building fire alarm system notifying occupants of a fire in the building
- Individuals are evacuating the building
- Stairwells occupied with individuals attempting to evacuate
- Elevators have recalled to the first floor

Charlie Side   **Delta Side**

2. Incident Priorities

What are your considerations for addressing the incident priorities of life safety, incident stabilization, and property conservation?

Life Safety

- Shelter in place individuals on lower floors
- Conduct a primary search on the fire floor and the floor above
- Identify an evacuation stairwell(s)
- Control the fire alarm panel to ensure the building is not circulating air on the fire floor

Incident Stabilization

- Deploy adequate resources to put in place a hoseline and a backup line
- Secure a water supply
- Check for extension

Property Conservation

- Remove or cover items on the fire floor and in areas below the fire floor
- If the sprinkler system has extinguished the fire, shut down the system

3. Operational Strategy

What are your considerations when selecting your operational strategy?

- High probability of survivable victims
- Resource availability/needs relating to the incident priorities

4. Command Options

What are your considerations when selecting a Command option: Investigation, Fast Attack, or Command?

- Strong Command presence requires a stationary Command post for this type of incident
- Multiple units responding
- A large building
- Potentially more Command and General Staff responsibilities will be needed
- High volume of potential victims
- High-profile incident

5. Tactics

What are your considerations, if any, for the tactics used to accomplish your incident priorities?

Rescue

- Sufficient amount of personnel needed to conduct primary search on fire floor and floor above
- Sufficient amount of personnel needed for sheltering in place and checking all stairwells
- Sufficient amount of personnel needed for occupant accountability outside of the structure

Exposure

- Internal exposures are the primary concern; maintain door control measures

Confinement

- Locate, isolate and extinguish the fire
- Gain control of the building's HVAC system

Extinguishment

- Initial response should focus only on extinguishment and search efforts

Overhaul

- Assign personnel to the floors directly below the fire and on the fire floor; consider implementing overhaul measures early in the incident if sufficient personnel are available and it will not comprise firefighter safety, fire attack, and search efforts

Ventilation

- Ensure stairwells are pressurized

Salvage

- Sprinkler control

- Quickly incorporate building management with salvage
- Control excess water runoff

6. Unusual Hazards or Conditions

What are any unusual hazards or conditions that must be addressed? How will you address those hazards and/or conditions?

- High fuel load
- Reliance on the building fire protection systems to assist with fire extinguishment
- Lot of occupants, making it difficult in ensuring adequate accountability
- Longer reflex time
- Need for additional personnel—two crews needed for every crew working
- Limited use of aerial ladder
- Lobby control is important

Notes

Courtesy of Chris Mickel, New Orleans (LA) F.D. Photo Unit.

Appendices

Appendix A
Chapter and Page Correlation to FESHE Requirements

FESHE Course Outcomes	Chapter References	Page References
1. Discuss fire behavior as it relates to strategies and tactics.	1, 4, 6	11-75, 164-165, 172-179, 258-259
2. Explain the main components of prefire planning and identify steps needed for a prefire plan review.	2	81-129
3. Identify the basics of building construction and how they interrelate to prefire planning and strategy and tactics.	2, 4, 6	102-129, 165, 186-199, 256-258
4. Describe the steps taken during size-up.	3, 4, 5, 6	135-142, 159-202, 211-217, 231-239
5. Examine the significance of fire ground communications.	3, 4, 5	146-148, 168-172, 183, 218-220
6. Identify the roles of the National Incident Management System (NIMS) and Incident Management System (ICS) as it relates to strategy and tactics.	3, 5, 6	142-146, 148-155, 207-221, 227-273
7. Demonstrate the various roles and responsibilities in ICS/NIMS.	3, 5, 6	142-146, 148-155, 207-221, 227-273

Glossary

Courtesy of Chris Mickel, New Orleans (LA) F.D. Photo Unit.

Glossary

A

After Action Reviews — Learning tools used to evaluate a project or incident to identify and encourage organizational and operational strengths and to identify and correct weaknesses.

Asphyxiation — Fatal condition caused by severe oxygen deficiency and an excess of carbon monoxide and/or other gases in the blood.

Autoignition — Initiation of combustion by heat but without a spark or flame (Reproduced with permission from NFPA 921-2011, *Guide for Fire and Explosion Investigations*, Copyright 2011, National Fire Protection Association).

Autoignition Temperature (AIT) — The lowest temperature at which a combustible material ignites in air without a spark or flame (Reproduced with permission from NFPA 921-2011, *Guide for Fire and Explosion Investigations*, Copyright 2011, National Fire Protection Association).

B

Backdraft — Instantaneous explosion or rapid burning of superheated gases that occurs when oxygen is introduced into an oxygen-depleted confined space. The stalled combustion resumes with explosive force; may occur because of inadequate or improper ventilation procedures.

Bowstring Truss — Lightweight truss design noted by the bow shape, or curve, of the top chord.

Buoyant — The tendency or capacity of a liquid or gas to remain afloat or rise.

C

Carbon-Based Fuels — Fuels in which the energy of combustion derives principally from carbon; includes materials such as wood, cotton, coal, or petroleum.

Carbon Dioxide (CO_2) — Colorless, odorless, heavier than air gas that neither supports combustion nor burns; used in portable fire extinguishers as an extinguishing agent to extinguish Class B or C fires by smothering or displacing the oxygen. CO_2 is a waste product of aerobic metabolism.

Carbon Monoxide (CO) — Colorless, odorless, dangerous gas (both toxic and flammable) formed by the incomplete combustion of carbon. It combines with hemoglobin more than 200 times faster than oxygen does, decreasing the blood's ability to carry oxygen.

Ceiling Jet — Horizontal movement of a layer of hot gases and combustion by-products from the center point of the plume, when a horizontal surface such as a ceiling redirects the vertical development of the rising plume.

Collapse Zone — Area beneath a wall in which the wall is likely to land if it loses structural integrity.

Combustion — A chemical process of oxidation that occurs at a rate fast enough to produce heat and usually light in the form of either a glow or flame. (Reproduced with permission from NFPA 921-2011, *Guide for Fire and Explosion Investigations*, Copyright 2011, National Fire Protection Association).

Combustion Zone — Area surrounding a heat source in which there is sufficient air available to feed a fire.

Compartmentation — The way that the arrangement of compartments creates or does not create a series of barriers designed to keep flames, smoke, and heat from spreading from one room or floor to another.

Concealed Space — Structural void that is not readily visible from a living/working space within a building, such as areas between walls or partitions, ceilings and roofs, and floors and basement ceilings through which fire may spread undetected; also includes soffits and other enclosed vertical or horizontal shafts through which fire may spread.

Conduction — Physical flow or transfer of heat energy from one body to another, through direct contact or an intervening medium, from the point where the heat is produced to another location, or from a region of high temperature to a region of low temperature.

Confinement — Fire fighting operations required to prevent fire from extending from the area of origin to uninvolved areas or structures.

Convection — Transfer of heat by the movement of heated fluids or gases, usually in an upward direction.

Crew Resource Management (CRM) — Training procedures intended to improve communications, leadership, and decision making to reduce human error.

D

Defensive Strategy — Overall plan for incident control established by the incident commander that involves protection of exposures, as opposed to aggressive, offensive intervention.

Direct Attack — (1) In structural fire fighting, an attack method that involves the discharge of water or a foam stream directly onto the burning fuel.

Draft Curtains — Noncombustible barriers or dividers hung from the ceiling in large open areas that are designed to minimize the mushrooming effect of heat and smoke and impede the flow of heat. *Also known as* Curtain Boards and Draft Stops.

E

End-of-Service-Time Indicator (ESTI) — Warning device that alerts the user that the respiratory protection equipment is about to reach its limit and that it is time to exit the contaminated atmosphere; its alarm may be audible, tactile, visual, or any combination thereof. *Also known as* Low Air Alarm.

Endothermic Reaction — Chemical reaction in which a substance absorbs heat.

Energy — Capacity to perform work; occurs when a force is applied to an object over a distance, or when a substance undergoes a chemical, biological, or physical transformation.

Entrain — To draw in and transport solid particles or gases by the flow of a fluid.

Entrainment — The drawing in and transporting of solid particles or gases by the flow of a fluid.

Exhaust — In terms of ventilation, the location where hot gases and the products of construction are leaving a structure.

Exothermic Reaction — Chemical reaction between two or more materials that changes the materials and produces heat.

Exposure — Structure surfaces or separate parts of the fireground to which a fire or products of combustion could spread.

Exposure Fire — A fire ignited in fuel packages or buildings that are remote from the initial fuel package or building of origin.

F

Fire — A rapid oxidation process, which is a gas phase chemical reaction resulting in the evolution of light and heat in varying intensities.

Fireground — Area around a fire and occupied by fire fighting forces.

Fire Point — Temperature at which a liquid fuel produces sufficient vapors to support combustion once the fuel ignites. The fire point is usually a few degrees above the flash point.

Fire Stop — Solid materials, such as wood blocks, used to prevent or limit the vertical and horizontal spread of fire and the products of combustion; installed in hollow walls or floors, above false ceilings, in penetrations for plumbing or electrical installations, in penetrations of a fire-rated assembly, or in cocklofts and crawl spaces.

Fire Tetrahedron — Model of the four elements/conditions required to have a fire. The four sides of the tetrahedron represent fuel, heat, oxygen, and self-sustaining chemical chain reaction.

Fire Triangle — Plane geometric model of an equilateral triangle that is used to explain the conditions/elements necessary for combustion. The sides of the triangle represent heat, oxygen, and fuel.

Fire Watch — Usually refers to someone who has the responsibility to tour a building or facility on at least an hourly basis, look for actual or potential fire emergency conditions, and send an appropriate warning if such conditions are found.

First Alarm Assignment — Initial fire department response to a report of an emergency; the assignment is determined by the local authority based on available resources, the type of occupancy, and the hazard to life and property.

Flammable (Explosive) Range — Range between the upper flammable limit and lower flammable limit in which a substance can ignite.

Flashover — Rapid transition from the growth stage to the fully developed stage.

Flash Point — Minimum temperature at which a liquid gives off enough vapors to form an ignitable mixture with air near the surface of the liquid.

Flow Path — The space between at least one intake and one exhaust outlet. The difference in pressure determines the direction of the flow of gases through this space. Heat and smoke in a high pressure area will flow toward areas of lower pressure.

Freelance — To operate independently of the incident commander's command and control.

Free Radical — Electrically charged, highly reactive parts of molecules released during combustion reactions.

Fuel — A material that will maintain combustion under specified environmental conditions (Reproduced with permission from NFPA 921-2011, *Guide for Fire and Explosion Investigations*, Copyright 2011, National Fire Protection Association).

Fuel-Limited — Fire with adequate oxygen in which the heat release rate and growth rate are determined by the characteristics of the fuel, such as quantity and geometry. *Also known as* Fuel-controlled (Reproduced with permission from NFPA 921-2011, *Guide for Fire and Explosion Investigations*, Copyright 2011, National Fire Protection Association).

Fuel Load — The total quantity of combustible contents of a building, space, or fire area, including interior finish and trim, expressed in heat units of the equivalent weight in wood.

G

Glue-Laminated Beam — (1) Wooden structural member composed of many relatively short pieces of lumber glued and laminated together under pressure to form a long, extremely strong beam. (2) Term used to describe wood members produced by joining small, flat strips of wood together with glue. *Also known as* Glued-Laminated Beam or Glulam Beam.

Green Roof — Roof of a building that is partially or completely covered with vegetation and a growing medium, planted over waterproof roofing elements. Term can also indicate the presence of green design technology including photovoltaic systems and reflective surfaces.

H

Heat — Form of energy associated with the motion of atoms or molecules in solids or liquids that is transferred from one body to another as a result of a temperature difference between the bodies, such as from the sun to the earth. To signify its intensity, it is measured in degrees of temperature.

Heat Flux — The measure of the rate of heat transfer to or from a surface, typically expressed in kilowatts per square meter (kW/m^2).

Heat of Combustion — Total amount of thermal energy (heat) that could be generated by the combustion (oxidation) reaction if a fuel were completely burned. The heat of combustion is typically measured in kilojoules per gram (kJ/g) or megajoules per kilogram (MJ/kg).

Heat Release Rate — Total amount of heat released per unit time. The heat release rate is typically measured in kilowatts (kW) or Megawatts (MW) of output.

Horizontal Smoke Spread — Tendency of heat, smoke, and other products of combustion to rise until they encounter a horizontal obstruction. At this point they will spread laterally (ceiling jet) until they encounter vertical obstructions and begin to bank downward (hot gas layer development).

Hydraulic Ventilation — Ventilation accomplished by using a spray stream to draw the smoke from a compartment through an exterior opening.

Hydrogen Cyanide (HCN) — Colorless, toxic, and flammable liquid until it reaches 79°F (26°C). Above that temperature, it becomes a gas with a faint odor similar to bitter almonds; produced by the combustion of nitrogen-bearing substances.

Hydrocarbon Fuel — Petroleum-based organic compound that contains only hydrogen and carbon; may also be used to describe those materials in a fuel load which were created using hydrocarbons such as plastics or synthetic fabrics.

I

Ignition — The process of initiating self-sustained combustion. (NFPA 921)

Ignition Source — Mechanism or initial energy source employed to initiate combustion, such as a spark that provides a means for the initiation of self-sustained combustion.

Immediately Dangerous to Life and Health (IDLH) — Description of any atmosphere that poses an immediate hazard to life or produces immediate irreversible, debilitating effects on health; represents concentrations above which respiratory protection should be required. Expressed in parts per million (ppm) or milligrams per cubic meter (mg/m³); companion measurement to the permissible exposure limit (PEL).

Incident Action Plan (IAP) — Written or unwritten plan for the disposition of an incident; contains the overall strategic goals, tactical objectives, and support requirements for a given operational period during an incident. All incidents require an action plan. On relatively small incidents, the IAP is usually not in writing; on larger, more complex incidents, a written IAP is created for each operational period and disseminated to all units assigned to the incident. When written, the plan may have a number of forms as attachments. *Also known as* Building Emergency Action Plan.

Incomplete Combustion — Result of inefficient combustion of a fuel; the less efficient the combustion, the more products of combustion are produced rather than burned during the combustion process.

Intake — In terms of ventilation, the location where air is being entrained toward a fire.

International Association of Fire Chiefs (IAFC) — Professional organization that provides leadership to career and volunteer chiefs, chief fire officers, and managers of emergency service organizations throughout the international community through vision, information, education, representation, and services to enhance their professionalism and capabilities.

Isolated Flames — Flames in the hot gas layer that indicate the gas layer is within its flammable range and has begun to ignite; often observed immediately before a flashover.

J

Jargon — The specialized or technical language of a trade, profession, or similar group.

Joule (J) — Unit of work or energy in the International System of Units (SI); the energy (or work) when a unit force (1 newton) moves a body through a unit distance (1 meter). Joules are defined in terms of mechanical energy. In terms of thermal energy, joules refer to the amount of additional heat needed to raise the temperature of a substance, such as the 4.2 Joules needed to raise the temperature of 1 gram of water 1 degree Celsius. Takes the place of calorie for heat measurement (1 calorie = 4.19 J).

K

Kinetic Energy — Energy possessed by a moving object because of its motion.

L

Laminar Flow — Movement of a liquid or gas at a low rate of speed and in a predictable direction.

Life Safety — Refers to the joint consideration of the life and physical well-being of individuals, both civilians and firefighters.

Life Safety, Incident Stabilization, and Property Conservation (LIP) — Three priorities at an incident, in order of importance.

Load-Bearing Wall — Wall that supports itself, the weight of the roof, and/or other internal structural framing components, such as the floor beams and trusses above it; used for structural support. *Also known as* Bearing Wall.

Lower Explosive (Flammable) Limit (LEL) — Lower limit at which a flammable gas or vapor will ignite and support combustion; below this limit the gas or vapor is too *lean* or *thin* to burn (lacks the proper quantity of fuel). *Also known as* Lower Flammable Limit (LFL).

M

Matter — Anything that occupies space and has mass.

Mayday — Internationally recognized distress signal.

Means of Egress — (1) Safe, continuous path of travel from any point in a structure to a public way. Composed of three parts: exit access, exit, and exit discharge. (2) Continuous and unobstructed way of exit travel from any point in a building or structure to a public way, consisting of three separate and distinct parts: exit access, exit, and exit discharge. (**Source:** NFPA 101, *Life Safety Code®*).

Mechanical Ventilation — Any means other than natural ventilation; may involve the use of fans, blowers, smoke ejectors, and fire streams.

Medical Surveillance — Rehabilitation function during an incident intended to monitor responders' vital signs and incident-stress levels.

Metal-Clad Door — Door with a metal exterior; may be flush type or panel type. *Also known as* Kalamein Door.

Miscible — Materials that are capable of being mixed in all proportions.

Mobile Data Computer (MDC) — Portable computer that, in addition to functioning as a Mobile Data Terminal, has programs that enhance the ability of responders to function at incident scenes.

N

National Incident Management System - Incident Command System (NIMS-ICS) — The U.S. mandated incident management system that defines the roles, responsibilities, and standard operating procedures used to manage emergency operations; creates a unified incident response structure for federal, state, and local governments.

Natural Ventilation — Techniques that use the wind, convection currents, and other natural phenomena to ventilate a structure without the use of fans, blowers, smoke ejectors, or other mechanical devices.

Negative-Pressure Ventilation (NPV) — Technique using smoke ejectors to develop artificial air flow and to pull smoke out of a structure. Smoke ejectors are placed in windows, doors, or roof vent holes to pull the smoke, heat, and gases from inside the building and eject them to the exterior.

Neutral Plane — Level at a compartment opening where there is an equal difference in pressure exerted by expansion and buoyancy of hot smoke flowing out of the opening and the inward pressure of cooler, ambient temperature air flowing in through the opening.

Noncombustible — Incapable of supporting combustion under normal circumstances.

Nonload-Bearing Wall — Wall, usually interior, that supports only its own weight. These walls can be breached or removed without compromising the structural integrity of the building. Also known as Nonbearing Wall.

O

Occupant Survival Profile — Type of incident size-up that evaluates whether an incident should be treated as a rescue or recovery.

Offensive Strategy — (1) In wildland fire fighting, a direct attack on the fire perimeter by crews, engines, or aircraft, or an aggressive indirect attack such as backfiring. (2) Overall plan for incident control established by the incident commander (IC) in which responders take aggressive, direct action on the material, container, or process equipment involved in an incident.

Open Burning — Description of a fire burning in the open with no restrictions to its oxygen supply.

Oriented Strand Board (OSB) — Wooden structural panel formed by gluing and compressing wood strands together under pressure. This material has replaced plywood and planking in the majority of construction applications. Roof decks, walls, and subfloors are all commonly made of OSB.

Oxidation — Chemical process that occurs when a substance combines with an oxidizer such as oxygen in the air; a common example is the formation of rust on metal.

Oxidizer — Any material that readily yields oxygen or other oxidizing gas, or that readily reacts to promote or initiate combustion of combustible materials. (Reproduced with permission from NFPA 400-2010, *Hazardous Materials Code*, Copyright 2010, National Fire Protection Association).

P

Personnel Accountability Report (PAR) — Roll call of all units (crews, teams, groups, companies, sectors) assigned to an incident. The supervisor of each unit reports the status of the personnel within the unit at that time, usually by radio. A PAR may be required by standard operating procedures at specific intervals during an incident, or may be requested at any time by the incident commander or the incident safety officer.

Piloted Ignition — Moment when a mixture of fuel and oxygen encounters an external heat (ignition) source with sufficient heat or thermal energy to start the combustion reaction.

Pitot Gauge — Instrument that is inserted into a flowing fluid (such as a stream of water) to measure the velocity pressure of the stream; commonly used to measure flow; functions by converting the velocity energy to pressure energy that can then be measured by a pressure gauge. The gauge reads in units of pounds per square inch (psi) or kilopascals (kPa). *Also known as* Pitot Tube.

Plan of Operation — Clearly identified strategic goal and the tactical objectives necessary to achieve that goal; includes assignments, authority, responsibility, and safety considerations.

Plenum — Open space or air duct above a drop ceiling that is part of the air distribution system.

Plot Plan — Architectural drawing showing the overall project layout of building areas, driveways, fences, fire hydrants, and landscape features for a given plot of land; view is from directly above.

Polar Solvents — Flammable liquids that have an attraction to water, much like a positive magnetic pole attracts a negative pole; examples include alcohols, esters, ketones, amines, and lacquers.

Positive-Pressure Attack (PPA) — The use and application of high volume ventilation fans before fire suppression which are intended to force heat and smoke toward desired exhaust openings.

Positive-Pressure Ventilation (PPV) — Method of ventilating a room or structure by mechanically blowing fresh air through an inlet opening into the space in sufficient volume to create a slight positive pressure within and thereby forcing the contaminated atmosphere out the exit opening.

Potential Energy — Stored energy possessed by an object that can be released in the future to perform work once released.

Power — Amount of energy delivered over a given period of time.

Preincident Planning — Act of preparing to manage an incident at a particular location or a particular type of incident before an incident occurs. *Also known as* Prefire Inspection, Prefire Planning, Preincident Inspection, Preincident Survey, or Preplanning.

Preincident Survey — Assessment of a facility or location made before an emergency occurs, in order to prepare for an appropriate emergency response. *Also known as* Preplan.

Pressure — Force per unit area exerted by a liquid or gas measured in pounds per square inch (psi) or kilopascals (kPa).

Primary Search — Rapid but thorough search to determine the location of victims; performed either before or during fire suppression operations. May be conducted with or without a charged hoseline, depending on local policy.

Products of Combustion — Materials produced and released during burning.

Protected Steel — Steel structural members that are covered with either spray-on fire proofing (an insulating barrier) or fully encased in an Underwriters Laboratories (UL) tested and approved system.

Pyrolysis — The chemical decomposition of a solid material by heating. Pyrolysis precedes combustion of a solid fuel.

Purlin — Horizontal member between trusses that support the roof.

R

Radiation — Transmission or transfer of heat energy from one body to another body at a lower temperature through intervening space by electromagnetic waves.

Rapid Intervention Crew or Team (RIC/RIT) — Two or more firefighters designated to perform firefighter rescue; they are stationed outside the hazard and must be standing by throughout the incident.

Reducing Agent — Fuel that is being oxidized or burned during combustion.

Rehabilitation — (1) Activities necessary to repair environmental damage or disturbance caused by wildland fire or the fire-suppression activity. (2) Allowing firefighters or rescuers to rest, rehydrate, and recover during an incident; also refers to a station at an incident where personnel can rest, rehydrate, and recover. *Also known as* Rehab.

Rescue — Saving a life from fire or accident; removing a victim from an untenable or unhealthy atmosphere.

Risk-Based Response — Method using hazard and risk assessment to determine an appropriate mitigation effort based on the circumstances of the incident.

Risk-Benefit Analysis — Comparison between the known hazards and potential benefits of any operation; used to determine the feasibility and parameters of the operation.

Rollover — Condition in which the unburned fire gases that have accumulated at the top of a compartment ignite and flames propagate through the hot gas layer or across the ceiling.

S

Salvage — Methods and operating procedures by which firefighters attempt to save property and reduce further damage from water, smoke, heat, and exposure during or immediately after a fire; may be accomplished by removing property from a fire area, by covering it, or by other means.

Search — Techniques that allow the rescuer to identify the location of victims and to determine access to those victims in order to remove them to a safe area.

Search and Rescue Operation — Emergency incident operation consisting of an organized search for the occupants of a structure or for those lost in the outdoors, and the rescue of those in need.

Secondary Search — Slow, thorough search to ensure that no occupants were overlooked during the primary search; conducted after the fire is under control by personnel who did not conduct the primary search.

Self-Heating — The result of exothermic reactions, occurring spontaneously in some materials under certain conditions, whereby heat is generated at a rate sufficient to raise the temperature of the material (Reproduced with permission from NFPA 921-2011, *Guide for Fire and Explosion Investigations*, Copyright 2011, National Fire Protection Association).

Shelter in Place — Having occupants remain in a structure or vehicle in order to provide protection from a rapidly approaching hazard, such as a fire or hazardous gas cloud. *Opposite of* evacuation. *Also known as* Protection-in-Place, Sheltering, *and* Taking Refuge.

Shielded Fire — Fire that is located in a remote part of the structure or hidden from view by objects in the compartment.

Situational Awareness — Perception of the surrounding environment and the ability to anticipate future events.

Size-Up — Ongoing evaluation of influential factors at the scene of an incident.

Smoke Explosion — Form of fire gas ignition; the ignition of accumulated flammable products of combustion and air that are within their flammable range.

Soffit — Lower horizontal surface such as the undersurface of eaves or cornices.

Solubility — Degree to which a solid, liquid, or gas dissolves in a solvent (usually water).

Span of Control — Maximum number of subordinates that that one individual can effectively supervise; ranges from three to seven individuals or functions, with five generally established as optimum.

Specific Gravity — Mass (weight) of a substance compared to the weight of an equal volume of water at a given temperature. A specific gravity less than 1 indicates a substance lighter than water; a specific gravity greater than 1 indicates a substance heavier than water.

Spontaneous Ignition — Initiation of combustion of a material by an internal chemical or biological reaction that has produced sufficient heat to ignite the material (Reproduced with permission from NFPA 921-2011, *Guide for Fire and Explosion Investigations*, Copyright 2011, National Fire Protection Association).

Stack Effect — (1) Tendency of any vertical shaft within a tall building to act as a chimney or "smokestack", by channeling heat, smoke, and other products of combustion upward due to convection. *Also known as* Stack Action. (2) Phenomenon of a strong air draft moving from ground level to the roof level of a building; affected by building height, configuration, and temperature differences between inside and outside air. *Also known as* Chimney Effect.

Structural Collapse — Structural failure of a building or any portion of it resulting from a fire, snow, wind, water, or damage from other forces.

Surface-To-Mass Ratio — Ratio of the surface area of the fuel to the mass of the fuel.

T

Tactical Ventilation — Planned, systematic, and coordinated removal of heated air, smoke, gases or other airborne contaminants from a structure, replacing them with cooler and/or fresher air to meet the incident priorities of life safety, incident stabilization, and property conservation.

Temperature — Measure of the average kinetic energy of the particles in a sample of matter, expressed in terms of units or degrees designated on a standard scale.

Thermal Conductivity — The propensity of a material to conduct heat within its volume. Measured in energy transfer over distance per degree of temperature.

Thermal Energy — Kinetic energy associated with the random motions of the molecules of a material or object; often used interchangeably with the terms *heat* and *heat energy*.

Thermal Equilibrium — The point at which two regions that are in thermal contact no longer transfer heat between them because they have reached the same temperature.

Thermal Imager — Electronic device that forms images using infrared radiation. *Also known as* Thermal Imaging Camera.

Thermal Layering — Outcome of combustion in a confined space in which gases tend to form into layers, according to temperature, gas density, and pressure with the hottest gases found at the ceiling and the coolest gases at floor level.

Thermoplastic — Plastic that softens with an increase of temperature and hardens with a decrease of temperature but does not undergo any chemical change. Synthetic material made from the polymerization of organic compounds that become soft when heated and hard when cooled.

Turbulent Flow — Movement of a liquid or gas at a high rate of speed and no definite pattern to the movement of the particles.

U

Upper Explosive (Flammable) Limit (UEL) — Upper limit at which a flammable gas or vapor will ignite; above this limit the gas or vapor is too *rich* to burn (lacks the proper quantity of oxygen). *Also known as* Upper Flammable Limit (UFL).

V

Vapor Density — Weight of pure vapor or gas compared to the weight of an equal volume of dry air at the same temperature and pressure. A vapor density less than 1 indicates a vapor lighter than air; a vapor density greater than 1 indicates a vapor heavier than air.

Vapor Pressure — The pressure at which a vapor is in equilibrium with its liquid phase at a given temperature; liquids that have a greater tendency to evaporate have higher vapor pressures at a given temperature.

Vaporization — Physical process that changes a liquid into a gaseous state; the rate of vaporization depends on the substance involved, heat, pressure, and exposed surface area.

Velocity — Rate of motion in a given direction; measured in units of length per unit time, such as feet per second (meters per second) and miles per hour (kilometers per hour).

Ventilation-Controlled — Fire with limited ventilation in which the heat release rate or growth is limited by the amount of oxygen available to the fire. (NFPA 921).

Ventilation-Limited — Fire with limited ventilation in which the heat release rate or growth is limited by the amount of oxygen available to the fire. *Also known as* Ventilation-controlled (Reproduced with permission from NFPA 921-2011, *Guide for Fire and Explosion Investigations*, Copyright 2011, National Fire Protection Association).

W

Watt (W) — The SI unit of power or rate of work equal to 1 joule per second (J/s).

Index

Courtesy of Ron Jeffers, Union City, NJ.

Index

A

AAR (After Action Review), 137
Abandoned building
    fire development in, 75
    life hazard to occupants, 198–200
Access to buildings
    blocked streets, 88
    size-up considerations, 186, 188
    street access size-up considerations, 187–188
Accountability. *See* Personnel accountability
Acronyms
    CAAN, 179
    CAN, 179
    CARA, 179
After Action Review (AAR), 137
AHJ. *See* Authority having jurisdiction (AHJ)
Air conditioning systems. *See* Heating, ventilation, and air conditioning systems (HVAC)
Air flow
    backdraft indicators, 55
    structure fire development, 59–61
Air management, size-up considerations, 192–193
Air monitoring research, 250
AIT. *See* Autoignition temperature (AIT)
Alarm activated doors, 95–96
Alterations, special hazards of, 190
Apartment building fire, 229
Apparatus
    Level I staging, 153
    Level II staging, 153
    pumpers used to boost supply line pressure, 99
    staging protocols, 153
Arched roofs, 127
Arcing, 25
Area of buildings, size-up considerations, 186
Arrival condition indicators, 200–202
    time of day, 200
    visual indicators, 202
    weather, 201–202
Arrival conditions for size-up
    arrival report, 168–169
    day of week, 164
    incident considerations, 167–172
    time of day, 164, 200
    visual indicators, 202
    weather, 201–202
Arson, 249, 250
Asphyxiation, 19
Assembly Occupancy
    life hazard to occupants, 197
    restaurants, 129
Atrium vents, 94
Atriums, 197
Attic, 115–116, 121
Authority having jurisdiction (AHJ)
    basement fires, 109
    overhaul, 249
    size-up considerations, 164, 168, 169
Autoignition, 16
Autoignition temperature (AIT)
    CO fuel gas, 52
    defined, 17
Automatic aid, 101–102

B

Backdraft, 50, 54–55
Barriers
    crew resource management (CRM), 184
    size-up considerations, 186
Basements
    defined, 114
    floor collapse risks, 108–109
    smoke fire hazards, 114
    stairwell flow of smoke, 114
    townhouse fire, 115
    ventilation considerations, 258, 269
Bearing wall, 104
Big box store fuel load, 257
Blowtorch effect, 254
Bowstring truss, 74, 127
Branch level of NIMS-ICS, 144
British thermal unit (Btu), 14
Btu (British thermal unit), 14
Building characteristics for size-up
    assembly occupancy, 197
    business occupancy, 194
    educational occupancy, 197–198
    egress, 188–189
    evaluation factors, 186–187
    industrial and storage occupancy, 196
    institutional occupancy, 196–197
    life hazard to occupants, 191–193
    mercantile occupancy, 194–195
    metal security doors and bars, 189–190
    mixed occupancy, 195
    residence occupancy, 193–194
    special hazards, 190–191
    street access, 187–188
    structure access, 188
    unoccupied, vacant, or abandoned structures, 198–200
Building codes
    code enforcement inspections, 82
    floor framing requirements, 108
    floor structural requirements, 108
    purpose of, 102
Building conditions noted in preincident surveys, 91–92
Building construction, 102–129
    building codes, 102
    Canadian construction, 110–111
    interior building arrangement, 111–125
    lightweight/engineered construction, 103
    occupancy types, 128–129
    reaction to fire. *See* Building construction and reaction to fire
    roof types and hazards, 125–128
    size-up considerations, 163, 165–166
    United States construction, 103–110
    ventilation considerations, 256–258
Building construction and reaction to fire, 63–75
    buildings under construction, 74–75
    changing conditions, 65
    collapse, 84
    compartment volume and ceiling height, 72
    compartmentation, 69–70
    dangerous building conditions, 63–64
    demolition hazards, 74–75
    elapsed time of structural integrity, 64–65
    fire resistance ratings, 64
    fuel load, 65–69

combustible roof materials, 68–69
contents, 66–67
defined, 65
determining, 66
exterior wall coverings, 68
furnishings and finishes, 68
location of the fire, 67
ventilation, 68
wood-frame buildings, 66
lightweight trusses and joists, 73–74
modern vs. legacy construction, 70–72
renovation hazards, 74–75
thermal properties, 72–73
Building Emergency Action Plan, 136
Building exterior, preincident survey information, 87–88
Building indicators
backdraft, 55
flashover, 52
Building interior, preincident survey information, 88–90
Building separation distances, 237
Buoyant liquids or gases, 22
Business Occupancy Classification
life hazard to occupants, 194
office building, 129

C
CAAN acronym, 179
CAN acronym, 179
Canada
construction types, 110–111
National Building Code of Canada (NBC), 110–111
Novoclimat standard, 111
Cancer of firefighters, 250
CARA acronym, 179
Carbon dioxide (CO_2)
defined, 21
health effects, 21
as product of combustion, 19, 21
Carbon monoxide (CO)
in black smoke, 174
characteristics, 19
defined, 19
health effects, 20
as product of combustion, 19
Carbon particles in black smoke, 174
Carbon-based fuels, 18, 19
Carcinogens in smoke, 250
Cedar Fire, 142
Ceiling height
fire development and, 72
size-up considerations, 186
Ceiling jet, 43
Cellars, 114
Celsius scale of measurement, 14, 23
Chemical energy, 23–24
Chemical flame inhibition. *See also* Extinguishment
defined, 40
extinguishing agents
Class A and B foam, 241
structure fire fighting operations, 63
structure fires, 63
Chemical reaction, self-sustained, 39–40
Chemicals
confinement tactics, 236
as fuel, 31
illegal drug lab, 190
smoke colors indicating, 175
Chimney effect, 269

Churches
life hazard to occupants, 197
open space construction, 72
Class A and B foam, 241
Cleveland vacant building fire, 182–183
CO. *See* Carbon monoxide (CO)
CO_2. *See* Carbon dioxide (CO^2)
Cocklofts, 115–116
Code enforcement inspections, 82
Code of Federal Regulations–Title 24: Housing and Urban Development; Chapter XX, Part 3280, 109
Codes. *See* Building codes
Collapse
brick house floor collapse, 92
combustible material construction, 84
curtain collapse, 120
fast-food restaurant roof, 88
floors, 108–109
lightweight/engineered construction, 103
during a search, 180
structural collapse
collapse zone, 117, 118, 120
contents of the structure, 121
defined, 117
factors determining potential for, 117
floor system collapse times, 123
grain silo, 121
indicators of potential for, 121–122
safety alert for firefighters, 124–125
snow and ice load, 118
Type I construction, 119
Type II construction, 119
Type III construction, 119
Type IV construction, 119
Type V construction, 120
water used to suppress the fire, 121
Collapse zone
defined, 117
equation for calculating, 118
factors determining size of, 120
reasons for establishing, 120
Combustible construction (Canada), 110
Combustion
combustion zone, 45
complete, 40
defined, 11
flaming, 17–18, 38, 40
heat of combustion, 13
incomplete, 18, 19, 40
nonflaming/smoldering, 17, 38
products of
carbon dioxide (CO_2), 19, 21
carbon monoxide (CO), 19–20
carbon-based fuels, 19
defined, 18
hydrocarbon fuels, 19
hydrogen cyanide (HCN), 20–21
smoke, 19
toxic effects, 20
Command organizational level of NIMS-ICS, 144
Command Post, 147
Command staff level of NIMS-ICS, 144
Commercial construction
bowstring truss, 74
fuel load, 71
life hazard to occupants, 192
open floor plans, 71

Communications
    crew resource management (CRM), 183
    Incident Command System (ICS), 147
    Incident Communications Plan, 147
    jargon, 147
    NIMS-ICS, 146-148
    purpose of, 146
    radio communications
        example, 147
        for incident information, 159, 160
        protocols, 148
Compartment fire development, 41-59
    decay stage
        described, 42
        fuel-limited conditions, 58
        ventilation-limited conditions, 59
    fully developed stage
        described, 41-42, 55-56
        fuel-limited conditions, 56
        ventilation-limited conditions, 56-57
    growth stage, 44-55
        backdraft, 55
        combustion zone, 45
        described, 41
        entrainment, 44-45
        flashover, 50, 54
        rapid fire development, 50-55
        thermal layering, 46-48
        transition from incipient stage, 46
        transition to ventilation-limited decay, 50
        ventilation-limited decay, 48-50
    incipient stage
        described, 41, 43-44
        transition to growth stage, 46
Compartmentalized structure, 113
Compartmentation
    defined, 69
    fire development and, 69-70
    open vs. compartmentalized floor plans, 69
    two rooms with a door between them, 70
    volume and ceiling height, 72
Complete combustion, 40
Comprehensive resource management, 151
Compression generating mechanical energy, 25, 26
Computer for size-up information, 159
Concealed spaces
    defined, 116
    fire spread through, 117
    preincident planning knowledge of, 93
    purpose of, 116
Concrete, steel used to strengthen, 104
Conduction, 26-27, 30
Confinement
    defined, 229
    high-rise buildings, 271
    HVAC considerations, 236
    objectives, 237
    options, 229
    tactics, 235-239
Construction. *See* Building construction
Construction types
    size-up considerations, 186
    Type I construction
        characteristics, 104-105
        fire resistance ratings, 64
        structural collapse, 119
    Type II construction
        characteristics, 105
        fire resistance ratings, 64
        structural collapse, 119
    Type III construction
        characteristics, 105-106
        fire resistance ratings, 64
        structural collapse, 119
    Type IV construction
        characteristics, 106-107
        fire resistance ratings, 64
        structural collapse, 119
    Type V construction
        characteristics, 107-109
        fire resistance ratings, 64
        structural collapse, 120
Contents
    fuel load of, 66-67
    furnishings as fuel load, 68, 165
    Mercantile Occupancy, 195
    preincident planning knowledge of, 92-93
    structural collapse and, 121
Contract for outside aid, 102
Convection, 27-28, 30, 61
Crawl spaces, 114
Crew resource management (CRM), 183-184
Crew resources, 146, 193
CRM (crew resource management), 183-184
Curtain boards, 94-95
Curtain collapse, 120
Customary system of measurement, 14, 23
Cutting torch, heat from a chemical reaction, 23

D

Day of the week, size-up considerations, 164
Daycare center occupancy life hazard to occupants, 196
Dead-end corridors, 189
Debris and fire development, 75
Decay stage of fire development
    described, 42
    fuel-limited conditions, 58
    ventilation-limited conditions, 59
Decision-making
    crew resource management (CRM), 183
    incident management, 135-142
    adjusting the plan, 141
    identify and prioritize problems, 139
    Incident Action Plan (IAP), 136
    indecision, result of, 141-142
    monitor results, 141
    Recognition Primed Decision Making (RPDM), 137
    solution determination, 139-140
    solution implementation, 140
    steps, 137
    size-up, 180
Defensive strategy, 214-216
Defensive tactics, 240
Demolition, hazards of, 74-75
Department of Housing and Urban Development (HUD), 109, 110
Direct attack, 241-242
Division level of NIMS-ICS, 144
Doors
    compartmentation, 70
    as means of egress, 188
    metal-clad (Kalamein door), 189-190
    smoke, heat, or alarm activated doors, 95-96
Draft curtains, 94
Draft stops, 94

E

Educational Occupancy
    elementary school, 129
    life hazard to occupants, 197–198
Egress, means of, 188–189
Electrical energy, 24–25
Electrical lines as safety hazard, 170
Electromagnetic wave, 29
Emergency response, preincident surveys, 85–86
End-of-Service-Time Indicator (ESTI), 192
Endothermic reaction, 14
Energy
    Celsius scale of measurement, 14, 23
    defined, 12
    endothermic reactions, 14
    exothermic reactions, 14
    Fahrenheit scale of measurement, 14, 23
    joule (J) unit of measure, 14
    kinetic, 13
    potential, 13
    radiated heat temperature measurements, 29
    thermal. *See* Thermal energy (heat)
    types, 14
    work, 12
Engineered construction, 103
Entrainment, 18, 44–45
ESTI (End-of-Service-Time Indicator), 192
Evacuation, preincident survey information, 90, 91
Exhaust, ventilation, 246
Exhaust flow of structure fire development, 59–61
Exothermic reaction, 14
Explosion
    investigations. *See* NFPA 921-2011, *Guide for Fire and Explosion Investigations*
    lower explosive limit (LEL), 38
    smoke explosion, 63
    upper explosive limit (UEL), 38
Explosive (flammable) range, 38
Exposures
    defined, 234
    examples, 235
    exposure fire, defined, 29
    external control, 239
    fire extension and confinement tactics, 237–238
    internal control, 239
    size-up considerations, 186
    tactics, 234–235
    ventilation considerations, 262
Exterior of buildings, preincident survey information, 87–88
Exterior wall coverings as fuel load, 68
External exposure control, 239
Extinguishment, 239–248
    chemical flame inhibition, 40, 63, 241
    extinguishing agents
        Class A and B foam, 241
        structure fire fighting operations, 63
    factors to consider, 239–240
    fire attack, 241–243
        direct attack, 241–242
        gas cooling, 242–243
        indirect attack, 242
    offensive to defensive attack, 240
    positive-pressure attack, 245–248
    risk vs. benefit, 241
    size-up considerations, 240
    transitional attack, 243–245

F

Factory Mutual (FM), 100
Factory-built homes, 109–110
Facts as size-up consideration, 163–164
Fahrenheit scale of measurement, 14, 23
Fast-attack or mobile command option, 219
Fast-food restaurant roof collapse, 88
Fatalities
    apartment building fire, 229
    line-of-duty death (LODD)
        deaths over time, 211
        improper ventilation LODD, 254–255
        structural collapse incident, 180
        The 10 Rules of Engagement for Structural Fire Fighting, 211
    NIOSH death prevention recommendation, 115, 209
FCSN (Firefighter Cancer Support Network), 250
Field sketch, 86, 87
Files in the mind, 137
Finishes as fuel load, 68, 165
Fire
    defined, 11
    investigations. *See* NFPA 921-2011, *Guide for Fire and Explosion Investigations*
Fire attack, 241–243
    direct attack, 241–242
    gas cooling, 242–243
    indirect attack, 242
Fire behavior
    indicators of, 228, 258–259
    size-up considerations, 164, 165
    ventilation considerations, 258–259
Fire Detail, 253
Fire detection and alarm systems, 95
Fire development. *See* Compartment fire development; Structure fire development
Fire dynamics, 11–75
    building construction reaction to fire, 63–75
        building compartmentation, 69–70
        construction, renovation, and demolition, 74–75
        construction features, 70–74
        construction type and elapsed time of structural integrity, 64–65
        fuel load, 65–69
        size-up considerations, 163
    compartment fire development, 41–59
        decay stage, 42, 58–59
        fully developed stage, 41–42, 55–57
        growth stage, 41, 44–55
        incipient stage, 41, 43–44, 46
    fuel, 31–37
        gases, 32–33
        liquids, 33–35
        solids, 35–37
    oxygen, 37–39
    science of fire, 11–22
        combustion, 17–21
        fire triangle and tetrahedron, 15
        ignition, 15–17
        physical science, 12–14
        pressure differences, 22
    self-sustained chemical reaction, 39–40
    structure fire development, 59–63
        fire fighting operations, 63
        flow path, 59–61
        smoke explosion, 63
        ventilation and wind considerations, 61–63
    thermal energy (heat), 22–30
        defined, 13
        heat release rate vs. temperature, 22–23

heat transfer, 26–30
sources of, 23–26
Fire escapes as means of egress, 189
Fire extinguishing systems, preincident planning knowledge of, 95. *See also* Extinguishment
Fire flow calculations, 99–100
Fire incidents
    basement fires
        floor collapse risks, 108–109
        townhouse, 115
        ventilation considerations, 258, 269
    building built on a hillside, 191
    Cedar Fire, 142
    Cleveland vacant building, 182–183
    garbage, 238
    grain silo, 121
    house fire, 138, 145
    Jersey City (NJ) three-alarm fire, 101
    New Orleans, 154
    Paradise Fire, 142
    separated firefighters at three-story apartment building, 229
    three-story apartment building, 229
    townhouse fire, 115
Fire point, 34, 35
Fire Protection Guide to Hazardous Materials, 39
Fire protection systems
    fire detection and alarm systems, 95
    fixed fire-extinguishing systems, 95
    preincident planning knowledge of, 95–97
    size-up considerations, 186
    smoke, heat, or alarm activated doors, 95–96
    standpipe systems
        operation during demolition, 74
        preincident planning knowledge of, 95
        preincident survey information, 87
        water supply requirements, 96
Fire spread and position of fuel, 36, 37
Fire stop, 106
Fire tetrahedron, 15
Fire triangle, 15
Fire wall, 85
Fire watch, 250
Firefighter Cancer Support Network (FCSN), 250
Firefighter survivability approaches, 181–185
    crew resource management, 183–184
    occupant survival profile, 181–183
    rules of engagement, 184–185
Fireground
    defined, 208
    size-up considerations, 185–202
        arrival condition indicators, 200–202
        building characteristics, 186–191
        life hazard to occupants, 191–200
        procedures, 166–167
Fireproof construction, 104
First alarm assignment, 139
First due or first in area, 83
Flame
    backdraft indicators, 55
    chemical flame inhibition, 40, 63, 241
    color of, 179
    flashover indicators, 52
    intermittent, 47
    isolated, 46, 47
Flaming combustion, 17–18, 38, 40
Flammable (explosive) range, 38
Flammable gases, characteristics, 33
Flash point, 34, 35

Flashover, 50–54
    defined, 50
    elements, 51–52
    growth stage, 54
    indicators, 52–53
    rollover indicator, 51
Flat roofs, 126
Floor joists, failure of, 73–74
Floor plan
    commercial structure open floor plan, 71, 74
    open floor plans
        business occupancy, 194
        commercial construction, 71, 74
        compartmentalized vs., 69
        Mercantile Occupancy, 72
        residence occupancy, 112, 257
        warehouse construction, 72
    open vs. compartmentalized, 69
    preincident survey information, 90
Floors
    brick house floor collapse, 92
    collapse risks, 108–109
    collapse times, 123
    firefighter safety, 124–125
    framing requirements, 108
Flow path
    defined, 46
    forced entry as new flow path, 243
    smoke, 176–177
    structure fire development, 59–61, 63
    thermal layering and, 47
FM (Factory Mutual), 100
Fog stream, 244
Frame construction, 107
Free burn, 42
Free radicals, 39
Freelance, 141–142, 218
Friction generating mechanical energy, 25, 26
Fuel, 31–37
    carbon-based, 18, 19
    chemical content, 31
    defined, 12
    gases, 32–33
    hydrocarbon, 18, 19
    inorganic, 31
    liquids, 33–35
    organic, 31
    reducing agent, 31
    removal, in fire fighting operations, 63
    solids, 35–37
    synthetic materials, 32
    vapor-to-air mixture, 38–39
Fuel load, 65–69
    big box stores, 257
    combustible roof materials, 68–69
    commercial structure, 71
    contents, 66–67
    defined, 65, 92
    determining, 66
    exterior wall coverings, 68
    furnishings and finishes, 68, 165
    location of the fire, 67
    lumber mill and sawdust, 92
    open floor plan office, 194
    plastic, 71
    preincident planning knowledge of, 92–93
    residence construction, 70–71
    siding, 68

ventilation, 68
wood-frame buildings, 66, 74
Fuel loading, 92
Fuel-controlled fire, 41
Fuel-limited decay, 58
Fuel-limited fire, 41, 56
Fully developed stage of fire development
    described, 41–42, 55–56
    fuel-limited conditions, 56
    ventilation-limited conditions, 56–57
Furnishings
    as fuel load, 68, 165
    ventilation considerations, 256–258

G

Garbage fire, 238
Gas cooling, 242–243
Gases
    flammable characteristics, 33
    flammable ranges, 39
    as fuel, 32–33
    vapor, 32
Gasoline, flash point vs. fire point, 34
General search, 232
General staff level of NIMS-ICS, 144
GIS (Global Information System), 87
Global Information System (GIS), 87
Global Positioning System (GPS), 87
Glued-laminated beam, 107
Glue-laminated beam, 107
Glulam beam, 107
GPS (Global Positioning System), 87
Graffiti as indication of occupancy, 164
Grain silo fire, 121
Green roof, 128, 129
Group level of NIMS-ICS, 144, 145
Growth stage of fire development, 44–55
    backdraft, 55
    combustion zone, 45
    described, 41
    entrainment, 44–45
    flashover, 50, 54
    rapid fire development, 50–55
    thermal layering, 46–48
    transition from incipient stage, 46
    transition to ventilation-limited decay, 50
    ventilation-limited decay, 48–50
Guide for Fire and Explosion Investigations. See NFPA 921-2011, *Guide for Fire and Explosion Investigations*
Gusset plates, 103

H

Hazardous materials
    Fire Protection Guide to Hazardous Materials, 39
    Hazardous Materials Code (NFPA 400-2010), 12
    pesticides, 85
    unified command structure for incidents, 148
HCN (hydrogen cyanide), 20–21
Heat
    alarm activated doors, 95–96
    backdraft indicators, 55
    defined, 11
    flashover indicators, 52
    indicators of, 178–179
    self-heating, 23–24
    thermal energy. *See* Thermal energy (heat)
    venting (NFPA 204), 94

Heat flux, 22, 23, 29
Heat of combustion, 13
Heat reflectivity and fire development, 73
Heat release rate (HRR)
    defined, 22
    temperature vs., 22–23, 29
    unconfined burning, 31
    watts, 32
Heat transfer, 26–30
    conduction, 26–27, 30
    convection, 27–28, 30, 61
    interaction among methods of, 30
    radiation, 27, 28–29, 30
Heating, ventilation, and air conditioning systems (HVAC)
    confinement tactics, 236
    plenum, 116
    preincident planning knowledge of, 93
    underfloor air distribution systems, 95
    ventilation considerations, 262
    window-mounted air conditioning units, 93
Heavy-timber construction, 106–107, 110
High-rise buildings
    collapse zone, 119
    fire confinement tactics, 271
    fire ventilation tactics, 269–271
    hazards, 119
    size-up considerations, 186
    standpipe operations during demolition, 74
    survey pattern, 89
Horizontal smoke spread, 270
Horizontal ventilation, 263–266
Hoseline
    attack lines, 210
    fog stream, 244
    long lay for adequate water supply, 98–99
    straight/solid stream, 244
    for transitional attack, 244
    use in heavy snow, 161
House fire, 138, 145
HRR. *See* Heat release rate (HRR)
HUD (U.S. Department of Housing and Urban Development), 109, 110
Humidity, arrival condition indicators, 201
Hurricane Katrina, 148
HVAC systems. *See* Heating, ventilation, and air conditioning systems (HVAC)
Hybrid modular homes, 110
Hydraulic ventilation, 263, 266
Hydrocarbon fuel, 18, 19
Hydrogen cyanide (HCN), 20–21

I

IAFC. *See* International Association of Fire Chiefs (IAFC)
IAP. *See* Incident Action Plan (IAP)
IARC (International Agency for Research on Cancer), 250
IC. *See* Incident Commander (IC)
ICC (International Code Council), 100
Ice
    arrival condition indicators, 201
    causing structural collapse, 118
ICS. *See* Incident Command System (ICS)
IDLH. *See* Immediately Dangerous to Life and Health (IDLH)
Ignition, 15–17
    autoignition, 16
    autoignition temperature, 17, 52
    ignition source, 173
    piloted, 16
    pyrolysis, 16
    spontaneous, 22, 23–24

vaporization, 16
Illegal drug lab chemical hazards, 190
Immediately Dangerous to Life and Health (IDLH)
    defined, 152
    personnel accountability, 152
    RIC/RIT requirements, 167
    two-in, two-out rule, 171–172
Incident Action Plan (IAP)
    adjusting as necessary, 141
    defined, 136
    development of, 217–218
    implementation of, 218–221
        command options, 218–220
        resource allocation, 220–221
    information included, 149
    NFPA 1021, *Incident Action Plan*, 149, 250
    personnel resource allocation, 100
    plan of operation, 180
    purpose of, 136, 149
    resource implementation, 149
    resource tracking, 153–154
    transfer of command, 149
    written IAP, 149
Incident Command Post (stationary command) option, 219–220
Incident Command System (ICS). *See also* National Incident Management System-Incident Command System (NIMS-ICS)
    department capabilities, 162
    Incident Action Plan (IAP), 149
    jargon for communications, 147
    model structure, 142
    organizational levels, 144
    radio communications, 147
    resource tracking, 154
    resources, 146
Incident Commander (IC)
    apparatus staging protocols, 153
    comprehensive resource management, 151
    decision-making skills, 135–136, 138, 141
    department capabilities, 162
    extinguishment responsibilities, 239–242, 245
    house fire decision errors, 138, 145
    Incident Action Plan (IAP), 141, 149, 218–221
    Incident Commander Rules of Engagement for Firefighter Safety, 184
    life hazard to occupants, 192
    life safety, 207–208
    occupant survival profile, 181–182
    operational strategies, 213–217
    personnel accountability, 151, 152
    personnel accountability report (PAR), 216
    preincident planning, 81–82, 92
    problem-solving skills, 139
    resource availability determination, 139
    resource tracking, 154
    responsibilities during ventilation-limited decay, 59
    search and rescue operation, 230
    size-up responsibilities, 166–167, 169, 185
    ventilation considerations, 255, 261
Incident Communications Plan, 147
Incident stabilization
    incident strategy, 210
    life safety, incident stabilization, and property conservation (LIP), 180, 207
Incipient stage of fire development
    described, 41, 43–44
    transition to growth stage, 46
Incomplete combustion, 18, 19, 40
Indecision, result of, 141
Indirect attack, 242

Industrial occupancy life hazard to occupants, 196
Industrialized housing, 109–110
Inorganic fuel, 31
Insulation and fire development, 72–73
Insurance Services Office (ISO), 100
Intake, ventilation, 246
Interior building arrangement, 111–125
    attics and cocklofts, 115–116
    basements, cellars, and crawl spaces, 114–115
    compartmentalization, 113
    concealed spaces, 116–117
    open floor plans, 112–113
    preincident survey information, 88–90
    structural collapse potential, 117–125
Intermittent flames, 47
Internal exposure control, 239
International Agency for Research on Cancer (IARC), 250
International Association of Fire Chiefs (IAFC)
    defined, 184
    Rules of Engagement, 184–185
    The 10 Rules of Engagement for Structural Fire Fighting, 211
International Building Code, 103
International Code Council (ICC), 100
International System of Units (SI), 14, 23, 32
Interstitial space, 115
Investigation option (nothing is showing), 219
ISO (Insurance Services Office), 100
Isolated flames, 46, 47

J

Jargon, 147
Jersey City (NJ) three-alarm fire, 101
Joists
    lightweight, failure of, 73–74
    wood I-joists, 108
Joule (J) unit of measure, 14

K

Kalamein door, 189–190
Kilo unit of measure, 32
Kilowatts (kW), 22, 32
Kilowatts per square meter (kW/m²), 22
Kinetic energy, 13

L

Ladders, 140, 141
Lamella roof, 127
Laminar flow of smoke, 178
Legacy construction, 70–72
LEL (lower explosive limit), 38
Level I apparatus staging, 153
Level II apparatus staging, 153
LFL (lower flammable limit), 38
Life safety. *See also* Safety
    area familiarization and, 83
    defined, 207
    firefighter safety criteria, 208
    firefighter survivability, 181–185
    preincident survey information, 90–93
        building conditions, 91–92
        building contents, 92–93
        evacuation considerations, 90, 91
        firefighter hazards, 91
        location of windows, 90
        occupant protection, 90
    strategy priorities, 207–209
    victims, 209

Life Safety Code (NFPA 101), 188
Life safety, incident stabilization, and property conservation (LIP), 180, 207
Lighting for nighttime incidents, 200
Lightweight construction, 103
Line-of-duty death (LODD)
    deaths over time, 211
    improper ventilation LODD, 254–255
    structural collapse incident, 180
    The 10 Rules of Engagement for Structural Fire Fighting, 211
LIP (life safety, incident stabilization, and property conservation), 180, 207
Liquids
    fire point, 34, 35
    flammable and combustible characteristics, 35
    flammable ranges, 39
    flash point, 34, 35
    as fuel, 33–35
    solubility, 34, 35
    specific gravity, 34, 35
    vapor pressure, 34, 35
Load-bearing wall, 104
LODD. *See* Line-of-duty death (LODD)
Low air alarm, 192
Lower explosive (flammable) limit (LEL), 38
Lower flammable limit (LFL), 38
Lumber mill fuel load, 92
LUNAR (location, united number, assignment, and resources needed), 229

M

Managing the incident, 135–155
    decision-making, 135–142
    adjusting the plan, 141
    identify and prioritize problems, 139
    Incident Action Plan (IAP), 136
    indecision, result of, 141–142
    monitor results, 141
    Recognition Primed Decision Making (RPDM), 137
    solution determination, 139–140
    solution implementation, 140
    steps, 137
    National Incident Management System-Incident Command System (NIMS-ICS), 142–155
    characteristics, 143
    communications, 146–148
    comprehensive resource management, 151
    defined, 142
    Incident Action Plan (IAP), 149
    modular concept, 143
    organizational levels, 144–145
    personnel accountability, 151–152
    purpose of, 142–143
    resource tracking, 152–155
    resources, 146
    scalable concept, 143
    span of control, 150–151
    stationary command (incident command post) option, 220
    unified command structure, 148
Manufactured homes, 109–110
Matter, defined, 12
Mayday, 180, 215, 229
MDC/T (Mobile Data Computer/Terminal), 159, 161
Means of egress, 188–189
Mechanical energy, 25–26
Mechanical ventilation, 263, 264–266
Medical surveillance, 248
Mega unit of measure, 32
Megawatts (MW), 22, 32

Mental "file," 137
Mercantile Occupancy
    contents, 195
    defined, 194
    open space construction, 72
    retail outlets, 129
Metal
    metal-clad doors, 189–190
    rust, 12
    security doors and bars, 189–190
    smoke colors and burning metal, 175
Methane, self-sustained chemical reaction, 40
Miscible, 34, 35
Mixed occupancy life hazard to occupants, 195
Mobile command option, 219
Mobile Data Computer/Terminal (MDC/T), 159, 161
Mobile homes, 109–110
Modern construction, 70–72
Modular homes, 110
Monitoring results of decisions, 141
Monitors, 94
Multijurisdictional incidents, 148
Multiple use occupancy, 129
Multistory buildings
    collapse zone, 119
    fire confinement tactics, 271
    fire ventilation tactics, 269–271
    hazards, 119
    size-up considerations, 186
    standpipe operations during demolition, 74
    survey pattern, 89
Mutual aid, 102

N

National Building Code of Canada (NBC), 110–111
National Fire Academy (NFA), 100
National Fire Protection Association (NFPA). *See also specific NFPA*
    construction types, 103
    Fire Protection Guide to Hazardous Materials, 39
    RIC/RIT requirements, 167
National Incident Management System-Incident Command System (NIMS-ICS), 142–155
    characteristics, 143
    communications, 146–148
    comprehensive resource management, 151
    defined, 142
    Incident Action Plan (IAP), 149
    modular concept, 143
    organizational levels, 144–145
    personnel accountability, 151–152
    purpose of, 142–143
    resource tracking, 152–155
    resources, 146
    scalable concept, 143
    span of control, 150–151
    stationary command (incident command post) option, 220
    unified command structure, 148
National Institute for Occupational Safety and Health (NIOSH)
    causes of line-of-duty death, 180
    death prevention recommendations, 115
    firefighter safety recommendations, 209
    improper ventilation LODD, 255
National Institute of Standards and Technology (NIST)
    floor system collapse times, 123
    furnishings as fuel load, 165
    offensive strategy, 213
    radiated heat temperature measurements, 29
Natural disasters, 148
Natural ventilation, 263–264

NBC (National Building Code of Canada), 110–111
Negative-pressure ventilation (NPV), 264
Neutral plane
    defined, 46
    rapid fire development indicators, 52–53
    reading smoke, 175
    thermal layering and, 47
New Orleans fire, 154
NFA (National Fire Academy), 100
NFPA. *See* National Fire Protection Association (NFPA)
NFPA 101, *Life Safety Code*, 188
NFPA 204, *Standard for Smoke and Heat Venting*, 94
NFPA 220-2015, *Standard on Types of Building Construction*, 64
NFPA 400-2010, *Hazardous Materials Code*, 12
NFPA 921, ventilation-controlled fire, 177
NFPA 921-2011, *Guide for Fire and Explosion Investigations*
    autoignition, 16
    autoignition temperature, 17
    combustion, 11
    fuel, 12
    fuel-limited fire, 41
    self-heating, 22
    spontaneous ignition, 22
    ventilation-limited fire, 41
NFPA 1021, I*ncident Action Plan*, 149, 250
NFPA 1142, *Standard on Water Supplies for Suburban and Rural Fire Fighting*, 100
NFPA 1500, personnel accountability, 152
NFPA 1561, S*tandard for Fire Department Incident Management System*, 149
NFPA 1620, *Standard for Pre-Incident Planning*, 82
NIMS-ICS. *See* National Incident Management System-Incident Command System (NIMS-ICS)
NIOSH. *See* National Institute for Occupational Safety and Health (NIOSH)
NIST. *See* National Institute of Standards and Technology (NIST)
Nomex®, 38
Nonbearing wall, 104
Noncombustible construction (Canada), 110
Noncombustible materials in Type I construction, 104
Nonflaming combustion, 17, 38
Nonload-bearing wall, 104
Nothing is showing (investigation option), 219
Novoclimat standard (Canada), 111
NPV (negative-pressure ventilation), 264

O

Occupancy
    Assembly Occupancy
        life hazard to occupants, 197
        restaurants, 129
    Business Occupancy Classification
        life hazard to occupants, 194
        office building, 129
    changes, hazards of, 190
    defensive strategy and, 215
    Educational Occupancy
        elementary school, 129
        life hazard to occupants, 197–198
    Mercantile Occupancy
        contents, 195
        defined, 194
        open space construction, 72
        retail outlets, 129
    multiple use, 129
    single use, 129
    size-up considerations, 163, 186
    size-up for life hazard to occupants, 191–200
        assembly occupancy, 197
        business occupancy, 194
        clues to hazards, 192
        educational occupancy, 197–198
        generally, 191–193
        industrial and storage occupancy, 196
        institutional occupancy, 196–197
        mercantile occupancy, 194–195
        mixed occupancy, 195
        residence occupancy, 193–194
        unoccupied, vacant, or abandoned structures, 198–200
Occupant survival profile, 181
Occupational Safety and Health Administration (OSHA)
    OSHA 1910.134, 152
    RIC/RIT requirements, 167
Offensive strategy, 212–213, 219
Offensive tactics, 240
Open burning, 42–43
Open floor plans
    business occupancy, 194
    commercial construction, 71, 74
    compartmentalized vs., 69
    Mercantile Occupancy, 72
    residence occupancy, 112, 257
    warehouse construction, 72
Open space hazards, 72, 191
Operational strategies, 212–217
    defensive strategy, 214–216
    defensive tactics, 240
    offensive strategy, 212–213, 219
    offensive tactics, 240
    transitioning strategies, 216–217
Operations Section Officer, 153
Organic fuel, 31
Orientation of fuel and fire spread, 36, 37
Oriented search, 232
Oriented strand board (OSB), 107, 108
OSB (oriented strand board), 107, 108
OSHA. *See* Occupational Safety and Health Administration (OSHA)
Outside aid, 102
Overcurrent/overload, 25
Overhaul, 248–253
    arson evidence, 249, 250
    facts to consider, 251
    fire watch, 250
    hazards of, 249–250, 252–253
    location for, 251–252
    medical surveillance during, 248
    procedures, 249
Oxidation
    defined, 12
    production of thermal energy, 24
    timeline, 13
Oxidizer
    common types, 11
    defined, 12
Oxygen
    concentration, effects of, 38
    exclusion, in fire fighting operations, 63
    fuel vapor-to-air mixture, 38–39
    needed for combustion, 37–39
    self-sustained chemical reaction, 40
Oxygen-limited environment, 254

P

Paint store life hazard to occupants, 195
Panelized homes, 110
Panelized roofing, 103
PAR (personnel accountability report), 151, 216
Paradise Fire, 142

Parapet, 74, 126
PASS (personal alert safety system), 180
Passaic (NJ) Fire Department use of hoselines, 99
PEL (permissible exposure limit), 152
Perceptions as size-up consideration, 164–165
Permissible exposure limit (PEL), 152
Personal alert safety system (PASS), 180
Personal protective equipment (PPE)
    IDLH environments, 171–172
    self-contained breathing apparatus (SCBA)
        air monitoring research, 250
        IDLH environments, 172
        mechanical energy, 25, 26
        radiated heat flux susceptibility, 29
        use during overhaul, 249
    temperature and heat transfer rate, 29, 30
Personnel accountability
    IDLH environments, 152
    personal alert safety system (PASS), 180
    purpose of, 151
    tracking board, 154, 155
    two-in, two-out rule, 152, 171–172
    types of systems, 152
Personnel accountability report (PAR), 151, 216
Personnel resource allocation, 221
Pesticides, preincident survey record of, 85
Phoenix (AZ) Fire Department SOP/G, 212
Photography for preincident surveys, 87
Photovoltaic (solar) panels, 128, 129
Physical science
    defined, 12
    endothermic reactions, 14
    energy, 12–14
    exothermic reactions, 14
    matter, 12
    oxidation, 12, 13
Piloted ignition, 16
Pitched roofs, 125, 126–127
Pitot gauge, 96
Pitot tube, 96
Plan of operation, 180
Planning. *See* Prefire planning; Preincident plan
Plastic, fuel load of, 71
Plenum, 116
Plot plan, 85, 88
Polar solvents, 34, 35
Polyurethane foam
    incipient stage of fire development, 44
    pyrolysis, 35, 36, 67
Position of fuel and fire spread, 36, 37
Positive-pressure attack (PPA), 245–248
Positive-pressure ventilation (PPV), 264–266
Potential energy, 13
Power, 32
PPA (positive-pressure attack), 245–248
PPE. *See* Personal protective equipment (PPE)
PPV (positive-pressure ventilation), 264–266
Precut homes, 110
Prefabricated homes, 109–110
Prefire inspection, 81
Prefire planning, 81–129
    building construction, understanding of, 102–129
        building codes, 102
        Canadian construction, 110–111
        interior building arrangement, 111–125
        lightweight/engineered construction, 103
        occupancy types, 128–129
        roof types and hazards, 125–128
        United States construction, 103–110

defined, 81
preincident survey preparation, 84–97
    building interior, 88–90
    emergency response, 85–86
    fire protection systems, 95–97
    information records, 86–88
    life-safety information, 90–93
    plot plan, 85, 88
    ventilation systems, 93–95
resources available, 97–102
    additional resources, 102
    automatic aid, 101–102
    fire flow calculations, 99–100
    mutual aid, 102
    water supply, 98–99
surveys, 81–84. *See also* Preincident survey
    area familiarization, 83–84
    defined, 82
    preincident planning, 81
    updating the plan, 82–83
Preincident inspection, 81
Preincident plan
    preincident planning, defined, 81
    property conservation, 272
    resource allocation, 221
Preincident survey, 84–97
    bowstring truss identification, 74
    during building construction, 102–103
    building interior, 88–90
    defined, 82
    emergency response considerations, 85–86
    fire protection systems, 95–97
    floor plan, 90
    information records, 86–88
        building exterior, 87–88
        field sketches, 86, 87
        global positioning, 87
        photography, 87
        survey form data, 86–87
        videography, 87
    life-safety information, 90–93
        building conditions, 91–92
        building contents, 92–93
        evacuation considerations, 90, 91
        firefighter hazards, 91
        location of windows, 90
        occupant protection, 90
    multistory buildings, 89
    pesticide contents, 85
    planning process, 81
    plot plan, 85, 88
    preparation for, 84–97
    ventilation systems, 93–95
        built-in devices, 94–95
        HVAC systems, 93
        underfloor air distribution systems, 95
Preplan, 82
Preplanning, 81
Presidential Directive 5, 142
Pressure
    atmospheric, 22
    defined, 22
    heated gases, 22
Primary response area, 83
Primary search, 231, 232–234
Probabilities as size-up consideration, 165–167
Problem identification and prioritization, 139
Products of combustion
    carbon dioxide (CO_2), 19, 21

carbon monoxide (CO), 19-20
carbon-based fuels, 19
defined, 18
hydrocarbon fuels, 19
hydrogen cyanide (HCN), 20-21
smoke, 19
toxic effects, 20
Projections as size-up consideration, 165-167
Property conservation
    incident priorities, 210
    life safety, incident stabilization, and property conservation (LIP), 180, 207
    operational tactics, 271-273
    waterproof covers for, 210
Protected steel, 104
Protection-in-place, 196
Pulsations of smoke, 177
Pumpers used to boost supply line pressure, 99
Purlin, 103
Pyrolysis
    black smoke indicators, 174-175
    defined, 16
    polyurethane foam, 35, 36, 67
    white smoke indicators, 173
    wood, 35, 36

R

Radiation, 27, 28-29, 30
Radio communications
    example, 147
    for incident information, 159, 160
    protocols, 148
Rain
    arrival condition indicators, 201
    causing structural collapse, 103
Rapid fire development, 50-55
    backdraft, 50, 54-55
    combustible furnishings and, 165
    evolution of, 51
    flashover, 50-54
Rapid Intervention Crew/Team (RIC/T)
    defined, 167
    purpose of, 138, 208
    size-up considerations for, 167
    uncertainty at a house fire, 145
Recognition Primed Decision Making (RPDM), 137
Reducing agent, 31
Reflective barriers for nighttime incidents, 200
Reflective surfaces, 129
Rehabilitation
    defined, 201
    incident personnel, 193
    temperature extremes and need for, 201
Relay pumping, 99
Renovation
    hazards of, 74-75
    updating prefire plans after, 82-83
Required fire flow, 99
Rescue, 227-230
    defined, 227
    Incident Commander responsibilities, 230
    search and rescue operation, defined, 228
Residence
    attic storage, 121
    building built on a hillside, 191
    factory-built, 109-110
    floor collapse at brick house, 92
    fuel load, 70-71

    life hazard to occupants, 193-194
    modern vs. legacy construction, 70
    open floor plans, 112, 257
    size-up considerations, 193-194
Resistance heating, 25
Resources
    comprehensive resource management, 151
    crew, 146, 193
    crew resource management (CRM), 183-184
    Incident Action Plan (IAP), 220-221
    preincident planning for, 97-102
        additional resources, 102
        automatic aid, 101-102
        fire flow calculations, 99-100
        mutual aid, 102
        water supply, 98-99
    single resources, 146
    size-up considerations, 164, 170-171
    strike team, 146
    task force, 146
    tracking, 152-155
    ventilation considerations, 263
Restaurant occupancy, 88, 129
Retention of heat and fire development, 73
RIC/RIT. *See* Rapid Intervention Crew/Team (RIC/T)
Risk vs. benefit
    extinguishment, 241
    strategy, 211-212, 215
Risk-benefit analysis, 199
Rollover, 51
Roof
    arched, 127
    automatic roof vents, 94
    bowstring truss, 74, 127
    failure of lightweight trusses, 73-74
    fast-food restaurant roof collapse, 88
    firefighter injury during ventilation, 209
    firefighter safety, 124-125
    flat, 126
    green roof, 128, 129
    hazards, 128
    materials as fuel load, 68-69
    panelized, 103
    photovoltaic (solar) panels, 128
    pitched, 125, 126-127
    prefire planning, 125-128
    rain roof hazards, 138
    rooftop garden, 128
    size-up considerations, 186
    tactical ventilation, 253
    trussless arched (lamella), 127
RPDM (Recognition Primed Decision Making), 137
Rules of Engagement for Firefighter Survival, 184-185
Rules of Engagement for Structural Fire Fighting - Increasing Firefighter Survival, 184, 211
Rust, 12

S

Safety
    electrical lines as hazard, 170
    Incident Commander responsibilities for, 136
    life safety. *See* Life safety
    search, 231
    structural collapse, 124-125
Salvage
    defined, 265
    operational tactics, 271-273
    purpose of, 271

Satellite imaging for preincident surveys, 87
Sawdust, fuel load of, 92
SCBA. *See* Self-contained breathing apparatus (SCBA)
Science of fire, 11–22
    combustion, 17–21
        flaming, 17–18
        nonflaming, 17
        products of, 18–21
    fire triangle and tetrahedron, 15
    ignition, 15–17
        autoignition, 16
        autoignition temperature, 17
        piloted, 16
        pyrolysis, 16
        vaporization, 16
    physical science, 12–14
    pressure differences, 22
Search
    conducting, 231–234
    defined, 227
    primary, 231, 232–234
    safety guidelines, 231
    secondary, 231, 234
    tactics, 227
    VEIS (vent, enter, isolate, search), 233
    victim removal, 234
Search and rescue operation, defined, 228
Secondary search, 231, 234
Section level of NIMS-ICS, 144
Security metal bars over windows, 163
Self-contained breathing apparatus (SCBA)
    air monitoring research, 250
    IDLH environments, 172
    mechanical energy, 25, 26
    radiated heat flux susceptibility, 29
    use during overhaul, 249
Self-heating, 22, 23–24
Self-sustained chemical reaction, 39–40
Shear walls, 103
Sheetrock®, 114
Shelter in place, 196, 227
Sheltering, 196
Shielded fire, 242
SI (International System of Units), 14, 23, 32
Siding as fuel load, 68
Single resources, 146
Single use occupancy, 129
Single-family construction, 70–71
Situational awareness
    crew resource management (CRM), 183
    defined, 164
    search and rescue operation, 228
Size-up, 159–202
    arrival condition indicators, 200–202
        arrival report, 168–169
        day of week, 164
        incident considerations, 167–172
        time of day, 164, 200
        visual indicators, 202
        weather, 201–202
    building characteristics, 186–191
        egress, 188–189
        evaluation factors, 186–187
        metal security doors and bars, 189–190
        special hazards, 190–191
        street access, 187–188
        structure access, 188
    confinement tactics, 236–237
    decision-making, 180
    defined, 159
    extinguishment, 240
    facts, 163–164
    firefighter survivability, 181–185
        crew resource management, 183–184
        occupant survival profile, 181–183
        rules of engagement, 184–185
    fireground factors, 185–202
        arrival condition indicators, 200–202
        building characteristics, 186–191
        life hazard to occupants, 191–200
        procedures, 166–167
    during the incident, 179
    life hazard to occupants, 191–200
        assembly occupancy, 197
        business occupancy, 194
        clues to hazards, 192
        educational occupancy, 197–198
        generally, 191–193
        industrial and storage occupancy, 196
        institutional occupancy, 196–197
        mercantile occupancy, 194–195
        mixed occupancy, 195
        residence occupancy, 193–194
        unoccupied, vacant, or abandoned structures, 198–200
    perceptions, 164–165
    plan of operation, 180
    projections and probabilities, 165–167
    reading smoke, 172–179
        black smoke, 174–175
        brown smoke, 173
        density of smoke, 176
        flame color, 179
        gray smoke, 173
        heat, 178–179
        neutral plane, 175
        pulsations, 177
        smoke movement, 177–178
        types of smoke activity, 176–177
        unusual colors, 175
        velocity, 177
        volume of smoke, 175–176
        white smoke, 173, 176
    360-degree survey
        on arrival, 170–171
        for decision-making, 180
        extinguishment, 243, 244
    weather considerations, 161, 164, 201–202
    while responding, 159–162
        department capabilities, 162
        time of day, 161
        weather conditions, 161
Skylights, 94, 260
Smoke
    alarm activated doors, 95–96
    backdraft indicators, 55
    basement fires, 114
    carbon dioxide (CO_2), 19, 21
    carbon monoxide (CO), 19–20
    carcinogens, 250
    flashover indicators, 52
    floating or hanging, 177
    flow path, 176–177
    health effects, 21
    horizontal smoke spread, 270
    laminar flow, 178
    lazy smoke, 177–178
    particle size, 172
    preincident survey record of smoke control system, 86

as product of combustion, 19
reading smoke, 172–179
 black smoke, 174–175
 brown smoke, 173
 density of smoke, 176
 flame color, 179
 gray smoke, 173
 heat, 178–179
 neutral plane, 175
 pulsations, 177
 smoke movement, 177–178
 types of smoke activity, 176–177
 unusual colors, 175
 velocity, 177
 volume of smoke, 175–176
 white smoke, 173, 176
size-up considerations, 162
smoke explosion, 63
turbulent flow, 178
velocity of, 162
venting (NFPA 204), 94
Smoldering combustion, 38
Snow
 arrival condition indicators, 201
 causing structural collapse, 103, 118
 hoseline advancement in snow, 161
Soffits, 268
Solar panels, 128, 129
Solids
 as fuel, 35–37
 position of fuel and fire spread, 36, 37
 surface-to-mass ratio, 35–37
Solubility, 34, 35
Solution, determination and implementation of, 139–140
SOP/SOG. *See* Standard Operating Procedures/Standard Operating Guidelines (SOP/SOG)
Span of control, 150–151
Sparking, 25
Specific gravity, 34, 35
Spontaneous ignition, 22, 23–24
Sprinkler system water supply requirements, 96
Stabilization of the incident
 incident strategy, 210
 life safety, incident stabilization, and property conservation (LIP), 180, 207
Stack action, 269
Stack effect, 269, 270
Staging protocols, 153
Stairs as means of egress, 189
Stairwell, flow of smoke from basement fires, 114
Standard for Fire Department Incident Management System (NFPA 1561), 149
Standard for Pre-Incident Planning (NFPA 1620), 82
Standard for Smoke and Heat Venting (NFPA 204), 94
Standard on Types of Building Construction (NFPA 220-2015), 64
Standard on Water Supplies for Suburban and Rural Fire Fighting (NFPA 1142), 100
Standard Operating Procedures/Standard Operating Guidelines (SOP/SOG)
 dispatch information, 160–161
 fireground operation factors, 185
 resource tracking, 152–154
 size-up considerations, 168, 169
 The 10 Rules of Engagement for Structural Fire Fighting model, 212
 two-in, two-out rule, 172
Standpipes
 operation during demolition, 74
 preincident planning knowledge of, 95
 preincident survey information, 87
water supply requirements, 96
Stationary command (Incident Command Post) option, 219–220
Steel
 protected, 104
 unprotected, 105
 used for concrete strength, 104
Storage
 attic storage, 121
 hazards of, 190
 life hazard to occupants, 196
Straight/solid stream, 244
Strategy
 defensive, 214–216
 Incident Action Plan (IAP), 217–221. *See also* Incident Action Plan (IAP)
 incident priorities, 207–210
  incident stabilization, 210
  life safety, 207–209
  property conservation, 210
 offensive, 212–213, 219
 risk vs. benefit, 211–212, 215
 transitioning, 216–217
Street access, 187–188
Strike team, 146
Strip mall occupancy
 life hazard to occupants, 195
 multiple use occupancy, 129
 size-up considerations, 187
Structural collapse
 collapse zone, 117, 118, 120
 contents of the structure, 121
 defined, 117
 factors determining potential for, 117
 floor system collapse times, 123
 grain silo, 121
 indicators of potential for, 121–122
 safety alert for firefighters, 124–125
 snow and ice load, 118
 Type I construction, 119
 Type II construction, 119
 Type III construction, 119
 Type IV construction, 119
 Type V construction, 120
 water used to suppress the fire, 121
Structural integrity and construction type, 64–65
Structure fire development, 59–63
 fire fighting operations, 63
 flow path, 59–61
 smoke explosion, 63
 ventilation, unplanned, 62
 wind conditions, 62–63
Surface-to-mass ratio, 35–37
Survey. *See* Preincident survey
Survivability approaches, 181–185
 crew resource management, 183–184
 occupant survival profile, 181–183
 rules of engagement, 184–185
Synagogues, life hazard to occupants, 197
Synthetic materials as fuel for fires, 32, 71

T

Tactical ventilation, 253–255
Tactical Work Sheet, 136, 154
Tactics, 227–273
 confinement, 229, 235–239
 exposures, 234–235
 extinguishment, 239–248
  fire attack, 241–243
  positive-pressure attack, 245–248

transitional attack, 243–245
overhaul, 248–253
property conservation, 271–273
salvage, 271–273
search and rescue, 227–234
tactical ventilation, 253–255
ventilation considerations, 255–263
    basement fires, 258
    building construction and furnishings, 256–258
    exposures, 262
    fire behavior indicators, 258–259
    location and extent of the fire, 259
    location for ventilation, 260
    occupant and firefighter risks, 256
    resources, 263
    type of ventilation, 259–260
    weather conditions, 261
ventilation types, 263–271
    basement fires, 269
    high-rise fires, 269–271
    horizontal, 263–266
    vertical, 267–268

Tag/passport systems, 152
Take no risk for no gain, 212
Taking refuge, 196
Task force, 146
Teamwork, crew resource management, 183
Temperature
    defined, 11
    extremes as arrival condition indicators, 201
    heat release rate vs., 22–23, 29
    reduction, in fire fighting operations, 63
The 10 Rules of Engagement for Structural Fire Fighting, 211
Thermal balance, 175
Thermal conductivity, 26, 27
Thermal energy (heat), 22–30
    defined, 13
    heat release rate vs. temperature, 22–23
    heat transfer, 26–30
        conduction, 26–27, 30
        convection, 27–28, 30, 61
        interaction among methods of, 30
        radiation, 27, 28–29, 30
    sources of, 23–26
        chemical energy, 23–24
        electrical energy, 24–25
        mechanical energy, 25–26
Thermal equilibrium, 26, 27
Thermal imager, 232, 243
Thermal imaging camera, 232
Thermal layering, 46–48
Thermal properties and fire development, 72–73
Thermoplastic, 94, 95
360-degree survey
    on arrival, 170–171
    for decision-making, 180
    extinguishment, 243, 244
Three-sided view of the structure for size-up, 187
Tie rods, 106
Time of day, size-up considerations, 161, 164, 200
Topography as size-up consideration, 186
Townhouse fire, 115
Tracking board, 154, 155
Traditional search, 232
Transitional attack, 213, 243–245
Transitioning strategies, 216–217
Trusses
    bowstring, 74, 127
    construction of, 108

firefighter safety, 124–125
lightweight, failure of, 73–74
lightweight/engineered construction, 268
parallel chord trusses, 108
web truss support collapse, 103
Trussless arched roof, 127
Turbulent flow of smoke, 178
Two-in, two-out rule, 152, 171–172
Type I construction
    characteristics, 104–105
    fire resistance ratings, 64
    structural collapse, 119
Type II construction
    characteristics, 105
    fire resistance ratings, 64
    structural collapse, 119
Type III construction
    characteristics, 105–106
    fire resistance ratings, 64
    structural collapse, 119
Type IV construction
    characteristics, 106–107
    fire resistance ratings, 64
    structural collapse, 119
Type V construction
    characteristics, 107–109
    fire resistance ratings, 64
    structural collapse, 120

U

UEL (upper explosive) limit, 38
UFAD (underfloor air distribution) system, 95
UFL (upper flammable limit), 38
UL. *See* Underwriters Laboratories Inc. (UL)
Underfloor air distribution (UFAD) system, 95
Underwriters Laboratories Inc. (UL)
    floor system collapse times, 123
    furnishings as fuel load, 165
    protected steel, 104
Unified command structure, 148
Unit level of NIMS-ICS, 144
United States construction. *See also specific Type of construction*
    Type I construction, 104–105
    Type II construction, 105
    Type III construction, 105–106
    Type IV construction, 106–107
    Type V construction, 107–109
    unclassified construction types, 109–110
United States Department of Housing and Urban Development (HUD), 109, 110
Unoccupied structures, life hazard to occupants, 198–200
Unprotected steel, 105
Upper explosive (flammable) limit (UEL), 38
Upper flammable limit (UFL), 38

V

Vacant buildings
    Cleveland fire, 182–183
    graffiti of occupancy, 164
    life hazard to occupants, 198–200
    size-up considerations, 198–200
Vapor, 32
Vapor density, 33
Vapor pressure, 34, 35
Vaporization, 16
VEIS (vent, enter, isolate, search), 233
Velocity of smoke, 162
Veneer damage from fire, 122

Ventilation
    atrium vents, 94
    basement, 258, 269
    black smoke indicators, 174
    blowtorch effect, 254
    building characteristic considerations, 71–72
    considerations for openings, 244
    coordinated, 177
    exhaust, 246
    firefighter injury during, 209
    flow path of structure fire development, 59–61
    forcing entry as new flow path, 243
    fuel load and, 68
    high-rise fires, 269–271
    horizontal, 263–266
    HVAC. *See* Heating, ventilation, and air conditioning systems (HVAC)
    hydraulic, 263, 266
    improper ventilation LODD, 254–255
    intake, 246
    mechanical, 263, 264–266
    natural, 263–264
    negative-pressure, 264
    positive-pressure, 264–266
    size-up consideration for, 166
    structure fire development and, 61–63
    tactical considerations, 255–263
        basement fires, 258
        building construction and furnishings, 256–258
        exposures, 262
        fire behavior indicators, 258–259
        location and extent of the fire, 259
        location for ventilation, 260
        occupant and firefighter risks, 256
        resources, 263
        type of ventilation, 259–260
        weather conditions, 261
    tactical ventilation, 253–255
    unknowingly creating openings, 256
    unplanned, 62
    VEIS (vent, enter, isolate, search), 233
    vertical, 267–268
    weather considerations, 261, 266
Ventilation systems
    built-in devices, 94–95
    preincident planning knowledge of, 93–95
        built-in devices, 94–95
        HVAC systems, 93
        underfloor air distribution systems, 95
Ventilation-controlled fire, 41, 177
Ventilation-limited decay, 48–50, 58–59
Ventilation-limited fire
    backdraft, 54
    defined, 41
    fuel loads and, 68
    fully developed fire, 56–57
Vertical ventilation, 267–268
Victim removal, 234
Videography for preincident surveys, 87
Visual arrival condition indicators, 202
Void space, 116. *See also* Concealed spaces

W

Walls
    automatic wall vents, 94
    case iron stars, 122
    collapse, 120
    fire wall, 85
    heat from a concealed fire, 251
    load-bearing, 104
    nonload-bearing, 104
    shear walls, 103
Warehouse construction
    fire wall, 85
    fuel load, 71
    open spaces, 72
Washington D.C. townhouse fire, 115
Water
    hydrants, 98
    lakes as supply of, 97
    long lay of hoseline, 98–99
    preincident planning for, 97–99
    solubility of substances, 34
    sprinkler or standpipe system requirements, 96
    structural collapse from, 121
Watt (W), 32
Weather
    arrival condition indicators, 201–202
        humidity, 201
        ice, 201
        rain, 201
        snow, 201
        temperature extremes, 201
        wind, 202
    ice
        arrival condition indicators, 201
        causing structural collapse, 118
    rain
        arrival condition indicators, 201
        causing structural collapse, 103
    size-up considerations, 161, 164, 201–202
    snow
        arrival condition indicators, 201
        causing structural collapse, 103, 118
        hoseline advancement in snow, 161
    ventilation considerations, 261
    wind
        arrival condition indicators, 202
        structure fire development and, 62–63
        ventilation, 266
        ventilation considerations, 261
Wide-area search, 232–233
Wind
    arrival condition indicators, 202
    structure fire development and, 62–63
    ventilation, 266
    ventilation considerations, 261
Windows
    as means of egress, 188–189
    metal bars for security, 163
    preincident survey noting location of, 90
Witnesses as sources of information, 213
Wood
    brown smoke as indicator of burning, 173
    I-joists, 108
    pyrolysis, 35, 36
    wood-frame construction characteristics, 107–109
    wood-frame construction fuel load, 66, 74
Work of energy, 12
Worship places, life hazard to occupants, 197